# Research Reports in Physics

## Research Reports in Physics

**Nuclear Structure of the Zirconium Region**
Editors: J. Eberth, R. A. Meyer, and K. Sistemich

**Ecodynamics** Contributions to Theoretical Ecology
Editors: W. Wolff, C.-J. Soeder, and F. R. Drepper

**Nonlinear Waves 1** Dynamics and Evolution
Editors: A. V. Gaponov-Grekhov, M. I. Rabinovich, and J. Engelbrecht

**Nonlinear Waves 2** Dynamics and Evolution
Editors: A. V. Gaponov-Grekhov, M. I. Rabinovich, and J. Engelbrecht

**Nonlinear Waves 3** Physics and Astrophysics
Editors: A. V. Gaponov-Grekhov, M. I. Rabinovich, and J. Engelbrecht

**Nuclear Astrophysics**
Editors: M. Lozano, M. I. Gallardo, and J. M. Arias

**Optimized LCAO Method** and the Electronic Structure of Extended Systems By H. Eschrig

**Nonlinear Waves in Active Media**
Editor: J. Engelbrecht

**Problems of Modern Quantum Field Theory**
Editors: A.A. Belavin, A.U. Klimyk, and A.B. Zamolodchikov

**Fluctuational Superconductivity of Magnetic Systems**
By M.A. Savchenko and A.V. Stefanovich

**Nonlinear Evolution Equations and Dynamical Systems**
Editors: S. Carillo and O. Ragnisco

**Nonlinear Physics**
Editors: Gu Chaohao, Li Yishen, and Tu Guizhang

**Nonlinear Waves in Waveguides** with Stratification
By S. B. Leble

A. V. Gaponov-Grekhov
M. I. Rabinovich
J. Engelbrecht (Eds.)

# Nonlinear Waves 3

## Physics and Astrophysics

Proceedings of the Gorky School 1989

With 109 Figures

Springer-Verlag
Berlin Heidelberg New York London
Paris Tokyo Hong Kong Barcelona

Professor Andrei V. Gaponov-Grekhov, Academician
Professor Mikhail I. Rabinovich
Institute of Applied Physics, ul. Ulyanova 46, SU-603600 Gorky, USSR

Professor Jüri Engelbrecht
Institute of Cybernetics, Estonian Academy of Sciences, Akadeemia tee 21,
SU-200108 Tallinn, Estonia, USSR

ISBN 3-540-52024-4 Springer-Verlag Berlin Heidelberg New York
ISBN 0-387-52024-4 Springer-Verlag New York Berlin Heidelberg



2157/3150-543210 – Printed on acid-free paper

*Dedicated to the memory of*
*Ya.B. Zeldovich*

# Preface

This volume is based on the tutorial reviews and short contributions presented at the 9th All-Union School on Nonlinear Physics held in Gorky in March 1989. The papers collected here give an up-to-date picture of the state of modern nonlinear physics in the Soviet Union. Thus it complements well the preceding two volumes on the 1987 School which also appeared in this series.

The subtitle "Physics and Astrophysics" may be a bit misleading, because large portions of the material deal, in particular, with quantum physics, solid state physics, dynamical chaos and self-organization.

Since the history of the Gorky School dates back to 1972 there already exists a certain tradition of it as a gathering place for Soviet "nonlinearists". Thus, the School is attended by a good mixture of well-known names and younger scientists, which ensures that tutorial introductions and reviews are included as well as shorter contributions on the particulars of the most recent developments.

Some keywords are general nonlinear dynamics, turbulence, structure of the universe, plasma dynamics, nonlinear stability, surface and internal waves, active media, superconductors, lasers, chemistry of solids, magnons, chaos, limit cycles, synergetics, cellular automata, pattern formation.

The editors are grateful to all the authors participating in this volume and look forward to their future cooperation in publishing the proceedings of the subsequent Schools. Special thanks are due to Ms. Eve Klement for her enormous efforts in technical editing and typing of the manuscripts.

Tallinn, *J. Engelbrecht*
April 1990

# To the Memory of Yakov Borisovich Zeldovich (1914–1987)

The All-Union School on Nonlinear Physics held in Gorky had a remarkable tradition: a driving force in the form of one person – Yakov Borisovich Zeldovich. This tradition was sadly broken at the Ninth School because he was not among us any more. In this volume the proceedings of the Ninth School are presented and we feel that we must share with the world of science our memories about this remarkable scientist.

Yakov Borisovich Zeldovich had an exceptionally wide spectrum of interests and worked in a variety of areas ranging from astrophysics to chemical technology. Of course, nonlinear physics and nonlinear dynamics have no right to "monopolize" Ya.B's fame. Nevertheless, when we turn to his papers or books, we cannot but agree that the nonlinear problems have been challenging him all his life, during both the early period of his activity and the last two decades, devoted mainly to the origin of galaxies and the large-scale structure of the Universe. When discussing these problems at the Congress on Self-Organization at Pushchino in 1983, he joked that synergetics had been his life-long interest but he had not known, until shortly before, the true name of his interest.

The papers of Ya.B. on exothermic chemical reactions and the theory of combustion date back to the early thirties and were actually among the earliest ones dealing with the theory of spatio-temporal structures, now one of the main branches of modern nonlinear dynamics. The radically nonlinear character of these problems (in particular, due to the temperature dependence of the rate of combustion reactions) required his exceptional ingenuity and versatility as well as the intuition of a brilliant physicist and the skill of a virtuoso mathematician. It was Ya.B. who developed the theory of nonlinear combustion and detonation waves and related (together with D.A. Frank-Kamenetsky) the flame propagation velocity and the combustion front stability to the real chemical kinetics of a system. These ideas also found their way into his classical investigations in the theory of shock waves.

Nonlinear problems, and the aspects of their specific formulation, asymptotic and qualitative methods of solution and simulation attracted him primarily because of his interest in quantum mechanics, the theory of particles, cosmology and, of course, astrophysics. In this short text we are unable to give even a brief description of his remarkable achievements in nonlinear physics. No doubt his efforts will be continued and reinforced by his colleagues and disciplines.

Yakov Borisovich took great pleasure in working with young scientists; he gave courses, supervised seminars and delivered lectures. In spite of being extremely busy, Ya.B. always participated, most willingly, in Gorky Schools on Nonlinear Physics. Due to his unique irresistible charm as a lecturer, which brought to life the excitement of discovery, his fantastic vigor, unfading with years, and his personal interest in the audience, he won the passionate affection of the listeners of our Schools. The style and method he adopted in research seem to be specially synthesized for nonlinear science. Indeed, as he said, a good command of the mathematical tools and an ability to overcome mathematical difficulties, bold ideas and physical intuition, a capable experiment and adequate simulation – any of these different but integral parts is equally important to ensure a keen insight into Nature [*Higher Mathematics for Beginners in Physics and Engineering* (in Russian) by Ya.B Zeldovich, I.M. Yaglom (Moscow, Nauka 1982)].

Many papers in this volume are written by the students or long-time followers of Ya.B. Zeldovich. We are sincerely grateful to each of them who found time to contribute to this edition.

Gorky,
April 1990

*A.V. Gaponov-Grekhov*
*M.I. Rabinovich*

## Contents

# Part I

# Physics and Astrophysics

# Magnetic Intermittency

*S.A. Molchanov*[1], *A.A. Ruzmaikin*[2], *and D.D. Sokoloff*[1]

[1]Moscow State University, Lenin Hills, 119899 Moscow, USSR
[2]Institute of Terrestrial Magnetism, Ionosphere and Radio Wave Propagation, USSR Academy of Sciences, 142092 Troitsk, Moscow Region, USSR

Random flows of electrically conducting fluid generate intermittent magnetic fields. This means that principal contribution of the mean density of magnetic energy comes from widely spaced concentrations of magnetic field rather than typical (most probable) values of magnetic field. The process of generation of the intermittent magnetic field distributions is discussed in terms of the theory of dynamic systems.

## 1. INTRODUCTION

Turbulent flows of electrically conducting fluids in planetary interiors, stars and galaxies are notable for invariably large magnetic Reynolds numbers, i.e. for the prevalence of inductive action of motions over Ohmic dissipation. This peculiarity is the key to understanding the origin of cosmical magnetic fields as a result of self-excitation of some weak initial field. ZELDOVICH has made pioneering contributions to this field of magnetohydrodynamics, one of which is the elementary self-excitation process known as the "figure-eight" dynamo, or rope dynamo.

In the limit of the evolution of a magnetic field with the infinitely large magnetic Reynolds number can be understood in easy terms: the field is frozen into the fluid and, consequently, the exponential growth of the separation between neighbouring fluid elements in the turbulent flow results in the exponential growth of the embedded magnetic field. The structure of the magnetic field, generated thereby, turns out to be intermittent in space and time: rare but intense field concentration makes a principal contribution to the mean magnetic energy and coexists with vast weakly magnetized regions.

In reality, the magnetic Reynolds number is large but finite. Physically, this means that the Ohmic dissipation and magnetic diffusion can also play a certain role. Even though magnetic diffusivity is small, its effect is essential because the growth of the magnetic field is accompanied by the exponential decrease of its scale and the appearance of strong gradients of field intensity. From the mathematical viewpoint, here we encounter a small coefficient at the highest-order derivative in the induction equation, i.e. at the Laplacian term. Nevertheless, the problem of field evolution in finitely-conducting medium also can be treated with the aid of the frozenness into a flow with random, Wiener trajectories rather than true streamlines.



Research Reports in Physics Nonlinear Waves 3
Editors: A.V. Gaponov-Grekhov · M.I. Rabinovich · J. Engelbrecht

On the other hand, the finiteness of the magnetic Reynolds number in turbulent flows gives rise to the enhanced dissipation associated with the reconnection of oppositely directed, closely spaced magnetic lines.

## 2. THE DYNAMO MAPS

Consider a random flow of infinitely conducting incompressible fluid with an embedded magnetic field. Let the initial field be localized in some region or regions and let it be so weak that its influence on the fluid motions can be neglected. For the sake of simplicity, we consider a closed magnetic flux tube with the field H, length L and cross-section S. An elementary process of magnetic field amplification proposed by ZELDOVICH is as follows. The flux tube is stretched to the one twice its initial size; then $L \to 2L$, $S \to S/2$ and $H \to 2H$ as follows from the incompressibility and the conservation of magnetic flux $F = HS$. Then the tube is twisted into a figure resembling number eight and folded so that the rings of the figure eight, which now have identically directed fields, overlap: $2L \to L$, $S/2 \to S$, $2H \to 2H$ and $F \to 2F$. As a result, both the magnetic field and the magnetic flux are doubled while the rope has restored its initial shape. Evidently, the magnetic flux is doubled not due to deviations from frozenness but rather due to the combination of two fluid volumes into a single one. When repeated, this procedure leads to the growth of the magnetic field with the doubling time $T/\ln 2$, where T is the period of the cycle.

The authors of /1/ have noticed that stretching at the first step of the cycle can be inhomogeneous so that, say, fraction $\mu$ ($\leq 1/2$) of the tube length is stretched up to the length L, as well as the remaining part $(1-\mu)L$ of the tube length. Then, after folding, the magnetic field is non-uniformly distributed across the tube: it is $\mu^{-1}H$ over the area $\mu S$ and $(1-\mu)^{-1}H$ over the remaining part. This transformation can be represented analytically as a generalized baker's map

$$x_{n+1} = \begin{cases} \mu x_n, y_n < \mu \\ \frac{1}{1-\mu} x_n + \mu, \ y_n > \mu \end{cases}$$

$$y_{n+1} = \begin{cases} y_n/\mu, \ y_n < \mu \\ (y_n-\mu)/(1-\mu), \ y_n > \mu \end{cases} \qquad (1)$$

$$H = H_y(x)$$

The rate of the exponential growth (the Lyapunov exponent) is given by

$$\gamma = \left[\mu \ln(1/\mu) + (1-\mu)\ln 1/(1-\mu)\right] T^{-1} \qquad (2)$$

and for $\mu = 1/2$ ZELDOVICH's result is recovered. Notice that the growing field is highly inhomogeneous and the most rapidly growing field concentraties with time on a

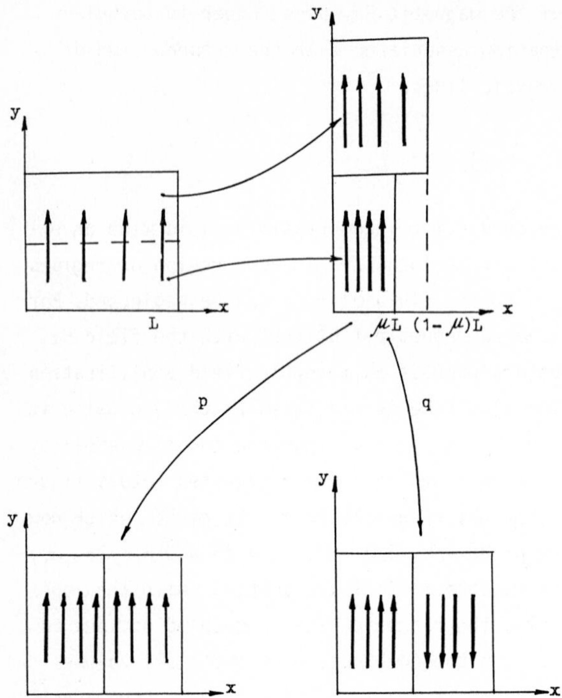


Figure 1. The discrete two-dimensional map with cutting of magnetic lines which illustrates the process in which magnetic flux is amplified with probability p and annihilates with probability q = 1-p.

fractal manifold of infinitely small measure. This fact reflects a real process of structuring and tangling of the magnetic field and necessitates the introduction of magnetic diffusion into (1), which can smear inhomogeneities of magnetic field with excessively small scales.

Note also that after stretching the tube can be folded with and without twisting. In the former case (the Zeldovich dynamo) the magnetic is doubled while in the latter case it becomes zero. Figure 1 illustrates this three-dimensional process with the help of a discrete map with cutting the magnetic lines. The magnetic lines are directed along the y-axis on the plane while their density varies along the x-axis. Thus, we can illustrate the three-dimensional process of magnetic field generation by a one-dimensional model (dependence on x) while the second coordinate y is required to account for the vectorial nature of magnetic fields.

Thus, the Zeldovich dynamo can be described in a surprisingly simple way in terms of maps while insurmountable difficulties prevent such a description in terms of magnetohydrodynamic equations. Of course, the results implied by dynamo maps are less rigorous than those following from direct analysis of MHD equations since the correspondence between the dynamo maps and the exact equations remains a problem. However, in applications, e.g. to astrophysics, we are not aware of all the details of fluid flows which can be drastically different from the turbulent laboratory flows. In such a situation we are forced to employ more or less reasonable models of turbulence and address a no less complicated problem of their adequacy; therefore, the rigourness of the direct analysis of MHD equations may be an illusion.

## 3. DYNAMO MAP WITH ENHANCED DISSIPATION

In this section we generalize the Zeldovich dynamo map in order to include a realistic effect of reconnection of oppositely directed magnetic lines passed together by a fluid flow. We recall that here we consider a weak magnetic field which does not produce noticeable effect on the flow. Within the framework of this approximation, annihilation of magnetic lines proceeds very rapidly, at the rate proportional to the magnetic Reynolds number /2/.

Thus, let us consider the flow in which during period T the magnetic flux is doubled with the same probability p and is annihilated with probability $q = 1-p$. This means that the magnetic ropes are twisted to figure eight shape with probability p and the ropes are folded without twisting with probability $q = 1-p$. After n cycles the magnetic field strength in an individual rope is given by

$$H_n = H_0 2^n \theta_1 \theta_2 \dots \theta_n , \tag{3}$$

where the independent random quantities $\theta_i$ are unities with probability p and zero with probability q. It is clear that for $p < 1$ every realization of the field $H_n$ (i.e. every particular product $\Pi\theta_i$) earlier or later becomes zero. Only one realization is exceptional which has $\theta_i = 1$ for any i, and the probability of this realization decreases with time. The average value of the magnetic field strength is $\langle|H_n|\rangle = 2^n p^n$ and the other statistical moments are given by $\langle|H_n|^k\rangle = 2^{nk} p^n$. Thereby, the growth rate of the k-th moment is

$$\gamma = \lim_{n\to\infty} \frac{\ln\langle|H_n|^k\rangle}{kn} = \ln 2 - \frac{|\ln p|}{k} . \tag{4}$$

Higher moments of magnetic field grow at greater rates than lower ones while a typical realization is decaying. Clearly enough, such a situation cannot be realized for a smooth overall distribution of the magnetic field and it consists of individual, widely separated peaks. With the growth of n these peaks occupy ever smaller and smaller volume. Roughly speaking, with every cycle the magnetic energy stored in the peaks is doubled. The available magnetic energy grows in expense of the kinetic energy. Fraction p of this energy preserves its magnetic form and survives to the subsequent cycle of evolution while the fraction q transforms into heat due to rapid dissipation.

In the limit $n \to \infty$ the magnetic field disappears at any given position (for $p < 1$) but a region the volume of which grows with time as $c^n$, with suitably chosen constant c, always contains a large number of magnetic peaks. To verify this, we consider this volume as a unit segment. Then the average magnetic flux is given by $c^n 2^n p^n$. Since the field strength within peaks is equal to $2^n$ and it determines the total flux even if occupying a small fraction of volume, the magnetic fluxes obtained by averaging over ensemble and over volume coincide if

$$c^n 2^n p^n = (c^n 2^n)^d ,$$

where d is the fractal (Hausdorff) dimension of the region occupied by the peaks. This yields

$$d = 1 - (\ln 1/p)/\ln(2c) .$$

For $c < (2p)^{-1}$ the Hausdorff dimension vanishes, i.e. the peaks eventually disappear in the considered volume. For $c \to \infty$ the dimension tends to unity which means that the considered volume increases so quickly that for a remote observer who observes the region with linear resolution C, the magnetic fields merge into a smooth averaged distribution. For $p = 1$ we have $d = 1$, i.e. for any value of C the field grows everywhere (all over the segment) because a typical realization becomes growing. Properties of intermittent random fields are discussed in more detail /3/.

When these results are applied to specific real dynamo systems, one can assume that at the nonlinear stage of the field evolution the rate of magnetic reconnection diminishes while a seed magnetic field is constantly reproduced by an external source. In this case the limiting distribution of the magnetic field takes the form of a widely spaced high concentration and the number n can be chosen in such a way that after n cycles the seed field grows up to the level when the influence of magnetic field on the flow becomes essential. The resulting magnetic concentrations have a relatively long lifetime because their decay occurs in a nonlinear regime. Such nonlinear models are discussed in /4/.

The considered model of magnetic field evolution does not include all physical effects expected in real MHD systems. However, it can be made slightly more complicated while preserving tractability. For instance, instead of random numbers $\theta_i$ which are either a unity or zero, one can consider random numbers $\tilde{\theta}_i$ equal to $1-\delta$ with probability p and $\delta$ with probability $q = 1-p$. This model accounts for the fact that annihilating magnetic lines are not exactly antiparallel. In this case a typical realization of magnetic field can be growing. The growth rate can be shown to be

$$\gamma = \ln 2 + p \ln(1-\delta) + q \ln \delta .$$

Furthermore, the stretching factor of flux loops can differ from two and the number of foldings can also be larger. At those positions where stronger stretchings prevail, thin long tubes with very strong magnetic field arise. Meanwhile, moderate magnetic field concentrated to very small regions arise at positions where foldings are unusually numerous. In both cases a weak magnetic diffusion leads to the formation of a skin-layer and restricts amplification of magnetic field and magnetic flux. Only in those cases when the stretching factor coincides with or is close to the folding factor occurs a considerable - formally infinite - amplification associated with the exact reproduction of the shape of the magnetic tube. These effects lead to even more pronounced intermittency of magnetic field.

## 4. THE STRUCTURE OF MAGNETIC FIELD

The dynamo maps are not well suited for the analysis of the spatial structure of magnetic concentrations. A direct analysis of MHD equations provides a more useful tool in this case, and this can be accomplished for a short-correlated random flow. Such an analysis also reveals two components in the space-time distribution of the magnetic field, viz. sparse regions with strong magnetic field (concentrations) and relatively weak extended background. Both components grow with time but the growth rates are different. The concentrations are actually thin widely spaced tubes of a large magnetic flux. They make a dominant contribution into the average magnetic energy and higher statistical moments of magnetic field /5/. Analysis of the second moment (the correlation function) sheds light on the geometric structure of the magnetic concentrations. The most easily excited and the most rapidly growing concentrations are simple magnetic loops of the size close to the turbulent correlation scale $l$ and thickness $lR_m^{-1/2}$, where $R_m$ is the magnetic Reynolds number. For larger magnetic Reynolds number appear helical loops of the kind of a winding of a 2-torus in which helical tubes of the diameter $lR_m^{-1/4}$ have closed or curved axes with typical scale $l$, and even more complicated configurations /6/.

The presence of intermittency in the magnetic field distribution is proved by the fact that the growth rates of the successive averages,

$$\gamma_k = \lim_{t\to\infty} \ln\langle|H|^k\rangle/(kt) \ ,$$

progressively grow with the order $k$ of the considered statistical moment (Fig.2). The limiting value $\gamma \equiv \gamma_k$ corresponds to the growth rate of a typical (unaveraged) realization of the magnetic field. Notably enough, the latter quantity is smaller than the growth rates of all moments, i.e. a typical realization of the magnetic field grows at smaller rate and is excited at larger $R_m$. Thereby, for moderate $R_m$ a typical realization of magnetic field can decay while higher moments, e.g. the magnetic energy, grow.

A typical realization corresponds, with a very high probability, to the background magnetic field (concentrations are rare events). This background cannot be described through ensemble averaging. First, the concentration should be filtered out. The easiest way to do this is through the consideration of the directions of the magnetic field,

$$l_i(t,x) = H_i(t,x)/|H(t,x)| \ .$$

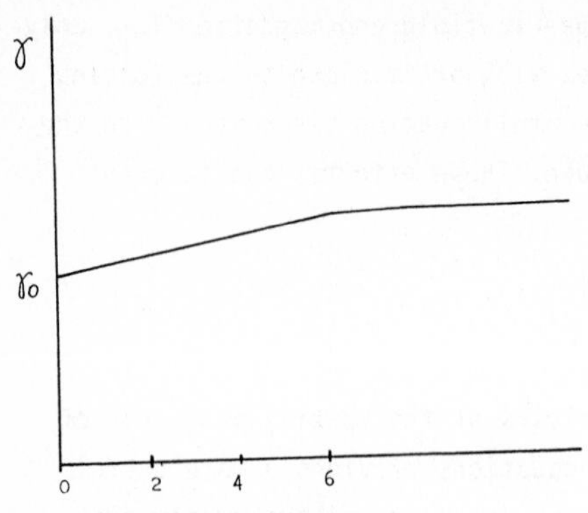


Figure 2. Qualitative dependence of the growth rate of a statistical moment of magnetic field on the order of the moment in the limit of large magnetic Reynolds number. The dependence for moderate orders can be evaluated explicitly in the short-correlation approximation /4/. For small $R_m$ the growth rate is negative, i.e. only sufficiently high-order statistical moments are growing.

This normalization makes the contribution of the concentrations and the background equal. Studies of the correlation function of the directions of the magnetic field show that the background magnetic field also consists of magnetic loops which, however, are more closely spaced and have a higher number of additional helical woundings of the type of ($\ln R_m$)-dimensional torus /7/. This implies that at arbitrary chosen position one would probably see a most complicated and tangled magnetic tube. This is the higher tangling of the background magnetic field which explains why it grows at smaller rate than stronger concentrations (quite similarly, the larger the number of zeros of an eigenfunction is, the higher is the corresponding energy level).

## REFERENCES

1. J.Finn, E.Ott. Chaotic flows and fast magnetic dynamo. Phys.Rev.Lett., 1988, 60, 760-763.
2. Ya.B. Zeldovich, S.A.Molchanov, A.A.Ruzmaikin, D.D.Sokoloff. Kinematic dynamo in the linear velocity field. J.Fluid Mech., 1984, 144, 1-11.
3. Ya.B.Zeldovich, S.A.Molchanov, A.A.Ruzmaikin, D.D.Sokoloff. Intermittency in a random medium. Sov.Phys.-Usp., 1987, 152, 1, 3-32.
4. Ya.B.Zeldovich, S.A.Molchanov, A.A.Ruzmaikin, D.D.Sokoloff. Self-excitation of a nonlinear scalar field in a random medium. Proc.Nat.Acad.Sci. USA, 1987, 84, 6323-6325.
5. S.A.Molchanov, A.A.Ruzmaikin, D.D.Sokoloff. A dynamo theorem. GAFD, 1984, 30, 241-259.
6. N.I.Kleeorin, A.A.Ruzmaikin, D.D.Sokoloff. Correlation properties of self-exciting fluctuative magnetic fields. Plasma astrophysics, ESA SP-251, Paris, 1986, 557-561.
7. R.F.Galeyeva, A.A.Ruzmaikin, D.D.Sokoloff. Typical realization of magnetic field in a random flow. Magnitnaya gydrodynamika, 1989, 4, 17-21 (in Russian).
8. S.A.Molchanov, A.A.Ruzmaikin, D.D.Sokoloff. Short-correlated random flow as the fast dynamo. Sov.Phys.-Dokl., 1987, 295, 3, 576-579.

# The Large Scale Structure of the Universe, the Burgers Equation and Cellular Structures

S. Gurbatov[1], A. Saichev[1], and S. Shandarin[2]

[1]Gorky State University, Prospect Gagarina 23, 603600 Gorky, USSR
[2]Institute of Physical Problems, Kosygina 2, 117334 Moscow, USSR

The problem of the large scale structure formation in the Universe is one of the most important ones in modern cosmology. It is closely connected to the origin of the primeval perturbations presumably at the inflationary stage and the nature of the so called Dark Matter. The most advanced theory to explain the formation of the large scale structure is based on the theory of gravitational instability in the expanding Universe. The nonlinear stage of the gravitational instability is shown to be approximately described in the frame of the "sticking" model which is based on the Burgers equation of the nonlinear diffusion.

The appearance of ordered structures is possible in nonlinear media without dispersion. These structures represent regions (domains, cells) with regular behaviour alternating with randomly located zones of dissipation. In the present work we consider the formation and evolution of such structures using the simplest equation of the theory of nonlinear waves - the Burgers equation and its generalization to the three-dimensional case. The basic properties of arising stochastic regime which is an example of strong turbulence, are discussed. The self-preserving nature of this regime is demonstrated.

The model three-dimensional Burgers equation is used to describe the nonlinear stage of gravitational instability in the expanding Universe. Together with the equation of continuity it provides an approximate description of the formation of nonlinear structures (pancakes, filaments and clumps of mass) as well as merging of the clumps. It is shown that at the late nonlinear stages the statistical characteristics of both density and velocity fields are determined by the statistical characteristics of gravitational potential fluctuations taken at the linear stage of the gravitational instability.

## 1. INTRODUCTION

The large scale structure of the Universe represents the inhomogeneities in the galaxy distribution on the scales roughly from 1 Mpc $\approx 3\cdot 10^{24}$cm to 100 Mpc. The galaxies concentrate around surfaces, lines or points forming "pancakes", filaments and clusters, respectively. There are giant voids of galaxies where bright galaxies have not been found /1/. It is generally assumed that the galaxy distribution on these large scales is similar to the mass distribution. The problem of the origin and evolution of the large scale structure is one of the fundamental issues of current cosmology. It is closely related to the problem of dark matter dominated in the mean density of the Universe as well as the problem of the origin of the primordial perturbations in the very early Universe.

In the modern cosmology there are few models proposed to explain the existence of the large scale structure. The most advanced one is based on the hypothesis of the

Research Reports in Physics Nonlinear Waves 3
Editors: A.V. Gaponov-Grekhov · M.I. Rabinovich · J. Engelbrecht

gravitational instability /2,3/. According to the gravitational instability model, both the galaxies and the large scale structure itself have been formed from small density fluctuations originated as quantum fluctuations during the very early stages when the Universe expanded exponentially. Let us remind that at present the Universe is likely expanding according to the power law: $a(t) \propto t^{2/3}$, where a(t) is the so-called scale factor characterizing the overall size of the Universe. At the later stages the density perturbation grew in the linear regime for a long time. Finally they reached the nonlinear stage ($\delta\rho/\rho \geq 1$). ZELDOVICH has shown that if the spectrum of the linear density perturbations satisfies the definite conditions, the first non-linear objects have shapes similar to pancakes /4/.

The direct N-body simulations performed in 2D and 3D systems of collisionless particles interacting only due to the gravity have confirmed the result by ZELDOVICH /5,6/. The model of the collisionless particles was chosen because most of the mass in the Universe is believed to exist in the form of hypothetical weakly interacting particles having nonzero mass like massive neutrinos or axions. The following investigation of the nonlinear gravitational instability by means of N-body simulations of various kinds showed that the formation of the first pancakes was followed by the formation of the cellular structure illustrated by Figs.1 and 2 taken from 2D simulations /7/. These figures show the particle distributions taken at the beginning of the nonlinear stage when

$$<(\delta\rho/\rho)^2>^{1/2}_{\text{lin.theor.}} \cong 1$$

(Fig.1) and at the much later stage when

$$<(\delta\rho/\rho)^2>^{1/2}_{\text{lin.theor.}} \cong 16$$

(Fig.2). One can easily see cellular structures in both figures. However, the typical sizes of the cells are clearly different.

Another feature to note the relative thinness of the "walls" separating the regions of low density of particles. The latter has become the reason to suggest the "sticking" model ( or the adhesion model) for the large scale structure formation /8-10/.

According to this model the motion of the selfgravitating collisionless matter in the expanding Universe can be approximately described as the motion of the cold ($T = 0$) viscous gas in common (non-expanding) space. Assuming that the coefficient of viscosity $\nu$ is infinitesimal, one can describe the motion of the gas as the motion of a pressureless sticky inert medium. Every particle of the medium moves with constant speed until it runs into another particle. Afterwards they stick and move together with the velocity corresponding to the conservation of the momentum. The comparison of this simple model with the direct 2D N-body simulation of the gravitational instability in the collisionless medium will be discussed.

Figure 1.

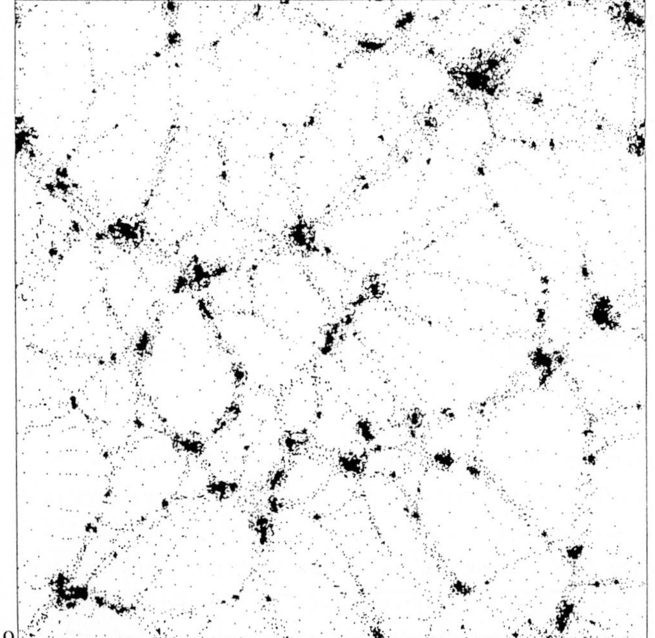


Figure 2.

The sticking model is formally based on the solution of the Burgers equation which will be discussed in detail in the following Section. Concerning its application to cosmology, it is worth stressing that it has been suggested for a quite restrictive purpose to *construct the "skeleton"* of the large scale structure. At the present form the sticking model *is not assumed* to be used for the description of the internal structure of the high density regions.

## 2. RANDOM NONLINEAR WAVES AND ONE-DIMENSIONAL TURBULENCE IN NONDISPERSIVE MEDIA

The extreme difficulty of analyzing nonlinear waves has given rise to another trend in the development of the theory of such waves: the transition from complicated equations of nonlinear random waves to a simpler model equation. One of such equations of strong turbulence (the turbulence with strong interaction of large number of coherent harmonic waves) is the Burgers equation /11/:

$$\partial u/\partial t + u\partial u/\partial x = \nu\partial^2 u/\partial x^2 \ , \qquad u(x,t) = u_0(x), \tag{1}$$

where $\nu$ is the dissipation coefficient, $u_0(x)$ is a random function if the initial field has a noise character. This equation describes two fundamental effects characteristic of any turbulence - the nonlinear transfer of energy over the spectrum and the damping of energy in the region of small scales. Equation (1) was proposed by BURGERS as a model of hydrodynamic turbulence. It was later shown that the description of one-dimensional acoustic waves is reduced to Eq.(1). The Burgers equation has found widespread application in nonlinear acoustics, in particular for the description of intense acoustic noise.

The evolution of the field u(x,t) is completely determined by the initial condi-io tions. Let $u_0$ and $\ell_0$ be the characteristic amplitude and spatial scale of the initial field. Then the dimensionless parameter $R_0 = u_0\ell_0/\nu$ (the Reynolds number) will characterize the relative influence of nonlinear and dissipative effects on the evolution of the field.

We assume that the initial noise spectrum has the form

$$g_0(k) = (1/2\pi)\int\langle u_0(x)u_0(x+\rho)\rangle \exp(ik\rho)d\rho = \alpha_n{}^2 k^n b_0(k). \tag{2}$$

Here $b_0(k)$ is a sufficiently rapidly decreasing function as $k \to \infty$ with the characteristic scale $k_* \cong 1/\ell_0$ and $b_0(0) \neq 0$. We shall below investigate how the asymptotic behaviour of the field depends on the exponent n in (2).

We consider the limiting case $R_0 >> 1$, and we shall show that the resultant nonlinear structure is stable at $n < 1$, i.e. it never departs from the linear regime. As $\nu \to 0$, the solution of the Burgers equation is rewritten in the form /11-13/

$$u(x,t) = (x-y(x,t))/t,$$

where y(x,t) is the coordinate of the absolute minimum of the function

$$s(x,y,t) = s_0(y) + (x-y)^2/2t; \quad s_0(y) = \int^y u_0(y)dy \,. \tag{3}$$

Here $s_0(y)$ is known as the action of the initial field. At $t \gg t_n = \ell_0/u_0$, function $y(x,t)$ is a discontinuous, piecewise constant function of x, while field u(x,t) transforms into a series of sawtooth pulses with the same slope $u_x' = 1/t$ /11-13/. Because of the merging of the discontinuities the characteristic distance $\ell(t)$ between them (external scale of the random sawtooth wave) increases. As an estimate for the external scale $\ell(t) \cong |x-y|$ we can take the condition that the parabola in (6) and the value of increment of the initial action are of the same order, i.e.

$$d_s(\ell(t))^{1/2} \cong \ell^2(t)/t$$

$$d_s(\rho) = \langle (s_0(x+\rho) - s_0(x))^2 \rangle \,,$$

where $d_s(\rho)$ is the structural function of the initial action. If $n > 1$, then the dispersion of $s_0(y)$ is limited and $d_s(\rho\to\infty) = 2\sigma_s^2$. If $n < 1$, then the structural function increases according to the power law $d_s(\rho) \approx \alpha_n^2 \rho^{1-n}$. Respectively, we have for the external scale $\ell(t)$ in these cases

$$\ell(t) = (\sigma_s t)^{1/2}, \qquad n > 1$$

$$\ell(t) = (\alpha_n t)^{2/(3+n)}, \qquad n < 1 \,. \tag{4}$$

At finite viscosity ($\nu \neq 0$) the width of the shock front $\delta \cong \nu t/\ell(t)$ will increase. For the effective Reynolds number we have here

$$R(t) = \ell(t)/\delta(t) = \ell^2/\nu t \cong \begin{cases} R_0 = \text{const}\,, & n < 1 \\ t^{(1-n)/(n+3)}, & n > 1\,. \end{cases}$$

We note that for the periodic signal, if the discontinuities were not to coalesce, then the Reynolds number would decrease according to the law $R(t) = \ell_0^2/\nu t$.

Thus, if $n < 1$, then the effective Reynolds number increases even in the nonlinear stage, i.e. the regime of strongly nonlinear sawtooth waves turns out to be structurally stable and never departs from the linear damping regime.

It is strongly shown in /4,5/ using the exact Hopf-Cole solution of (1) that in case $n < 1$ the establishment of a strongly nonlinear sawtooth regime actually takes place regardless of the value of the initial Reynolds number. In this case the correlation function and the energy spectrum are asymptotically self-similar and can be represented in the form of

$$B(\rho,t) = \frac{\ell^2(t)}{t^2} \tilde{R}(\rho/\ell(t)); \quad g(k,t) = \frac{\ell^3(t)}{t^2} \tilde{g}(k\ell(t)) \,, \tag{5}$$

where $\ell(t)$ is determined by (4). In the region of large scale ($k\ell \ll 1$) the nonlinearity and the dissipation cannot change the form of the spectrum at $n < 1$, and $g(k,t) = \alpha_n^2 k^n b_0(0)$.

The appearance of the shocks leads to a universal, shortwave asymptotics of the energy spectrum of the field:

$$g(k,t) \cong k^{-2}, \quad \text{if } k\ell(t) >> 1.$$

For the acoustic noise the case $n \geqq 2$ is more typical. The most rigorous estimate shows that the effective Reynolds number decreases but because of the effect of the merging of the shocks, this increase is logarithmically slow /13,14/.

$$R(t) = \ell^2/\nu t = R_0/(\ell n\ t/t_n)^{1/2} . \tag{6}$$

In this case, if $R_0 >> 1$, all the statistical characteristics of the field $u(x,t)$ are self-similar over the interval $t_n < t < t_1$. Here $t_n \cong \ell_0/u_0$ is the characteristic time of formation of sawtooth structure, while $t_1$ is the time emergence into the linear regime and can be estimated from the condition $R(t_1) \cong 1$ (6). Because of the effect of the merging of the shocks, this time is extremely large: $t_1 \cong t_n \exp(R_0)$. We note that for the periodic signal $t_1 \cong t_n R_0$.

For $n \geqq 2$ and $R_0 >> 1$ the one- and two-point probability densities, correlation function and energy spectrum are found on the basis of the solution (2), (3) /13, 14/. They are also self-similar and depend only on the external scale

$$\ell(t) = \ell_0(t/t_n)/(\ell n\ t/t_n)^{1/4} ,$$

i.e. on the average distance between the shocks. The energy spectrum $g(k,t)$ has a universal behaviour:

$$g(k,t) \cong k^2 t^{1/2} \quad \text{if } k\ell << 1; \qquad g(k,t) \cong k^{-2} t^{-3/2} \quad \text{if } k\ell >> 1 .$$

According to (5), the maximum of the spectrum is displaced in the direction of small wave numbers proportional to $1/\ell(t)$. Physically, this picture of the evolution of the spectrum is due to the following: the appearance of discontinuities (shocks) in the wave leads to the dissipation of the energy of the wave and to the appearance of slowly decaying components of the spectrum in the region of large wave number. The growth of the low-frequency components is connected with the growth of the external scale because of the merging of the shocks.

The initial stage of appearance of self-similar regime was observed in the experiments connected with propagation of intense acoustic noise in the tube /18,19/.

## 3. THE FORMATION OF STRUCTURES IN GRAVITATIONALLY UNSTABLE MEDIUM

In this section the three-dimensional Burgers equation is used to describe the nonlinear stage of gravitational instability. We discuss only the evolution of mass density inhomogeneities leaving aside a very complicated process of galaxy formation itself. Neither de we intend to study the internal structure of large-scale inhomogeneities. Instead of that our aim is to consider the global geometry of the Large Scale Structure (LSS) and some of its statistical properties, such as typical masses of the clusters and the typical size of cells. The suggested model is a generalization of a very well known Zeldovich approximate solution /4/ describing the beginning of the nonlinear stage of gravitation instability in the expanding Universe.

a) Basic Equations

In case of the expanding Universe it is useful to introduce peculiar velocity

$$\vec{V} = \vec{u} - \vec{r}\dot{a}/a = a\, dx/dt , \tag{7}$$

where $\vec{u}$ is the physical velocity, $a = a(t)$ is the scale factor describing the mean (Hubble) expansion of the Universe and $\vec{X} = \vec{r}/a$ are coordinates comoving to the Hubble expansion of the Universe. By using these quantities, one excludes insignificant effects of the general expansion. In these variables the equation of density $\rho(\vec{x},t)$ and velocity inhomogeneities are as follows /2/

$$\partial\vec{V}/\partial t + (\vec{V}\nabla_x)\vec{V}/a + \vec{V}\dot{a}/a = -\nabla_x\psi/a \tag{8}$$

$$\Delta\psi = 4\pi G a^2(\rho(x,t) - \overline{\rho})$$

$$\partial\rho/\partial t + 3\rho\dot{a}/a + \nabla_x(\rho\vec{V})/a = 0 ,$$

where $\psi$ is the perturbation of gravitational potential; $\overline{\rho}(t)$ is the mean density of the Universe.

In the linear approximation describing the evolution of small density fluctuations, a system of equations (8) has two solutions: one increasing and the other decreasing with time. Both modes are dispersion free which means that perturbation to every scale grows with the same rate. In a linear approximation density perturbation grows in the increasing mode as $\delta = (\rho-\overline{\rho})/\overline{\rho} \approx b(t)$, where $b(t)$ is the growing solution of equation $a\ddot{b} + 2\dot{a}\dot{b} + 3\ddot{a}b = 0$. In particular, in Einstein-de Sitter Universe with $\Lambda = 0$ and $\Omega = 1$ both $a(t) \cong t^{2/3}$ and increasing mode $b(t) \cong t^{2/3}$ (here $\Lambda$ is a cosmological term and $\Omega$ is a dimensionless mean density of the Universe: at $\Omega < 1$ the Universe is open and expands forever while at $\Omega > 1$ it is closed and at some time it stops to expand, it will collapse).

Hereafter we shall discuss the increasing mode only as the LSS arise due to the growth of density perturbations.

The Zeldovich approximate solution (mentioned above) /11/ relates the Eulerian coordinate $x_i$ of every particle at time t with the Lagrangian on $\vec{q}$

$$x_i = a_i + b(t)s_i(\vec{q}) , \tag{9}$$

where $\vec{s} = \nabla\Phi_0$ is a potential vector field defined by density perturbations at the linear stage. It is easy to find the relations between $\Phi_0(\vec{q})$ and $\delta \equiv (\rho-\bar{\rho})/\bar{\rho}$ and in the linear regime

$$\delta = -b\nabla^2\Phi_0, \quad \psi = 3\ddot{a}ab\Phi_0 \,. \tag{10}$$

The Zeldovich approximation assumes that solution (9) is also correct at the beginning of the nonlinear stage when $\delta > 1$. One can find a detailed discussion of the validity of the Zeldovich approximation in reviwes /4,14/. Let us note only that the Zeldovich solution (9)-(10) is exact in case of one-dimensional perturbations until the particle trajectories interact.

In the framework of the Zeldovich solution density is calculated as Jacobian of the transformation to Eulerian from the Lagrangian coordinates

$$\eta \equiv \rho a^3 = \eta_0/|\partial x_i/\partial q_j| \,. \tag{11}$$

Using (4), one easily finds

$$\eta = \eta_0/|\delta_{ik}+bd_{ik}| \,, \qquad d_{ik} = \partial^2\Phi_0/\partial q_i\partial q_j \,. \tag{12}$$

Approximation (9) also permits to calculate the acceleration of every particle

$$d\vec{V}/dt = (\dot{a}/a + \ddot{b}/\dot{b})\vec{V} \tag{13}$$

Since solution (9) is approximately selfconsistent, equation (13) can exclude gravitational potential $\psi$ from system (8) and reduce the number of equations. Substituring new variables $b$, $\eta$, $\vec{v}$

$$\eta = \rho a^3 \,, \quad \vec{v} = \vec{V}/a\dot{b} \tag{14}$$

in (8), (13) instead of $t$, $\rho$ and $V$, one gets

$$\partial\vec{v}/\partial b + (\vec{v}\nabla_x)\vec{v} = 0, \quad \partial\eta/\partial b + \nabla_x(\eta\vec{v}) = 0 \tag{15}$$

Thus in these variables the first equation describes "inertial" motion of matter, and (9) and (12) are the solution of (15) in terms of Lagrangian coordinates.

Solutions (9), (12) predict the formation of highly anisotropic concentrations of mass, i.e. pancakes /3-6/. However, the comparison of this approximation with numerical simulations has shown that it predicts too fast growth of the thickness of pancakes /5,6,14/. In reality, the thickness of pancakes quickly becomes approximately stable because of the action of gravity. The particles which run into pancakes begin oscillating about the central plane. At the same time they keep moving along the plane which results in formation of filaments and knots in the density distribution. As we do not intend to describe the mass distribution inside the concentrations, we shall roughly describe this effect by adding a viscosity term into the right hand side of the first equation (15). Thus, instead of (15) we suggest using three-dimensional Burgers equation which in our notations is as follows /8,10,15/:

$$\partial\vec{v}/\partial b + (\vec{v}\nabla)\vec{v} = \nu\,\Delta\vec{v} \tag{16}$$

b) Solution of the basic equations and formation of structures

It is well known that one-dimensional Burgers equation has exact analytic solution. This permits us to investigate the velocity fields satisfying this equation from the statistical point of view in many cases /1-4/. Similar analysis is also possible in the three-dimensional case. For potential velocity fields the exact solution of equation (16) is as follows (here we assume that $\nu \to 0$)

$$\vec{V} = (\vec{x}-\vec{q}(\vec{x},b))/b \ , \tag{17}$$

where q are the coordinates of the absolute minima of function

$$G(\vec{x},\vec{q},b) = \Phi_0(\vec{q}) + (\vec{x}-\vec{q})^2/2b \ . \tag{18}$$

Here $\Phi_0$ is the potential of the initial velocity field: $\nu \equiv s = \nabla\Phi_0$. It is evident that $\vec{q}$ is one of the roots of equation (16) and therefore $\vec{q}(\vec{x},b)$ are the Lagrangian coordinates of the particles which came into point $\vec{x}$ at "time" b. At the linear stage when solution (9) is the only one, (17) is the evident solution of (15) and the velocity field is continuous. However, later on when equation (16) has several roots, solution (17) "selects" only those particles which have not yet run into singularities. At this stage the velocity field becomes discontinuous.

There is a very convenient graphic technique describing solution (17). One can easily see that the coordinates of the absolute minimum of $G(\vec{x},\vec{q},b)$ coincide with the coordinates of the point where paraboloid $P(\vec{x},\vec{q},b) = -(\vec{x}-\vec{q})^2/2b + h$ is tangential to the hypersurface of $\Phi_0(\vec{q})$ at the smallest h. Of course, the type touching P and $\Phi_0$ depends on the relation between the curvature of P (which is of order of b) and that of $\Phi_0$ (which is of $b_0 \approx \ell_0/\sigma_0$ where $\ell_0$ and $\sigma_0$ are the typical scale and the amplitude of the initial velocity field).

Let us proceed to the final conclusions. At $b \gg b_0$ coordinate of absolute minimum $\vec{q}$ becomes a step-like function of $\vec{x}$: $\vec{q}(\vec{x},b)$ is practically constant in separate regions $\Sigma_i$ with discontinuities on their borders. According to equation (17), the velocity field inside every region (cell) has a universal structure.

$$\vec{v} = (\vec{x}-\vec{q}_i)/b, \qquad x \in \Sigma_i$$

where $\vec{q}_i$ are coordinates of deep minima of $\Phi_0(\vec{q})$. If $\Phi_0(\vec{q})$ is a random function, then these cells are approximately polyhedrons randomly distributed in space. And as b grows, so do the cells with the deeper minima of $\Phi_0$. Figure 1 shows the typical cellular structure of the velocity field in the two-dimensional case.

However, our main aim is to study the distribution of density. Inside the cells one can calculate the density using equation (12), as the particles inside the cells have not experienced collisions. To find the density inside the borders, ribs and apices of the cells, let us assume that at finite $\nu$ function

$$q_i(\vec{x},b) = x_i - v_i(\vec{x},b)b \tag{19}$$

is still a Lagrangian coordinate, and the density of matter is determined from equation (12). This definition of density results in additional term in the equation of continuity. For example, in one-dimensional case the system of basic equations becomes

$$\partial \upsilon/\partial b + \upsilon \partial \upsilon/\partial x = \nu \partial^2 \upsilon/\partial x^2 ; \quad \partial \eta/\partial b + \partial(\eta \upsilon)/\partial x = \nu \partial^2 \eta/\partial x^2 . \tag{20}$$

The appearance of the diffusion term is justified by the conservation of momentum.

One can infer from (9) and (19) that in the course of time density distribution is changing in the cellular structure: matter concentrates into borders of cells with even higher concentration into ribs (or filaments). The highest concentration of mass is in apieces (or knots). Finally, all the mass practically concentrates in knots. The The density inside cells decreases as $\eta \approx b^{-3} \, \ell n \, b^{-3/2}$ /15/.

Typical mass of the clumps in the knots of the cellular structure grows due to their merging. This describes the hierarchical clustering which is well known in cosmology /9,10,12/. Assuming the large scale asymptote of the initial density perturbation spectrum to be a power law $\delta_k{}^2 \sim k^n$, one can easily find the asymptote of the potential spectrum $\Delta_k{}^2 \approx k^{-4}\delta_k{}^2 \approx k^{n-4}$.

Let us also assume that the spectrum has a short wave cut off at $k > \ell_0{}^{-1}$. Then the typical size of the cell $\ell(b)$ can be estimated from the condition that the absolute minimum in (17) and (18) is in the region where the increments of both P and $\Phi_0$ are of the same order. This results in the following equation for (b)

$$\ell^2(b) \approx 2b\{D[\ell(b)]\}^{1/2} , \tag{21}$$

$$D(x) = \langle(\Phi_0(x) - \Phi_0(0))^2\rangle .$$

As it follows from (19), the increase of $\ell$ depends on the dispersion of $\Phi_0$. There is a critical index $n_\star$ when the growth of $\ell$ changes qualitatively

$$\ell(b) \sim \begin{cases} \ell_0(b/b_0)^{1/2} , & n > 1 \\ b^{2/(n+3)} , & -1 < n < 1 . \end{cases} \tag{22}$$

Using simple relation $M_\ell \sim \eta\ell^3$, one can estimate the typical mass of the clumps.

Finally, let us briefly discuss the comparison of the model under consideration with simple one-dimensional numerical simulations of gravitational instability /16/. If $\delta_k{}^2(k,b)$ and $\upsilon_k{}^2(k,b)$ are the spectra of density perturbations and velocity field, respectively, then equation (20) gives rise to the following relation /17/

$$\delta_k{}^2(k,b) = k^2 b^2 \upsilon_k{}^2(k,b) . \tag{23}$$

Let us restrict ourselves to the case when $\langle\Phi_0\rangle$ is less than infinity and $n > 4$. Using the known results about spectrum $\upsilon_k{}^2$ /13,14/, one can find from (23) that nonlinear effects result in the formation of the universal spectrum $\delta_k{}^2 \sim k^4$ growing at the linear stage ($b \ll b_0$) as $\delta_k{}^2 \sim b^4$ and at the nonlinear stage ($b \gg b_0$) as $\delta_k{}^2 \sim b^{5/2}$. In addition, at $b \gg b_0$ the spectrum of density perturbations acquires a selfsimilar character and depends on the single scale $\ell(b) \sim b^{1/2}$. These conclusions are in a good agreement with the results of numerical simulations modelling the gravitational instability in collisionless medium /16/. The only difference is observed in the shortwave part of the spectrum: the theory predicts $\delta_k^2 \sim$ const at $k > \ell^{-1}$ and the numerical simulations give $\delta_k^2 \sim k^{-1}$. This is explained by the poor description of the internal structure of the mass concentrations in collisionless medium by the Burgers equation.

c) The comparison of the sticking model with N-body simulations

In this section we shall briefly discuss the results of the comparison of the sticking model (at $\nu \to 0$) with the direct N-body simulations performed in 2D case /20/.

The N-body simulations using the particle-mesh code were performed with 64x64 particles on equivalent mesh with periodic boundary conditions. The initial perturbation was a random Gaussian firld with the spectrum of density perturbations proportional to $k^2$ with sharp cutoffs at the longest possible wave corresponding to the size of the region and at $\Lambda_{min} = \Lambda_{max}/8$. The results of the simulations are illustrated in Figs. 3a, 3b and 3c. At the stage when $\langle(\delta\rho/\rho)^2\rangle^{1/2}_{\text{lin.theor.}} \approx 6$, three pictures are shown: Fig.3a illustrates the particle distribution calculated in the N-body simulation; Fig.3b shows the skeleton of the structure constructed by means of a special geometrical technique solving the Burgers equation (at $\nu = 0$) with the same initial conditions as in the N-body simulation (large points show the positions of the pancakes and circles show the positions of the clumps of mass); Fig.3c represents the combination of both pictures constructed for easier comparison of the

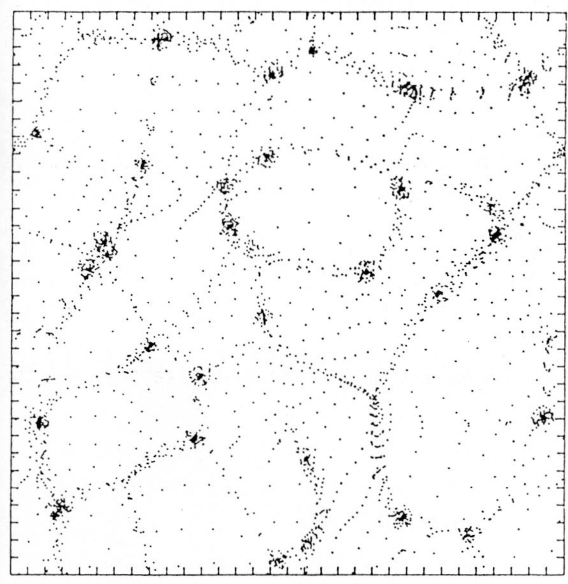

Figure 3a.

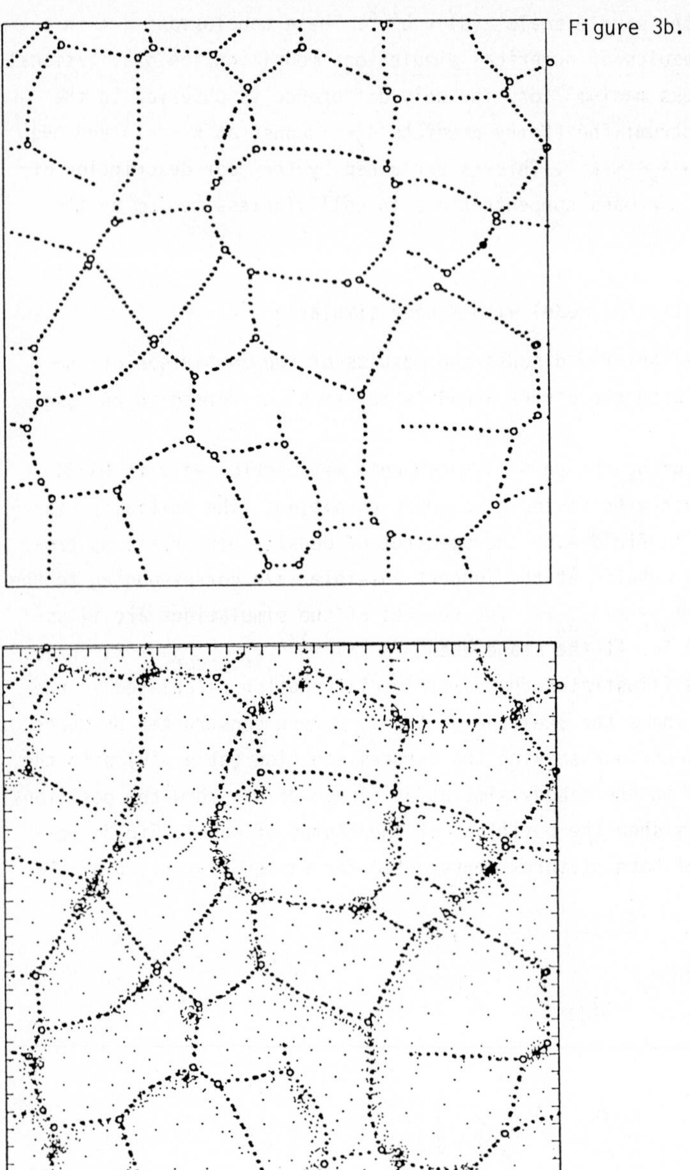

Figure 3b.

Figure 3c.

picture. Slight differences between two pictures are within the mesh size that limits the accuracy of bothe calculations.

## REFERENCES


1. J.H.Oort. Ann.Rev.Astron.Astrophys., 1983, 21, 373.
2. P.J.E.Peebles. The Large Scale Structure of the Universe. Princeton Univ. Press, Princeton, 1980.
3. S.F.Shandarin, Ya.B.Zeldovich. Rev.Mod.Phys., 1989, 61, April 1.
4. Ya.B.Zeldovich. Astron.Astrophys., 1970, 5, 84.
5. A.G.Doroshkevich, E.V.Kotok, I.D.Novikov, A.N.Polyudov, S.F.Shandarin, Yu.S.Sigov. Mon.Not.R.Astr.Soc., 1980, 192, 321.
6. A.A.Klypin, S.F.Shandarin. Mon.Not.R.Astr.Soc., 1983, 204, 891.
7. A.Melott, S.Shandarin. Astrophys.J. August issue (to be published).
8. S.N.Gurbatov, A.I.Saichev, S.F.Shandarin. Sov.Phys.Dokl., 1985, 30, 921.
9. L.A.Kofman, S.F.Shandarin. Nature, 1988, 334, 129.
10. S.N.Gurbatov, A.I.Saichev, S.F.Shandarin. Mon.Not.R.Astr.Soc., 1989, 236, 385.
11. I.M.Burgers. The Nonlinear Diffusion Equation. Drodrecht, Reidel, 1974.
12. Sh.Kida. J.Fluid Mech., 1979, 93, 2, 337.
13. S.N.Gurbatov, A.I.Saichev, I.G.Yakushev. Sov.Phys.Usp., 1983, 26(10), 857.
14. S.N.Gurbatov, A.N.Malachov, A.I.Saichev. Nonlinear Random Waves in Nondispersive Media. Moscow, Nauka 1990 (also Manchester University Press).
15. S.N.Gurbatov, A.I.Saichev. Izvestiya VUZov-Radiophysics. 1984, 27, 456.
16. E.V.Kotok, S.F.Shandarin. Sov.Astron., 1989.
17. S.N.Gurbatov, A.I.Saichev. Izvestiya VUZov-Radiophysics, 1988, 31.
18. F.M.Pestorius, D.T.Blackstock. Propagation of finite amplitude noise. Proc. 1973 Symp. Finite wave effects in fluids. Copenhagen, 1974, p.22-24.
19. L.Bjørnø, S.N.Gurbatov. Sov.Phys.-Acoustics, 1985, 31, 3.
20. L.Kofman, D.Pogosyan, S.Shandarin. Mon.Not.R.Astr.Soc., 1989 (submitted for publication).

# Generation of Observation Turbulence Spectrum and Solitary Dipole Vortices in Rotating Gravitating Systems

*V.V. Dolotin and A.M. Fridman*

Council of Astronomy, USSR Academy of Sciences, Pyatnitskaya 48, 109017 Moscow, USSR

The nonlinear equation describing the dynamics of perturbations with a small but finite amplitude in a uniformly rotating gas system with inhomogeneous density is obtained. In a special case of short-wave perturbations (with wavelength $\lambda$ much less than the Jeans wavelength $\lambda_J$, $\lambda \ll \lambda_J$), the obtained equation can be reduced to the known nonlinear equation for the Rossby waves with the scalar and vector nonlinearity. Neglecting the scalar nonlinearity, we obtain the equation of the same type as in the opposite limit of long-wave perturbations, $\lambda \gg \lambda_J$ (the Charney-Obukhov equation or the Hasegava-Mima equation). The weak turbulence spectrum of these equations leads to the known observable relations between the main parameters of the galactic gas population, in particular, to the Salpeter's stellar mass spectrum found in 1955. The stationary solution of these equations (named after Larichev and Reznik) in the form of a solitary dipole vortex (modone) with the circular separatrix is obtained. In conclusion, the possible observable manifestations of modones (double galaxies, galactic nuclei and stars) are discussed.

## 1. INTRODUCTION

An analogy between the main equations for particles interacting by the Coulomb law in the magnetic field (the Boltzmann-Vlasov equations) and the main equations for particles interacting by the Newtonian law (the kinetic equation in the rotating coordinate system and the Poisson equation) was the basis for creating the stability theory of gravitating systems /1,2/. For this purpose the methods of the theory of collective processes in plasma physics are widely used. On the one hand, the principles of linear stability theory of gravitating systems are known. Besides the theory of stability of solitary figures of rotation /1-6/, there has also been some progress in the construction of a linear theory of involved figures of rotation /7/. On the other hand, the foundation of the theory of nonlinear waves /8/ and turbulence /9/ began to be formulated not long ago. Nonlinear theory was developed only for the model of an infinitely thin gravitating disk, i.e. for the case of large-scale perturbations with sizes much exceeding the thickness of the disk. In these first papers the Jeans perturbations of the finite amplitude were investigated and for the first time they were shown to spread as soliton envelopes /8/. In a similar way /10/, a theory of weak Jeans turbulence /9/ was constructed and $E_k \sim k^{-7/4}$ spectrum coinciding practically with the Kolmogorov's spectrum "proper" /11,12/ was obtained. As follows from /107, the Kolmogorov's spectrum is understood as any power turbulent spectrum. The Kolmogorov's spectrum "proper" is the spectrum



Research Reports in Physics Nonlinear Waves 3
Editors: A.V. Gaponov-Grekhov · M.I. Rabinovich · J. Engelbrecht

$E_k \sim k^{-5/3}$ obtained in /11, 12/ from dimensional estimates using the assumption of the constancy of the energy flow along spectrum.

In a number of recent observations of the Galaxy's gas component (see, for instance, /13/), the difference of the turbulence spectrum from the Kolmogorov's spectrum propoer is noted. Thus, the dependence of the velocity of mass centrum of gaseous condensation v on its size l proves to be /13/:

$$v_l \sim l^{0.5} \tag{1}$$

and, correspondingly, for gas density $\rho$ the observations give /13/

$$\rho_l \sim l^{-1} . \tag{2}$$

Deviation from the Kolmogorov's spectrum proper ($v \sim l^{1/3}$) exceed the observational errors so much that observers tend to speak with confidence about the spectrum which differs from that obtained in /11, 12/. Moreover, in some theoretical papers, instead of the hypothesis about the constancy of energy flow along the spectrum /11,12/ other hypotheses are suggested. Thus, in /14/ two fundamental hypotheses are put forward: 1) the constancy along the spectrum of rotational moment of gaseous condensations which results in the constancy of pressure, $P(l) \equiv \rho(l)v_l^2 = \text{const}$ along with the hypothesis that the characteristic time of motion of gaseous condensations coincides with the Jeans time. Here the authors identify /14/ the observed dependence (1) with the spectrum $E_k \sim k^{-1}$ and note it to be "more sloping than Kolmogorov's", $E_k \sim k^{-5/3}$. Without regarding "two fundamental hypothesis" of the authors mentioned above /14/, let us note that their last conclusions are erroneous, since instead of the relation $E_k \sim v_k^2$ used in /14/, the following relation should be written /15/ $\int_k^\infty E_k dk \sim v_k^2$ /16/ from which we have the spectrum

$$E_k \sim k^{-2} , \tag{3}$$

which is steeper than the Kolmogorov's spectrum proper /11,12/. Below we show how the spectrum may be obtained from the initial equations of hydrodynamics for a rotable gravitating compressed system within the frame of the theory of weak turbulence. Neglecting the terms of the third order in perturbation amplitude which is considered infinitesimal, we shall confine ourselves to deriving the nonlinear equation from which in two limit cases ($\lambda \gg \lambda_J$, $\lambda \ll \lambda_J$, where $\lambda_J$ is the Jeans perturbation scale), we obtain nonlinear equations for gravitational potential $\Psi$ and density $\rho$, correspondingly.

First of all, let us discuss the nonlinear equation for gravitational potential $\Psi$. It contains the so-called vector non-linearity /17/ and is known in hydrodynamics as the Charney-Obukhov equation /18-20/ and in plasma physics as the Hasegawa-Mima equation /21/. In the linear approximation this equation describes the *Rossby gravitational waves* /22/. We want to call them "gravitating", because unlike their analogs in hydrodynamics, i.e. the Rossby waves /23-25/ and those in plasma physics, i.e. the drift waves /27,28/, the existence of the Rossby gravitating waves owes to the perturbed gravitational potential. The Rossby waves were investigated by MARGULIS in 1893 and later by other authors before Rossby's paper was published (see references in /26/). For the first time one of the authors of the present paper, AMF, heard about one and the same law of dispersion for the Rossby waves and drift waves from Prof. LEONTOVICH in August 1965 long before the publication of papers by Japanese scientists (see also /26/).

The equation governing the Rossby nonlinear gravitating waves describes both turbulent pulsations and stationary solitary dipole vortexes. The Charney-Obukhov (or Hasegawa-Mima) equation which is similar to ours, was investigated in /30/ by a well-known method of the theory of weak turbulence (i.e. with the help of kinetic equations for waves, see, for example, /29/) and a turbulence spectrum was obtained. We used this spectrum here showing that it corresponds to spectrum (3) and determines not only the observed correlations (1), (2), but also the Salpeter's stellar mass spectrum /31/:

$$n \sim m^{-1.5}, \tag{4}$$

where n is a number of stars in the unit volume with masses from m to $\infty$. The stationary solution of the nonlinear equation for the Rossby gravitating waves obtained here describes solitary dipole vortexes. This solution is similar to the one obtained in /32/ for dipole vortexes in incompressible liquid in $\beta$-plane. Below we prove the uniqueness of the solution obtained in /32/ under conditions which are usually satisfied in real systems and suggesting the separatrix of the dipole vortex to have a circular form (see also /40/). The formation of solitary dipole vortexes in a medium described by the equation with one vector nonlinearity is natural. An exception is the case of one vortex analyzed in /33/, when its centre is at the line of tangential velocity break in the presence of two opposite flows. As correctly noted by the authors in /33/, the direction of rotation in a half of vortex should also be reversed, if the direction of one of the flows is reversed in such a way that the resulting flow becomes homogeneous (according to /32/). As a result, we shall obtain a dipole vortex /32/. One can find a remark on possible observational manifestations of dipole vortex structures in astronomical objects in the Conclusion.

The second nonlinear equation for density $\rho$ derived in the present paper, differs from the first one (for $\psi$ potential) in the additional term - the so called scalar nonlinearity /17/. With the dominating role of the latter over vector nonlinearity, the stationary solution of such an equation, as it is known, describes solitary vortexes /17,26/.

Here the following remark should be made. According to /15/, waves in shallow water correspond to two-dimensional perturbations in the gaseous medium with a plane adiabatic exponent $\gamma_{\perp} = 2$. The equation for the Rossby nonlinear waves in shallow water /26/ looks just similar to nonlinear equation which is obtained here for a solid rotating compressible gaseous medium with an arbitrary (three-dimensional) adiabatic exponent $\gamma$ /29/. As seen from this equation, however, the scalar nonlinearity becomes zero when $\gamma = 2$. Therefore, in that special case when $\gamma = \gamma_{\perp} = 2$ /4/, it is impossible to speak about an analogy between nonlinear gas dynamics of small, but finite two-dimensional perturbations, and the shallow water equations, though linearized equations, are obviously equivalent.

The plan of the paper is the following. In Sect.2 the authors formulate the problem. The derivation of the main nonlinear equations is given in Sect.3. In Sect.4 stationary solutions for two nonlinear equations obtained from the main equation in limit cases $\lambda \gg \lambda_J$ and $\lambda \ll \lambda_J$ are presented. After finding the solution of the second nonlinear equation for $\rho$, we shall be interested only in the dominating role of the vector nonlinearity. In this case both equations (for $\Psi$ and $\rho$) are proved to be coinciding, which simplifies the analysis. However, neglecting the scalar nonlinearity we have imposed the limit on the system parameters. These limits slightly differ from the known limitations for the case with shallow water /26/ which is the cause of a certain specific character of astrophysical objects. This difference makes easier to fulfill the condition that the scalar nonlinearity is much less than the vector nonlinearity. The spectrum of weak turbulence is analyzed in Sect.5. Possible astrophysical applications are discussed in Sect.6.

## 2. MODEL. BASIC EQUATIONS

Let us consider a solid, rotating ($\Omega_0$ = const) and inhomogeneous in density ($\rho_0' \neq 0$) gravitating system. We denote stationary values by index "0" and the differentation with respect to radial coordinate r by "prime". The central parts of galaxies and many polytropic rotating gaseous configurations are arranged just in such a way /34/.

The basic equations are: the equations of hydrodynamics in the coordinate system /15/ rotating with the angular velocity $\Omega_0$ and the Poisson equation:

$$\frac{d}{dt}\vec{v}_{\perp} = 2\,\vec{v}_{\perp},\Omega_0 - \nabla_{\perp}\tilde{\chi}, \qquad (5)$$

$$\frac{d\rho}{dt} + \rho\,\mathrm{div}\,\vec{v}_{\perp} = 0\,, \qquad (6)$$

$$\Delta_{\perp}\Psi = 4\pi G\rho\,, \qquad (7)$$

where

$$\frac{d}{dt} \equiv \frac{\partial}{\partial t} + (\vec{\tilde{v}}_{\perp},\nabla_{\perp}); \quad \Delta_{\perp} \equiv \frac{1}{r}\frac{\partial}{\partial r}r + \frac{1}{r}\frac{\partial}{\partial\phi}\,. \qquad (8)$$

Function $\Psi$ is assumed to stand of two parts

$$\Psi(r,\phi,t) = \Psi_0(r) + \tilde{\Psi}(r,\phi,t) \ . \tag{9}$$

Deriving Eq.(5), the following equilibrium condition of the system along r was used (assuming $\vec{\Omega}_0 || 0z$):

$$\Omega_0^2 r = \frac{\partial \chi}{\partial r} \tag{10}$$

Function $\chi$ is introduced in the following way (the rest of the notions are generally accepted /15/):

$$\frac{\partial \chi}{\partial x_i} = \frac{\partial \Psi}{\partial x_i} + \frac{1}{\rho}\frac{\partial P}{\partial x_i} \ . \tag{11}$$

Assuming that the gas is barotropic, $\rho = \rho(P)$, the last addendum in (11) may be presented as

$$\frac{1}{\partial(P)}\frac{\partial P}{\partial x_i} = \frac{\partial f(P)}{\partial x_i} \ , \tag{12}$$

where function f(P) is determined from the equation $df(P)/dP = 1/\rho(P)$. Finally we have from (11), (12)

$$\chi = \Psi + f(P) \ . \tag{13}$$

As follows from the recorded form of the equations, we have confined ourselves to the analysis of two-dimensional perturbations, lying in $(r,\phi)$ plane, i.e. perpendicular to the axis of rotation.

## 3. DERIVATION OF THE MAIN NONLINEAR EQUATIONS

Let us multiply Eq.(5) by $\vec{1}_z \equiv \vec{\Omega}_0/\Omega_0$. We obtain then

$$\vec{\tilde{v}}_\perp = \vec{\tilde{v}}_\chi + \vec{\tilde{v}}_I{}' \ , \tag{14}$$

where

$$\vec{\tilde{v}}_\chi \equiv \frac{1}{2\Omega_0}\left[\vec{e}_z, \vec{\nabla}_\perp \chi\right] \ , \tag{15}$$

$$\vec{\tilde{v}}_I{}' \equiv \frac{1}{2\Omega_0}\left[\vec{e}_z, \frac{d\vec{\tilde{v}}_\perp}{dt}\right] . \tag{16}$$

We substitute (14) in (16) for $\vec{\tilde{v}}_\perp$:

$$\vec{\tilde{v}}_I{}' \equiv \frac{1}{4\Omega_0^2}\left[\vec{e}_z, \frac{d}{dt}\left[\vec{e}_z, \vec{\nabla}_\perp \tilde{\chi}\right]\right] + \frac{1}{4\Omega_0^2}\left[\vec{e}_z, \frac{d}{dt}\left[\vec{e}_z, \frac{d\vec{\tilde{v}}_\perp}{dt}\right]\right].$$

Substituting expression (14) into the last term of this equation, we see that it turns out to be $\sim\Omega_0^{-3}$. We omit this term, since we are interested only in "slow" motions. Let us note that

$$|\vec{\tilde{v}}_I{}'/\vec{\tilde{v}}_\chi| \sim \frac{d}{\Omega_0 dt} \ll 1 . \tag{17}$$

Hence

$$\frac{d}{dt} \cong \frac{\partial}{\partial t} + (\vec{\tilde{v}}_\chi, \vec{\nabla}_\perp) \equiv \frac{d_0}{dt} , \tag{18}$$

from which

$$\vec{\tilde{v}}_I = \frac{1}{4\Omega_0^2}\left[\vec{e}_z, \frac{d_0}{dt}\left[\vec{e}_z, \vec{\nabla}_\perp\tilde{\chi}\right]\right] . \tag{19}$$

Using (15), (19) we calculate div $\vec{\tilde{v}}_\chi$ and div $\vec{\tilde{v}}_I$:

$$\text{div } \vec{\tilde{v}}_\chi = 0 , \tag{20}$$

$$\text{div } \vec{\tilde{v}}_I = -\frac{1}{4\Omega_0^2}\left(\frac{\partial}{\partial t} + \frac{1}{2\Omega_0}\left[\vec{\nabla}_\perp\tilde{\chi}, \vec{\nabla}_\perp\right]_z\right)\Delta_\perp\tilde{\chi} . \tag{21}$$

Introducing (20), (21) into continuity equation (6) and using the Poisson equation (7), we obtain finally the *main nonlinear equation*:

$$\left(\frac{\partial}{\partial t} + \frac{1}{2\Omega_0}\, \vec{\nabla}_\perp\tilde{\chi}, \vec{\nabla}_\perp\, {}_z\right)\Delta_\perp\tilde{\Psi} - \frac{\omega_0^{2}{}'}{2\Omega_0}\frac{1}{r}\frac{\partial\tilde{\chi}}{\partial\phi} -$$

$$-\frac{\omega_1^2}{4\Omega_0^2}\left(\frac{\partial}{\partial t} + \frac{1}{2\Omega_0}\,\vec{\nabla}_\perp\tilde{\chi}\, ,\vec{\nabla}_\perp\, {}_z\right)\Delta_\perp\tilde{\chi} = 0 , \tag{22}$$

where $\omega_0^2 \equiv 4\pi G\rho_0$ is the square of the Jeans frequency, and $\omega_0^{2}{}' \equiv d\omega_0^2/dr$ .

Let us consider successively two limit cases: 1) $\tilde{\Psi} \gg \tilde{f}$, 2) $\tilde{\Psi} \ll \tilde{f}$ .

1) $\tilde{\Psi} \gg \tilde{f}$. This case corresponds to the presence of long-wave perturbations $\lambda \gg \lambda_J$ ($\lambda_J$ being the Jeans wave length) which is possible only for sufficiently flattened configurations along the axis of rotation, $h \ll R$ (where h, R are the thickness and the radius of the system, correspondingly), because $\lambda_J \sim h$ /2/.

Under the above mentioned condition Eq.(22) has the form

$$\left(\frac{\partial}{\partial t} + \frac{1}{2\Omega_0}\left[\vec{\nabla}_\perp\tilde{\Psi}, \vec{\nabla}_\perp\right]_z\right)\Delta_\perp\tilde{\Psi} + \frac{1}{\alpha - 2}\frac{\omega_0^{2}{}'}{\Omega_0}\frac{1}{r}\frac{\partial\tilde{\Psi}}{\partial\phi} = 0 , \tag{23}$$

where

$$\alpha \equiv \frac{\omega_0^2}{2\Omega_0^2} . \tag{24}$$

Let us note, by the way, that $\alpha$ = I /1/ in the special case of a cold cylinder with homogeneous density which rotates with angular velocity $\Omega_0$.

Let us analyze the linear approximation of Eq.(23). Assuming

$$\tilde{\Psi}(r,\phi,t) \sim \exp\left[i(k_r r + m\phi - \omega t)\right] , \tag{25}$$

we obtain

$$\omega = - \frac{1}{\alpha-2} \frac{\omega_0^{2\,\prime}}{\Omega_0} \frac{k_\phi}{k_\perp^2} , \quad k_\phi \equiv m/r , \quad k_\perp^2 \equiv k_r^2 + k_\phi^2 , \tag{26}$$

i.e. the expression for the frequency of the Rossby gravitating waves. Since in real systems $\omega_0^{2\,\prime} \sim \rho_0' < 0$ the direction of azimuthal component of phase velocity of the Rossby gravitating waves depends on the sign of quantity $(\alpha-2)$: at $\alpha - 2 > 0$ in the direction of rotation (towards the "east"), and at $\alpha - 2 < 0$ towards the "west". The necessary condition (17) is fulfilled either in the case of anisotropic spectrum $k_\phi \ll k_r$ , or when the gravitating disk is on the border of gravitational instability (this is usually fulfilled for galactic protoplanetary disks /1,35/, and some planetary rings /6/).

2) $\tilde{\Psi} \ll \tilde{f}$. This condition is more universal than the first one, because $\lambda \ll \lambda_J$ is valid for short-wave pulsations that exist in systems of any kind of geometry. It is obvious that only turbulence in gas clouds may form perturbations with $\lambda \ll \lambda_J$. The main role in the creation of turbulent viscosity may be attributed to such perturbations $\lambda \lesssim h \sim \lambda_J$ in the $\alpha$-model the accretion disk /36,37/.

Equation (22) under the condition (2) takes the form

$$\left(\frac{\partial}{\partial t} + \frac{1}{2\Omega_0} \left[\vec{\nabla}_\perp \tilde{f}, \vec{\nabla}_\perp\right]_z\right) \Delta_\perp \tilde{f} + 2\Omega_0 \frac{\rho_0'}{\rho_0} \frac{1}{r} \frac{\partial \tilde{f}}{\partial \phi} = 0 . \tag{27}$$

In the linear approximation, taking $\tilde{f}$ as (25), we obtain from (27)

$$\omega = -2\Omega_0 \frac{k_\phi}{k_\perp^2} \ln' \rho_0 , \tag{28}$$

which is the expression for the frequency of the Rossby waves again. Let us note that since in real object $\ln' \rho_0 < 0$, the azimuthal component of phase velocity of the Rossby (shortwave) waves in rotating gravitating systems is directed towards rotation (i.e. towards the "east").

Let us come from function $\tilde{f}$ to $\tilde{\rho}$ in Eq.(27). We use the adiabatic curve from which we have $P/P_0 = (\rho/\rho_0)^\gamma$ yielding

$$P = P_0\left[1 + \gamma \frac{\tilde{\rho}}{\rho_0} + \frac{\gamma(\gamma-1)}{2} \left(\frac{\tilde{\rho}}{\rho_0}\right)^2\right] .$$

Note that this expression is bounded to the terms of the second order in finitesimal amplitude $\tilde{\rho}/\rho_0$.

Then, with the help of (12) we have, correspondingly,

$$\frac{1}{r} \frac{\partial \tilde{f}}{\partial \phi} = \frac{1}{r} \frac{\partial}{\partial \phi} \tilde{\zeta}\left(1 - \frac{\gamma-2}{2} \frac{\tilde{\zeta}}{c_0^2}\right), \quad \tilde{\zeta} \equiv c_0^2 \frac{\tilde{\rho}}{\rho_0} , \quad c_0^2 \equiv \frac{\partial P_0}{\partial \rho_0} .$$

Finally we have

$$(\frac{\partial}{\partial t} + \frac{1}{2\Omega_0}\left[\vec{\nabla}_\perp\tilde{\zeta},\vec{\nabla}_\perp\right]_z)\Delta_\perp\tilde{\zeta} + (2-\gamma)\frac{\Omega_0}{c_0^2}\frac{\rho_0'}{\rho_0}\frac{1}{r}\frac{\partial\tilde{\zeta}^2}{\partial\phi} + 2\Omega_0\frac{\rho_0'}{\rho_0}\frac{1}{r}\frac{\partial\tilde{\zeta}}{\partial\phi} = 0 . \qquad (29)$$

It is seen that besides the term with vector nonlinearity (proportional to vector product), Eq.(29) contains the scalar nonlinearity which is absent in Eq.(23). Let us estimate the weight of the terms with scalar and vector nonlinearity:

$$\frac{\text{scalar nonlinearity}}{\text{vector nonlinearity}} \sim \frac{a^2}{\rho^2}\frac{a}{R} ; \qquad \rho \equiv \frac{c_0}{\Omega_0} . \qquad (30)$$

Comparing (30) with the corresponding relation for nonlinear Rossby waves in shallow water /26/, we see that there is an extra multiplier $\frac{a}{R} < I$ in Eq.(30), which includes the contribution of vector nonlinearity. In the present paper we shall confine ourselves only to those astrophysical applications /67/ for which vector nonlinearity is dominating. Therefore, below we shall be interested only in the solution of Eq.(29) where the scalar nonlinearity is neglected:

$$(\frac{\partial}{\partial t} + \frac{1}{2\Omega_0^2}\,\vec{\nabla}_\perp\tilde{\zeta},\vec{\nabla}_\perp\,{}_z)\Delta_\perp\tilde{\zeta} + 2\Omega_0\frac{\rho_0'}{\rho_0}\frac{1}{r}\frac{\partial\tilde{\zeta}}{\partial\phi} = 0 . \qquad (31)$$

In this case, Eqs.(23) and (31) derived from the main equation (22) in the two limit cases ($\lambda >> \Lambda_J$ and $\lambda << \lambda_J$) coincide within the accuracy up to coefficients.

## 4. SOLITARY DIPOLE VORTEXES

Let us perform a sequency of the transformations of independent variables in Eqs.(23), (31). First of all, let us move over to the local Descartes coordinate system

$$\frac{\partial}{\partial r} \equiv \frac{\partial}{\partial x} , \qquad \frac{1}{r}\frac{\partial}{\partial\phi} \equiv \frac{\partial}{\partial y}$$

Let in the system of three variables x, y, t the last two prove to be mutually related by the following relationship: $y = \eta + ut$. In such a system of two independent variables (let it be x and $\eta = y - ut$) Eqs.(23), (31) will have the form

$$\hat{D}(\xi_i)\Delta_\perp\xi_i = \Lambda_i\frac{\partial\xi_i}{\partial\eta} , \qquad i = 1,2 . \qquad (32)$$

Here we have introduced the following notations:

$$\hat{D}(\xi_i) \equiv -\frac{\partial}{\partial\eta} + \frac{1}{V_i}\,\nabla_\perp\xi_i,\nabla_\perp\,{}_z ;$$

$$\xi_1 \equiv \tilde{\Psi}, \quad V_1 \equiv 2u\Omega_0, \quad \Lambda_1 \equiv -\frac{2\omega_0^{2\,\prime}}{(\alpha-2)V_1} ; \qquad (33)$$

$$\xi_2 \equiv \tilde{\zeta}, \quad V_2 \equiv 2u\Omega_0, \quad \Lambda_2 \equiv -\frac{4\Omega_0^2}{V_2}\frac{\rho_0'}{\rho_0} .$$

It is not difficult to understand that Eq.(32) may be rewritten in the form

$$J(\xi_i - V_i x, \Delta_\perp \xi_i + V_i \Lambda_i x) = 0 , \tag{34}$$

where J is the Jacobi determinant. The last equation yields

$$\Delta_\perp \xi_i + \Lambda_i V_i x = F(\xi_i - V_i x) , \tag{35}$$

where F is an arbitrary function.

We are interested only in localized solutions, i.e. $\xi_i \to 0$ when $\eta \to \infty$ at any x. In the vicinity of an infinitely remote point along $\eta$, $\eta \to \infty$, we have

$$F(-V_i x) = \Lambda_i V_i x$$

at any x. Therefore, function F must be a linear function in the region which is sufficiently remote from the point $x = \eta = 0$. Let us choose this linear function in all the plane $(x,\eta)$:

$$\Delta_\perp \xi_i + V_i \Lambda_i x = -k^2 (\xi_i - V_i x), \quad r < a ; \tag{36}$$

$$\Delta_\perp \xi_i + V_i \Lambda_i x = p^2 (\xi_i - V_i x), \quad r > a , \tag{37}$$

where $k > 0$, $p > 0$. The reason for the division of $(x,\eta)$-region into the internal and external parts of the circle of radius a with the centre at point $x = \eta = 0$ will be explained below.

Introducing polar coordinates $r$, $\phi(x = r\cos\phi,\ \eta = r\sin\phi)$ instead of (36) and (37), we shall obtain

$$\left(\frac{\partial^2}{\partial r^2} + \frac{1}{r}\frac{\partial}{\partial r}\right) + \frac{1}{r^2}\frac{\partial^2}{\partial\phi^2} + k^2)\xi_i = V_i (k^2 - \Lambda_i) r \cos\phi, \quad r < a ; \tag{38}$$

$$\left(\frac{\partial^2}{\partial r^2} + \frac{1}{r}\frac{\partial}{\partial r} + \frac{1}{r^2}\frac{\partial^2}{\partial\phi^2} - p^2\right)\xi_i = 0 , \quad r > a . \tag{39}$$

Equation (39) proves to be homogeneous, because we must demand $\xi_i \to 0$ at $r \to \infty$ From this condition we find

$$p^2 = -\Lambda_i . \tag{40}$$

We look for the solution of Eq.(38) in the form $\xi = \tilde{\xi} + \xi^*$ (here and below index i will be omitted) where $\xi^*$ is the particular integral of the nonhomogeneous equation and $\tilde{\xi}$ is the general solution of the homogeneous equation which is well-known:

$$\tilde{\xi}(r,\phi) = \sum_{m=0}^{\infty} (A_m \cos m\phi + B_m \sin m\phi) J_m(kr) , \quad r < a .$$

We look for a particular integral of the nonhomogeneous equation (38) in the form

$$\xi^* = r(a \sin\phi + b \cos\phi) . \tag{41}$$

Substituting (41) into (38) we obtain

$$a = 0, \quad b = V(1-\Lambda/k^2) . \tag{42}$$

Finally we find the solution of Eq.(38):

$$\xi_I = \sum_{m=0}^{\infty} (A_m \cos m\phi + B_m \sin m\phi) J_m(kr) + V(1-\Lambda/k^2) r \cos\phi , \quad r < a , \tag{43}$$

where $J_m(z)$ is the Bessel function of real argument.

The general solution of Eq.(39) has the form

$$\xi_{II} = \sum_{m=0}^{\infty} (C_m \cos m\phi + D_m \sin m\phi) K_m(pr), \quad r > a , \tag{44}$$

where $K(z)$ is the McDonald function.

The equations of motion and continuity include first derivatives of $\rho$ and v with respect to r, which at r=a must be continuous. It follows from (5)-(8) and (15) that $\xi$ and its first two derivatives must also be continuous. (Here lies the difference of the gravitating compressible medium with $\Delta_\perp \Psi = 4\pi G\rho$ from the case of incompressible nongravitating liquid /32/, where the second derivative, i.e. the vertex, may be discontinuous). For this reason Eqs.(36), (37) may be written in the following way

$$F(\xi, \xi_r', \xi_{rr}'', \xi_{\phi\phi}'', r, \phi) = \begin{cases} -k^2 f(r,\phi) , & r < a \\ p^2 f(r,\phi) , & r > a \end{cases} , \tag{45}$$

so that there are similar functions F and f, continuous at point r=a. So, subtracting the first equation from the second one at point r=a we shall obtain

$$(p^2+k^2) f(a^2,\phi) = 0 . \tag{46}$$

Since earlier we have demanded $p > 0$, $k > 0$, then from (46), (36), (37) we obtain

$$(\xi - Vx)\Big|_{r=a} = 0,$$

or

$$\xi_I(a,\phi) = \xi_{II}(a,\phi) = V_a \cos\phi .$$

Substituting solutions (43), (44) into these expressions, we find that $B_m = D_m = 0$ for all m and $A_m = C_m = 0$ for all m except m = I:

$$A_1 = V\Lambda a/k^2 J_1(ka) , \quad C_1 = Va/k_1(pa) .$$

So, finally we have

$$\xi = \begin{cases} \Omega_0 a u \left[\frac{2}{a} + \frac{\Lambda}{k^2} \left(\frac{J_1(kr)}{J_1(ka)} - \frac{r}{a}\right)\right] \cos\phi , & r < a \\ \Omega_0 a u \frac{K_1(pr)}{K_1(pa)} \cos\phi, & r > a \end{cases} \tag{47}$$

Here we have used the fact that according to (33) $V_2 = u$ (thus, (47) describes $\tilde{\xi}_2$ while $\tilde{\xi}_1$ differs by multiplier $(\alpha-2)/(\alpha-1)$).

From (14) we have: $\vec{v}_\perp = \vec{v}_\chi + \vec{v}_I{}'$ and from (17): $|\vec{v}_\chi| \gg |\vec{v}_I{}'|$. Hence, $v_\perp \approx v_\chi$. Further, from (15): $\vec{v}_\chi = \left[\vec{e}_z, \vec{\nabla}_\perp \chi\right]/2\Omega_0$, i.e.

$$v_r = -\frac{1}{2\Omega_0 r}\frac{\partial\tilde{\chi}}{\partial\phi}, \quad v_\phi = \frac{1}{2\Omega_0}\frac{\partial\tilde{\chi}}{\partial r} \tag{48}$$

Let R be the characteristic scale of changing of stationary parameters (the size of the system) in the direction perpendicular to the rotation axis, and $a/R \ll I$. Then in solution (47) quantity $\Lambda$ may be considered as a constant with the accuracy up to $a/R \ll I$ and, consequently, from (48), (47) we obtain

$$2v_r = \begin{cases} \left[1 - \frac{\Lambda}{k^2}\left(1 - \frac{a}{r}\frac{J_1(kr)}{J_1(ka)}\right)\right] u \sin\phi, & r < a \\ \frac{a}{r}\frac{K_1(pr)}{K_1(pa)}\, u \sin\phi, & r > a \end{cases} \tag{49}$$

$$2v_\phi = \begin{cases} \left[1 - \frac{\Lambda}{k^2}\left(1 - ka\frac{J_1{}'(kr)}{J_1(ka)}\right)\right] u \cos\phi, & r < a \\ pa\frac{K_1{}'(pr)}{K_1(pa)}\, u \cos\phi, & r > a \end{cases} \tag{50}$$

As it follows from (49), the continuity of $v_r$ at $r=a$ is fulfilled automatically. The continuity of $v_\phi$ at $r=a$ is fulfilled at points $z_0 \equiv ka$, which are the roots of the following equation (let us remember that p is fixed: $py = -\Lambda$):

$$\frac{s_0}{z_0}\left(1 - z_0\frac{J_1{}'(z_0)}{J_1(z_0)}\right) = s_0\frac{K_1{}'(s_0)}{K_1(s_0)} - 1 .$$

Here $s \equiv pa$. After elementary transformations we obtain /17/:

$$s_0/z_0 \cdot K_1(s_0)J_r(z_0) = -J_1(z_0)K_2(s_0) , \tag{51}$$

or

$$K_1(s_0)J_3(z_0) + J_1(z_0)K_3(s_0) = 0 .$$

This equation has a numerable set of positive roots $z_{0n} = z_{0m}(s_0)$, $n = 1,2,\ldots$ . The first three roots ($n = 1,2,3$) of the dependence $z_{0n} = z_{0n}(s_0)$ are shown in Fig.1 /38/.

Let us analyze the structure of vector field $\vec{v}_\perp$. With this aim we shall calculate $\mathrm{rot}\,_z\vec{v}_\perp$ at $r < a$. Let us substitute into expression

$$\mathrm{rot}\,_z\vec{v}_\perp = \frac{1}{r}\frac{\partial}{\partial r}(rv_\phi) - \frac{1}{2}\frac{\partial v_r}{\partial\phi}$$

the expressions for $v_r$ and $v_\phi$ from (49), (50) at $r < a$. As a result we obtain

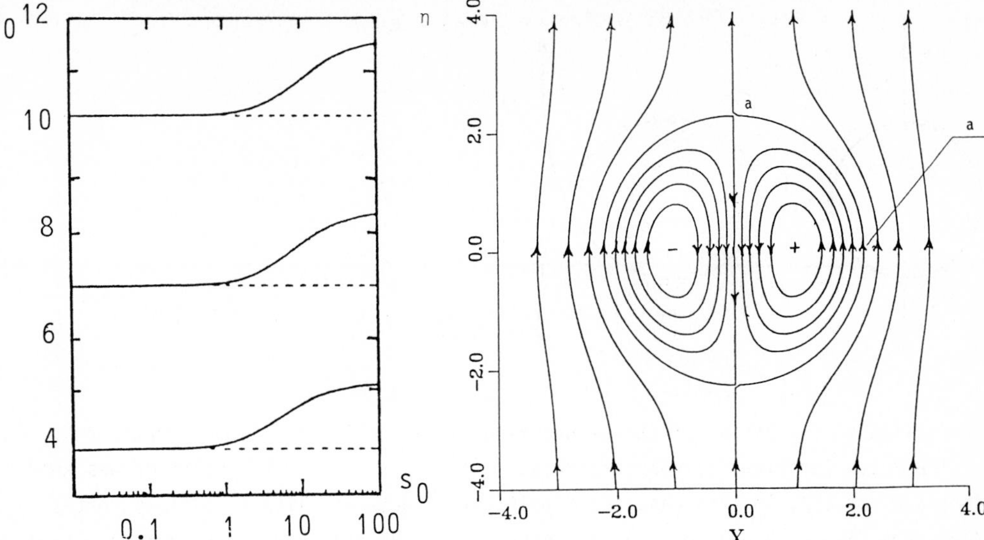


Figure 1. Figure 2.

$$\mathrm{rot}_z \vec{v}_\perp = \frac{\Lambda a}{2J_1(ka)} \quad J_1''(kr) + \frac{1}{kr} J_1'(kr) - \frac{1}{k^2 r^2} J_1(kr) \; u \cos\phi$$

or after a certain simplification

$$\mathrm{rot}_z \vec{v}_\perp = -\frac{1}{2} \Lambda a \frac{J_1(kr)}{J_1(ka)} u \cos\phi \,, \quad r < a \,. \tag{52}$$

It follows from (52) that $\mathrm{rot}_z v_\perp$ changes its sign when there is a change in $r$ inside the circle of radius a, $r < a$, in the points where $J_1(kr) = J_1(z) = 0$. The roots of the last equation (39) are $z_n \approx 4,7,10,\ldots$ ($n = 1,2,3,\ldots$). As $r < a$, then $z < z_0$. Analysis of the stability of the obtained solutions (see references in /38/) have shown that the first root of (51) is the most stable one. It is seen from Fig.1 that $z_{01} < 7$, therefore, for $J_1(kr)$ when $r < a$, only one change of sign may be realized. This corresponds to the structure shown in Fig.2. Let us note that consistency of Eq.(40) and of condition $p > 0$ is possible for (i) $\Lambda_i = \Lambda_1$ if $u > 0$ and $\alpha < 2$, and if $u < 0$ and $\alpha > 2$; and for (ii) $\Lambda_i = \Lambda_2$ if $u < 0$. Under these conditions, the structure shown in Fig.2 has a form of a solitary dipole vortex (modon). It is also called isolated since the amplitude of this vortex falls exponentially when $r > a$, because it is described by the MacDonald function $K_1$. In order to consider opposite cases: (i) $\Lambda_i = \Lambda_1$ if $u < 0$ and $\alpha < 2$, and if $u > 0$ and $\alpha > 2$; and (ii) $\Lambda_1 = \Lambda_2$ if $u > 0$, the sign of the r.h.s. in Eq.(37) should be changed:

$$\Delta_\perp \xi + V \Lambda x = - p^2 (\xi - Vx) \,, \quad r > a \,. \tag{37a}$$

The solution of this equation now depends not on the McDonald function $K_1$, but on

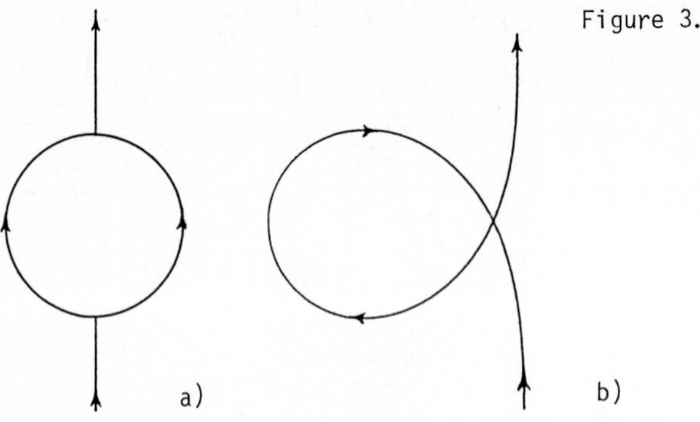

Figure 3.

the cylindrical function of the second kind (it is called the Neuman function $N_1$ or the Veber function $Y_1$), decreasing as $\sim r^{-1/2}$ at $r \to \infty$. Such a solution was known long ago /24,25/; in this case the dipole vortex acts not like a solitary vortex /26/, but like a long-range "force centre" (cf. with dipole Hertz vibrator).

Let us make some remarks to end this Section. First, let us note that the obtained solution (47) is unique for (36), (37). It was obtained from solutions (43), (44) by matching them at the circumference of radius a under the conditions of continuity of functions and of their first derivatives. For the first time, uniqueness of the solution for a two-layer flow was considered in /40/. The case when separatrix is not a circle r=a (Fig.3a), but a loop(Fig.3b), is shown in Fig.4 /41/. The cases of discontinuous conditions on the separatric are considered in /42/. Secondly, the reason for the division of the whole plane $(x,\eta)$ into two regions: I $(r < a)$ and II $(r > a)$ is now clear. Linear behaviour of function $\xi_i - Vx$ and $\Delta_\perp \xi_i + V\Lambda_i x$ proved /38/ from the condition of finiteness of solution determined

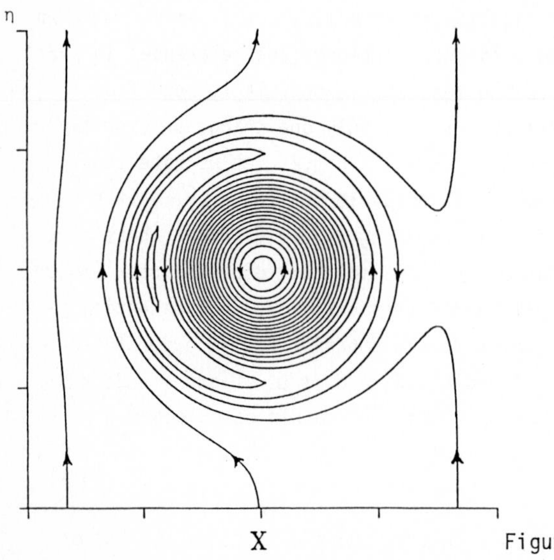

Figure 4.

an equation for cylindrical functions. Since the general solution of this equation grows into infinity either at point $r = 0$, or at $r \to \infty$, and we need the bounded solution in the whole plane, then we have to divide the plane into two parts, in each of which the solution is formed by cylindrical functions being finite there. Hence two equations (36) and (37) are linear. Thirdly, comparing our Eq.(34) with Eq.(4) from /32/ which describes dipole vortexes in uncompressible liquid in the $\beta$-plane with a solid cover, their complete similarity is easily seen.

## 5. SPECTRUM OF WEAK TURBULENCE

It is shown in /30/ that the equations line (23) and (31) describe the Kolmogorov powerlike spectrum of weak turbulence, for the first time obtained by ZAKHAROV in 1965 for weak dispersive compressible fluids (see references in /10/). Assuming perturbations to be one-dimensional, $k_x >> k_y$, the following spectral dependences of energy density have been found /30/ (see the Appendix):

$$W_k^{(1)} \sim k_y^{-3/2} k_x^{-2}, \quad W_k^{(2)} \sim k_y^{-3/2} k_x^{-3} . \tag{53}$$

A question arises about the dependence of turbulence spectrum of the Rossby waves $W_k$ on $k_z$. When the Rossby waves occur in the shallow water, the absence of such a dependence due to the shallow water approximation seems natural. Let us note that drift waves in plasma being much larger in size along z than the cross dimension a, $L >> a$, have a similar dispersion law. It follows from the theory of drift waves /48/ and from experiments /49-51/ that i) $k_z << k_y$ (let us remember that we assumed $k_y << k_x$); ii) the dependence of turbulent diffusion coefficient on z coordinate is is absent. Let us explain the latter conclusion, because in the given case it is rather important. To reveal the diffusion mechanism across the magnetic field in /51/ oscillograms were made of oscillations density in a plasma pillar and gauge current. In the absence of instability (small amplitudes of oscillations) the gauge current was small in value and constant in time. At the developed instability, the gauge current had a peak appearance correlated with density oscillations, and the current maxima were observed in phase with the density maxima. Thus, plasma flow across the field had a character of ejections. The ejection occurred practically simultaneously along the whole length of plasma cord (i.e. the phase shift was absent in different points of the axis) and was running together with the wave by azimuth. Since the diffusion coefficient was close to the Bohm's coefficient, turbulent diffusion was observed.

Thus, according to /48-50/ (at the limit of applicability of the theory, $k_x \sim k_y \sim k_\perp$) we get

$$W_k \sim k_\perp^{-(\alpha_1+\alpha_2)} \cdot k_z^{-\alpha_3} ,$$

where $\alpha_3 << \alpha_2, \alpha_1$. After transforming to spherical coordinates in k-space, $k_\perp \hat{=}$

$= k \sin \theta$, $k_z = k \cos \theta$, we obtain

$$W_k \sim k^{-(\alpha_1+\alpha_2+\alpha_3)} f(\theta) \sim k^{-(\alpha_1+\alpha_2)} f(\theta) ,$$

where

$$f(\theta) = (\sin \theta)^{-(\alpha_1+\alpha_2)} \cdot (\cos \theta)^{-\alpha_3} .$$

Next we integrate $W_k$ over angle $\theta$ from $\theta_0$ to $\pi/2$, where $\theta_0$ is determined from the condition of applicability of drift wave theory /48/: $\mathrm{Actg}\,\theta_{0min} \cong k_\perp/k_{zmax}$, $(k_z)_{max} \sim \omega_*/V_{Ti}$, $V_{Ti}$ being the thermal velocity of ions, $\omega_*$ the drift frequency. This results in

$$W_k \sim k^{-(\alpha_1+\alpha_2)} .$$

According to (53)

$$W_k^{(1)} \sim k^{-3.5} ; \quad W_k^{(2)} \sim k^{-4.5} . \tag{54}$$

Let us note that spectra (53) coincide with two of the three spectra found in /44-46/ for the case of the Rossby short waves. The explanation why the third spectrum obtained in /44-46/ cannot be realized, is given in /30/. Earlier the second spectrum (54) was obtained from dimensional estimations for the Rossby waves in the -plane /47/. Formulae (54) should be understood as $W_k \sim k^{-\gamma}$, where $\gamma \in (3.5;4.5)$, i.e. $\gamma \simeq 4$. Indeed, numerical solutions of the equation of the same type as those obtained in the present paper (23), (31), have shown the form of the spectrum to be close to $\gamma \simeq 4$ /43/.

## 6. ASTROPHYSICAL APPLICATIONS

### 6.1. Observational Correlations between the Parameters of Gas Structures and Salpeter Mass Spectrum /16/

Let us determine function E(k) from the condition

$$E(k)dk = W_k k^2 dk . \tag{55}$$

According to /15/

$$\int_k^\infty E_k dk \sim v^2 , \tag{56}$$

where $\Lambda \sim 1/k$. Since $W_k \sim k^{-4}$, then $E(k) \sim k^{-2}$ and from (56) we obtain /16/

$$\rho_\lambda \sim \lambda^{-1} . \tag{58}$$

Comparing (58) with (2), we can see that the observed dependence (2) for gas density of scale $\lambda$ /13/ is obtained. The Navier-Stokes equation results in the following estimations $v\nabla v \sim \nabla P/\rho$, or $v_\lambda^2/\lambda \sim P_\lambda/(\rho_\lambda \lambda)$. Using (57), (58), we obtain $P_\lambda \equiv$

$\equiv n_\lambda m_\lambda v_\lambda{}^2$ = const ($m$ is the mass of gas elements of scale $\lambda$, $n_\lambda$ is a number of such masses in a unit volume). From the last equation, with regard to $m \sim \rho\lambda^3 \sim \lambda^2$, i.e. $\lambda \sim m^{1/2}$, we obtain:

$$n \sim m^{-1.5} \, , \tag{59}$$

which coincides with the observed Salpeter mass spectrum (4) /31/. With the help of theoretical spectra (54) we can also explain deviations from average masses (i.e. $m \sim m_\theta$) distributions of the number of stars (4) for the region of large ($m \gg m_\theta$) and small ($m \ll m_\theta$) masses (for more detail see /16/).

## 6.2. About Possible Observational Manifestations of Modons Generation

Using the results of Sect.4, let us compare the direction of velocity $u_M$ of modons (solitary dipole vortexes) with y-th component of the phase velocity of the Rossby waves $u_{ph}$ in the two limiting cases: 1) $\lambda \gg \lambda_J$, 2) $\lambda \ll \lambda_J$. In calculations we shall take the fact that in all observed systems density decreases with radius into account, i.e.

1) $\lambda \gg \lambda_J$

$$u_{phy} \equiv \frac{\omega}{k_\phi} = -\frac{1}{\alpha-2}\frac{\omega_0{}^{2\,\prime}}{k_\perp{}^2} = \begin{cases} > 0 \, , & \alpha > 2 \, ; \\ < 0 \, , & \alpha < 2 \, ; \end{cases} \tag{60}$$

$$u_M = \begin{cases} < 0 \, , & \alpha > 2 \\ > 0 \, , & \alpha < 2 \end{cases} \tag{61}$$

2) $\lambda \ll \lambda_J$

$$u_{phy} \equiv \frac{\omega}{k_\phi} = -2\Omega_0 \frac{\ln'\rho_0}{k_\perp{}^2} > 0 \, ; \tag{62}$$

$$u_M < 0. \tag{63}$$

It is seen from (60)-(63) that in both limit cases

$$u_M/u_{phy} < 0 \, . \tag{64}$$

The physical sense of the last condition is the following. Being stationary structures, modons cannot spend their energy on the Rossby waves radiation due to the Cherenkov resonance. This is guaranteed by condition (64) /26/.

So far astrophysical literature has dealt with monopole vortexes /26,52-56/ the observable discovery of which has either stimulated the development of the theory (as, e.g. /52/), or has been itself the result of the test of theoretical investigations (such is the recent paper /56/).

Meanwhile, during the last three decades at least, astrophysical literature gives many examples of observations of pairs of close objects. The hierarchy of these objects is extremely broad: pairs of galaxies /57,59/, double cores of galaxies /58,60-62/, and binaries /31/. The determination of the direction of the rotation of spectral binaries is a problem that can be hardly solved in the near future. A similar problem for eclipsed binaries is now planned jointly with the Special Astrophysical Observatory of the USSR Acad. Sci. and the Sternberg Astronomical Institute of Moscow University. Let us note that the most active stars are close binaries. A good example of such a pair is SS 433 object /63/. It is noted in /66-68/ that properties of SS 433 resemble in miniature the active nuclei of galaxies. As for the determination of the rotation for the pairs of galaxies and for double cores, in cases where those directions are determined, they turn to be opposite more often. This is the case of the Markarian galaxy 266 with two nuclei, rotating into opposite directions /60/. Two-dimensional analogs of the objects mentioned above may be dipole solitary vortexes in laboratory experiments in shallow water /26,69/ and ocean.

In conclusion we express our gratitude to Drs. V.D.Larichev, A.B.Mikhailovsky, S.S.Moiseev and M.V.Nezlin for theoretical discussions and to A.I.Ginsburg, B.V.Komberg, A.P.Peterosian, A.V.Tutukov, E.Ye.Khachikian and B.M.Shustov for discussion on astrophysical applications.

## APPENDIX

Let us assume a dispersion law to be

$$\omega_k \sim k_y + \mu\Omega_k , \tag{A.1}$$

where $\mu$ is a small parameter and

$$\Omega_k \sim |k_y|^a |k_x|^b |k_z|^c \operatorname{sgn} k_y . \tag{A.2}$$

Here we used the well-known form of kinetic equation /29/

$$\frac{\partial N_k}{\partial t} \sim \int u(k,k_1,k_2)\Big[N_{k_1}N_{k_2} - N_k N_{k_1} \cdot \operatorname{sgn}(\omega_k \omega_{k_r}) -$$

$$- N_k N_{k_2} \operatorname{sgn}(\omega_k \omega_{k_1})\Big]\delta(\omega_k - \omega_{k_1} - \omega_{k_2})\delta(k-k_1-k_2)dk_1 dk_2 , \tag{A.3}$$

where $N_k$ is a "number of quanta". Further

$$u(k,k_1,k_2) \equiv |V(k,k_1,k_2)|^2 \tag{A.4}$$

The scale invariance of V is proposed:

$$V(\varepsilon_y k_y, \varepsilon_x k_x, \varepsilon_z k_z; \varepsilon_y k_{1y}, \varepsilon_x k_{1x}, \varepsilon_z k_{1z}; \varepsilon_y k_{2y}, \varepsilon_x k_{2x}, \varepsilon_z k_{2z}) =$$

$$= \varepsilon_y^u \varepsilon_x^v \varepsilon_z^w V(k_y, k_x, k_z; k_{1y}, k_{1x}, k_{1z}; k_{2y}, k_{2x}, k_{2z}) \quad \text{(A.5)}$$

Let us also assume

$$N_k \sim |k_y|^\alpha |k_x|^\beta |k_z|^\gamma \quad .$$

Our aim is to find the values of the coefficients $\alpha$, $\beta$ and $\gamma$ for a stationary spectrum when $\partial N_k/\partial t = 0$.

It is easy to see that

$$\partial N_k/\partial t \sim I_1 + I_2 + I_3 , \quad \text{(A.6)}$$

where

$$I_j = \int_0^\infty dk_{1y} dk_{2y} \int_{-\infty}^\infty dk_{1x} dk_{2x} dk_{1z} dk_{2z} G , \quad j = 1,2,3$$

$$G \begin{smallmatrix}1\\2\\3\end{smallmatrix} \equiv U \begin{smallmatrix}1\\2\\3\end{smallmatrix} (N_1 N_2 \mp NN_1 \mp NN_2) \delta(k \mp k_1 \mp k_2) \delta(\Omega \mp \Omega_1 \mp \Omega_2)$$

where for the sake of brevity $u_1 \equiv u(k,k_1,k_2)$, $u_2 \equiv u(k_2,k_1,k)$, $u_3 \equiv u(k_1,k,k_2)$ and $N$, $N_1$, $N_2$, $\Omega$, $\Omega_1$, $\Omega_2$ mean correspondingly $N_k$, $N_{k_1}$, $N_{k_2}$, $\Omega_k$, $\Omega_{k_1}$, $\Omega_{k_2}$. The set of $\Omega$, $\Omega_1$, $\Omega_2$ is considered to be positive, without restriction of generality.

Let us introduce dimensionless arguments

$$p_i = k_{iy}/k_y , \quad h_i = k_{ix}/k_x , \quad q_i = k_{iz}/k_z , \quad i = 1,2 .$$

Then

$$k_i = (k_y p_i, k_x \sigma_{hi} h_i, k_z \sigma_{qi} q_i) , \quad i = 1,2, \quad \sigma = \pm 1 . \quad \text{(A.7)}$$

In a similar way, $\nu_i = \Omega_i/\Omega$ , $n_i = N_i/N$, $i = 1,2$. So

$$I_1 = \frac{N_k}{|\Omega_k|} |k_y|^{2u+1} |k_x|^{2v+1} |k_z|^{2w+1} \cdot \sum_\sigma J_1(\sigma_{h_1}, \sigma_{q_1}; \sigma_{h_2}, \sigma_{q_2}) , \quad \text{(A.8)}$$

where

$$J_1(\sigma) = \int dp_1 dp_2 dh_1 dh_2 dq_1 dq_2 \hat{u}_{1,\sigma} (n_1 n_2 - n_1 - n_2) \cdot$$

$$\cdot \delta(1-p_1-p_2) \delta(1-\nu_1-\nu_2) \delta(1-\sigma_{h_1} h_1 - \sigma_{h_2} h_2) \cdot \delta(1-\sigma_{q_1} q_1 - \sigma_{q_2} q_2) , \quad \text{(A.9)}$$

$$u_{1,\sigma} \equiv u(\vec{1}; p_1, \sigma_h, h_1, \sigma_{q_1} q_1; p_2 \sigma_{h_2} h_2, \sigma_{q_2} q_2) , \quad \vec{1} \equiv (1,1,1) .$$

In such a manner we may also rewrite $I_2$ and $I_3$, and besides $I_2(\sigma)$, $I_3(\sigma)$ may be reduced to the form of $J_1(\sigma)$ by some algebra. As a result

$$\frac{\partial N_k}{\partial t} \sim \frac{N_k^2}{|\Omega_k|} |k_y|^{2u+1} |k_x|^{2v+1} |k_z|^{2w+1} \cdot I , \quad \text{(A.10)}$$

$$I = \int dp_1 dp_2 n_1 n_2 KRQ_1 \delta(1-\nu_1-\nu_2)\delta(1-p_1-p_2) \tag{A.11}$$

$$k = 1 - \lambda_1 - \lambda_2 , \quad R = 1 - n_1^{-1} - n_2^{-1} ,$$

$$\lambda_i = p_1^{a-2(1+u+\alpha)} h_1^{b-2(1+v+\beta)} q_i^{c-2(1+w+\gamma)} , \quad i = 1,2 ;$$

$$Q_1 = \sum_\sigma u_{1,\sigma} \delta(1-\sigma_{h_1} h_1 - \sigma_{h_2} h_2)\delta(1-\sigma_{q_1} q_1 - \sigma_{q_2} q_2) \tag{A.12}$$

$$d\vec{p}_1 \equiv dp_1 dh_1 dq , \quad d\vec{p}_2 \equiv dp_2 dh_2 dq_2 .$$

The condition of stationarity $\partial N_k/\partial t = 0$ is now reduced to $k = 0$.

Let us assume that, for example, $b \neq 0$ in (A.2). By making a substitution $h_i = (\nu_i p_i^{-a} q_i^{-c})^{1/b}$, $i = 1,2$ , we rewrite $k = 0$ as

$$1 - p_1^{\xi} \nu_1^{\eta} q_1^{\zeta} - p_2^{\xi} \nu_2^{\eta} q_2^{\zeta} = 0 \tag{A.13}$$

where

$$\xi = 2\, a(1+v+\beta)/b - (1+u+\alpha) ,$$
$$\eta = 1 - 2(1+v+\beta)/b , \quad \zeta = 2\, c(1+v+\beta)/b - (1+w+\gamma) .$$

Condition (A.13) is fulfilled when

$$\xi = 0 , \quad \eta = 1 , \quad \zeta = 0 \tag{A.14}$$

or

$$\xi = 1 , \quad \eta = 0 , \quad \zeta = 0 .$$

As for $\alpha$, $\beta$, $\gamma$ the following two sets are obtained

$$\alpha_0^{(1)} = -(1+u) , \quad \beta_0^{(1)} = -(1+v) , \quad \gamma_0^{(1)} = -(1+w) \tag{A.15}$$

$$\alpha_0^{(2)} = a/2 - (3/2+u), \quad \beta_0^{(2)} = b/2 - (1+v), \quad \gamma_0^{(2)} = c/2 - (1+w) \tag{A.16}$$

The additional spectra found in /44-46/ correspond to the case when the only terms not vanishing in (A.12) are those with $\sigma_{h_1} = \sigma_{h_2} = 1$ or $\sigma_{q_1} = \sigma_{q_2} = 1$ .

Let us represent the sought solution of our equations (23) or (27) in the form $\tilde{\Psi}$ or

$$\tilde{f} = \int \phi_k(t) e^{-i\omega_k t} e^{ikr} d^2 r , \tag{A.17}$$

where two-dimensional vectors k and r have x and y-components and $\omega_k = (1/(\alpha-2))\cdot (\omega_0^{2\prime}/\Omega_0)(k_y/k_\perp^2)$ or $\omega_k = 2\Omega_0(\rho_0'/\rho_0)(k_y/k_\perp^2)$ for equations (23) and (27), corres-respondingly. $\omega_k$ is an eigenfrequency of the Rossby waves. Then in terms of $\phi_k$ the Fourier representation of (23) and (27) gives

$$\frac{\partial \phi_k}{\partial t} \sim \sum_{k_1+k_2=k} k_1, k_2\, t\, \frac{k_{2\perp}^2 - k_{1\perp}^2}{k_\perp^2} \phi_{k_1} \phi_{k_2} \cdot \exp(-i(\omega_{k_1} + \omega_{k_2} - \omega_k)t) . \tag{A.18}$$

As (see /70/)

$$W_k \sim k_\perp^2 |\phi_k|^2 \,, \tag{A.19}$$

then

$$N_k \sim k_\perp^2 |\phi_k|^2 / |k_y| \quad . \tag{A.20}$$

Using a normalized potential

$$c_k \sim (1+k_\perp^2 \rho_0^2)\phi_k / |k_y|^{1/2} \tag{A.21}$$

and taking into account the decay condition $\omega_{k1} + \omega_{k2} - \omega_k \simeq 0$, we may represent (A.18) as

$$i \frac{dc_k}{dt} = \sum_{k_1+k_2=k} V(k,k_1,k_2) c_{k_1} c_{k_2} \exp(-i(\omega_{k_1}+\omega_{k_2}-\omega_k)t) \tag{A.22}$$

with

$$V(k,k_1,k_2) \sim |k_y k_{1y} k_{2y}|^{1/2} \cdot \left(\frac{1}{k_{1x}} + \frac{1}{k_{2x}} - \frac{1}{k_x}\right) \,. \tag{A.23}$$

According to (A.2) and (A.5), this corresponds to

$$\alpha = 1 \,, \quad b = -2, \quad u = 3/2 \,, \quad v = -1 \,. \tag{A.24}$$

Then the system (A.23), (A.24) yield

$$W_k^{(1)} \sim k_y^{3/2} k_x^{-2} \,, \quad W_k^{(2)} \sim k_y^{3/2} k_x^{-3} \,. \tag{A.25}$$

## REFERENCES


1. V.L.Polyachenko, A.M.Fridman. Equilibrium and Stability of Gravitating Systems. Nauka, Moscow, 1976.
2. A.M.Fridman, V.L.Polyachenko. Physics of Gravitating Systems. Vol.2. Springer, New York, Berlin et al, 1984.
3. V.L.Polyachenko, A.M.Fridman. Zh.Eksp.Teor.Fiz., 1988, 94, 1.
4. I.I.Pasha, A.M.Fridman. Sov.Phys.-JETP (in press).
5. A.M.Fridman. Sov.Phys.-JETF, in press.
6. N.N.Gor'kavij, A.M.Fridman. Sov.Phys.-Usp., in press.
7. M.G.Abramian. Doctorate thesis. Byurakan Astrophysical Observatory, 1986.
8. A.B.Makhailovskiy, V.I.Petriashvili, A.M.Fridman. Pis'ma v JETP, 1977, 26,129; 341. Astron.Zh., 1979, 56, 279 (in Russian).
9. S.M.Churilov, I.G.Shukhman. Astron.Zh., 1981, 58, 260; 1982, 59, 1093 (in Russian).
10. V.E.Zakharov. In: Plasma Physics Foundations. Eds. A.A.Galeev, P.Sudan, Vol.2, Energoatomizdat, Moscow, 1984, 48 (in Russian).

11. A.N.Kolmogorov. Sov.Phys.-Doklady, 1941, 30, 299 (in Russian).

12. A.M.Obukhov. Izvestiya AN SSSR, Geography and Geophysics, 1941, 4-5, 453 (in Russian).

13. P.C.Myers. Astrophys.J., 1983, 270, 105.

14. R.N.Henriksen, B.E.Turner.Astrophys.J., 1984, 287, 200.

15. L.D.Landau, E.M.Lifshitz. Hydrodynamics. Nauka, Moscow, 1986 (in Russian).

16. A.M.Fridman. Astronomicheskij Tsirkular, 1988, 1532, 25 (in Russian).

17. A.B.Mikhailovskij. Nonlinear Phenomena in Plasma and Hydrodynamics. Ed. R.Z. Sagdeev. Mir, Moscow, 1986 (in Russian).

18. I.G.Charney. Geophys.Publ.Kosjones Vors.Videtship Acad. Oslo, 1948, 17, 3.

19. A.M.Obukhov. Izvestiya AN SSSR, Geography and Geophysics, 1949, 13, 4, 281 (in Russian).

20. A.M.Obukov. Turbulence and Dynamics of Atmosphere. Gidrometeoizdat, Leningrad, 1988, pp.409 (in Russian).

21. A.Hasegawa, K.Mima. Phys.Fluids, 1978, 21, 87.

22. A.G.Morozov. Astron.Zh., 1985, 62, 209 (in Russian).

23. C.G.Rossby. J.Marine Res., 1939, 2, 38.

24. G.K.Batchelor. An Introduction to Fluid Dynamics. Mir, Moscow, 1973 (in Russian).

25. J.Pedlosky. Geophysical Hydrodynamics. Mir, Moscow, 1984 (in Russian).

26. M.V.Nezlin. Sov.Phys.-Usp., 1986, 150, 3 (in Russian).

27. L.I.Rudakov, R.Z.Sagdeev. Sov.Phys.-Dokl., 1961, 138, 581.

28. A.B.Mikhailovskij. Theory of Plasma Instabilities, Vol.2. Atomizdat, Moscow, 1977 (in Russian).

29. A.A.Galeev, R.Z.Sagdeev. In: Problems of Plasma Theory, Vol.7. Ed.. M.A.Leontovich. Atomizdat, Moscow, 1973 (in Russian).

30. A.B.Mikhailovskij, S.B.Novakovskij, V.P.Lakhin, S.B.Makurin, E.A.Novakovskaya, O.G.Onishchenko. Preprint of IKI, No.1356, 1988.

31. A.G.Massevitch, A.V.Tutkov. Stellar Evolution: Theory and Observations. Nauka, Moscow, 1988 (in Russian).

32. V.D.Larichev, G.M.Reznik. Sov.Phys.-Doklady, 1976, 231, 1077.

33. R.Z.Sagdeev, V.D.Shapiro, V.I.Shevchenko. Pis'ma v Astron.Zh., 1981, 7, 505.

34. J.-L.Tassoul. Theory of Rotating Stars. Mir, Moscow, 1982.

35. A.V.Zasov. Doctorate thesis. Moscow Univ., 1988.

36. N.I.Shakura. Astron.Zh., 1972, 49, 945.

37. N.I.Shakura, R.A.Sunayev. Astron. and Astrophys., 1973, 24, 337.

38. J.D.Meiss, W.Horton. Phys.Fluids, 1983, 26, 990.

39. E.Janke, F.Ende, F.Lösch. Special Functions. Nauka, Moscow, 1964.

40. G.R.Flier, V.D.Larichev, J.C.McWilliams, G.M.Reznik. Dyn.Atmos.Oceans, 1980, 5, 1.

41. V.D.Larichev, G.M.Reznik. Oceanology, 1976, 16, 961.

42. J.Nycander. Preprint of Inst. of Technol. of Uppsala Univ., VPTEC 8768 R., Aug.1987.

43. A.Hasegawa, C.G.Maclennan., Y.Kodama. Phys.Fluids, 1979, 22, 2122.

44. A.G.Sazontov. In: Fine Structure and Synoptic Changeability of Seas. Tallinn, 1980, 147.

45. A.G.Sazontov. Preprint IPF Acad.Sci., 30, Gorki, 1981.

46. A.S.Monin, L.I.Piterbarg. Sov.Phys.-Dokl., 1987, 295, 86.

47. E.N.Pelinovskij. Oceanology, 1978, 17, 192.

48. A.A.Galeev, S.S.Moiseev, R.Z.Sagdeev. Atomic Energy, 1963, 15, 451.

49. N.S.Buchelnikova. Universal Instability in Potassium Plasma. Preprint IYaF, Acad.Sci., Siberian Branch, Novosibirsk, 1963.

50. N.D'Angelo, R.W.Motley. Phys.Fluids, 1963, 6, 422.

51. N.S.Buchelnikova, R.A.Salimov. Preprint No.54, IYaF Acad.Sci.USSR, Siberian Branch, Novosibirsk, 1966.

52. A.V.Zasov, G.A.Kyazumov. Pis'ma v Astron.Zh., 1981, 7, 131.

53. M.V.Nezlin, E.N.Snezhkin, V.L.Polyachenko, A.S.Trubnikov, A.M.Fridman. Pis'ma v Astron.Zh., 1986, 12, 504.

54. P.V.Baev, Yu.N.Makov, A.M.Fridman. Pis'ma v Astron.Zh., 1987, 13, 964.

55. V.I.Korchagin, V.I.Petviashvili, A.D.Ryabtsev. Pis'ma v Astron.Zh., 1988, 14, 317.

56. V.L.Afanasyev, S.N.Dodonov, J.Boulesteix, F.Bonnarel. Pis'ma v Astron.Zh., in press.

57. V.A.Ambartsumian. Nauchn.Trudy, P.298, Erevan, Armenian Acad.Sci., 1960.

58. V.A.Ambartsumian, ibid, 291.

59. D.A.Verner. Pis'ma v Astron.Zh., 1985, 11, 664; 1987, 13, 95; Astron.Tsirk, 1966, 1466, 1.

60. A.R.Petrosian, K.A.Saakian, A.E.Khachikian. Astrofizika, 1979, 15, 62, 210, 373; 1980, 16, 621.

61. Yu.N.Koroviakovskij. A.R.Petrosian, K.A.Saakian, A.E.Khachikian. Astrofizika, 1981, 17, 231.

62. M.Fraux, G.D.Illingworth. Astrophys.J., 1988, 327, Letters, 55.

63. B.Margon, S.A.Grandy, R.Dawnes. Astrophys.J., 1980, 241, 306.

64. A.C.Fabian, M.J.Rees. Month.Not.RAS, 1979, 187, 13.

65. B.V.Komberg. Preprint IKI, Acad.Sci.USSR, No.834, 1983.

66. V.M.Luty, A.M.Cherepashchuk. Astron.Zh., 1986, 63, 897.

67. A.M.Fridman. Astron.Tsirk., 1988, 1533, 25.

68. A.M.Fridman. Sov.Phys.-Dokl., in press.

69. A.I.Ginzburg, A.G.Kostianoy, A.M.Pavlov, K.N.Fedorov. Izvestiya Acad.Sci. USSR, Physics of Atmosphere and Ocean, 1987, 23, 170.

70. W.Horton. In: Handbook of Plasma Physics, Vol.II. North Holland, Amsterdam et al, 1984.

# Nonlinear Dynamics of Particle-Like Solutions of Inhomogeneous Fields

*I.S. Aranson, K.A. Gorshkov, A.S. Lomov, and M.I. Rabinovich*

Institute of Applied Physics, USSR Academy of Sciences,
46 Ulyanov Str., 603600 Gorky, USSR

Stable localized structures of two- and three-dimensional fields are investigated. A theory describing the interaction of the "elementary particles" and the formation of their bound states is constructed for scalar real and scalar complex fields. Three types of elementary structures: a ball, a torus and a baseball are revealed for the real fields. Localized spirals and toroidal rolls are the elementary structures of the complex fields.

## 1. INTRODUCTION

In the recent five-ten years there has appeared quite a number of papers dealing with various not one-dimensional spatial patterns in nonlinear media: periodic and quasi-periodic lattices, localized (particle-like) solutions, spiral structures, vortices, etc. /1-5/. As a rule, unified methods are applied for the analysis of the structures of different origins and the obtained results can be used in different fields of physics, and not only physics. This means that we witness the advent of a new field of nonlinear dynamics - the theory of structures in nonlinear media. Many problems of this theory were posed and solved before, in the theory of nonlinear waves in the first place. However, they were, as a rule, related (except, perhaps, hydrodynamics) to one-dimensional patterns: solitons, shock waves, dissipative structures (combustion fronts, etc.). The investigation of multi-dimensional structures that was necessitated by astrophysics, the physics of atmosphere and ocean, the nonlinear field theory, microelectronics and biophysics, gave new formulations of the problems. Among these problems are, for example, the study of pattern bifurcation, i.e. the off-beat variation of their topology when the governing parameter passes the critical value; the spatial interaction of structures, that is, the formation of bound spatial states, including "planet-like" states; their regular or chaotic spatio-temporal dynamics explaining, in particular, many manifestations of nonlinear field turbulence. The latter is also known as spatio-temporal chaos of structures, or topological turbulence. The main problems of the multi-dimensional theory are the mechanisms of the spatial localization of structures, their stability and the interaction of such particle-like patterns. All these problems are considered in this paper.



Research Reports in Physics **Nonlinear Waves 3**
Editors: A.V. Gaponov-Grekhov · M.I. Rabinovich · J. Engelbrecht

## 2. THE FORMULATION OF THE PROBLEMS

The particle-like solutions, i.e. the localized solutions of a nonlinear field, are the solutions decreasing rather fast from the localization maximum to the periphery. A more accurate and formal definition is based on the convergence of some integrals.

In order to work out the approach to the construction of the theory of localized structures, we shall consider laboratory and computer experiments.

In spite of the diversity of the media and experimental situations in which localized structures are observed, these structures are remarkably universal and the origin of their universality is, yet, to be elucidated. However, even now it is clear that the space dimensions, the form of nonlinearity and the type of spatial dispersion are most essential for pattern formation. Note that identical particle-like solutions are possible even in the fields with different dynamics, in particular, in conservative and nonequilibrium dissipative fields, if their nonlinear characteristics and dispersion features are balanced.

Let us put aside for the present the localization mechanisms and consider some peculiarities of the individual dynamics and interaction of structures. As a rough approximation based on various experiments one can arrive at a conclusion that there are "strong" and "weak" interactions of particle-like solutions. When the interaction is weak, the localized structures (dislocations, solitons, etc.) are located rather far from one another and the field of one structure at the centre of the localization of another can be considered to be weak. This facilitates the construction of a consistent theory of such adiabatic interactions. Strong coupling is the interaction during which particle-like solutions change qualitatively, appear and die "colliding" with each other. The result of strong interaction depends on the particle prehistory and field characteristics. We believe that in the description of the nonlinear dynamics of ensembles of structures strong interactions can be replaced by the rules according to which mutual conversions occur with subsequent transition to weak interactions that can be described approximately taking into account the conservation of structure of the localized object.

Apparently, not every nonlinear dynamics of localized structures can be described within the interaction scheme presented above ("adiabatic" - weak interaction and "bifurcation" - strong interaction). However, experiment furnished ample examples Therefore it is essential to formulate the general models describing local experimental situations, on the one hand, and allowing for a fairly complete analytic and computer investigation on the other hand. As such models we shall take generalized gradient systems, i.e. systems in the form

$$\hat{T}(\partial_t)\psi = - \frac{\delta F}{\delta\psi} + \varepsilon G(\psi,\vec{r},t), \quad \varepsilon \ll 1 . \tag{1}$$

Here $\psi$ is a set of physical variables; F is a functional having the sense of the free energy of the system (medium, field); G is the nonlinear operator which takes into account the external field effect, the deviation of (1) from the potential system and other factors.

For $\varepsilon = 0$, the dynamics of the particle-like solutions of (1) is determined by the form of the operator $T(\partial_t)$. If $T = \partial_t$, Eq.(1) is a conventional gradient system where all solutions are statical, as $t \to \infty$, and correspond to the local minima of the functional F (in this case F is the Lyapunov functional $dF/dt = -\int|\partial\psi/\partial t|^2 dr \leq 0$). From here it follows that the localized structures in such media either become fixed or go to the infinity, or die, for example, as a result of merging. However, the transition on the way to this final result may be rather versatile and multistaged. Consider as an example a result of a computer experiment with a two-dimesnional Swift-Hohenberg model /6/:

$$F = \int\{\frac{1}{2}\alpha\psi^2 - \frac{1}{3}\beta\psi^3 + \frac{1}{4}\psi^4 + \frac{1}{2}\left[(1+\nabla^2)\psi\right]^2\}d\vec{r} , \qquad (2)$$

where the dynamics of defect-localized structures was investigated on the background of a period lattice. Figure 1 which plots F(t) demonstrates that the transition to the state $F = F_{min} = const$ is, indeed, a sequence of adiabatic stages separated by fast motion to which the defect merging corresponds.

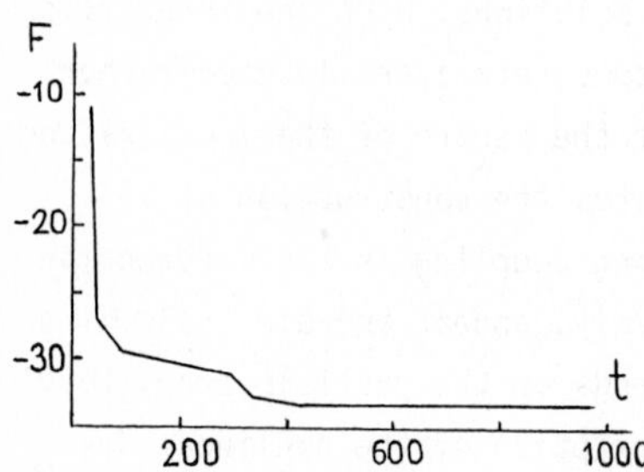


Figure 1. The free energy functional F versus time for model (2).

The existence of steady-state localized structures does not, apparently, depend on the form of the operator $\hat{T}(\partial_t)$ (they are the solutions of the equation $\delta F/\delta\psi = 0$), therefore it is not surprising that conservative media ("Hamiltonian" fields and non-equilibrium dissipative media) may have the same particle-like solutions. However, if these solutions do not exactly correspond to the minimum F, they will not tend to statical solutions in the conservative case, instead they will oscillate, perhaps even irregularly, like a ball rolling without friction over the bottom of a pit.

When the perturbations are small ($\varepsilon << 1$), particle-like structures will not be statical in dissipative media either, they will interact with external fields, walk chaotically, change their shape gradually, etc.

It will be shown below that the identification and investigation of models of the form (1) contribute to a better understanding of localized pattern formation and interaction.

## 3. TOPOLOGY OF MULTI-DIMENSIONAL STRUCTURES. EXAMPLES

3.1. It seems to be impossible to classify the topological properties of localized structures when the theory of such nonlinear patterns is only on the verge of its formation. Therefore we shall restrict ourselves to some examples which, however, demonstrate vividly the topological diversity of possible localized structures. The first example refers to the well-known generalized Ginzburg-Landau equation describing many physical models of two-dimensional nonequilibrium media:

$$\frac{\partial\psi}{\partial t} = \psi f(|\psi|^2) + \Delta\psi + \varepsilon G(\psi,\psi^*,x,y,t) \ . \tag{3}$$

Here $\psi = u + iv$ is a complex-valued function, f is a nonlinear function, and G is the operator taking into account the external field and the deviation (when $\varepsilon \neq 0$) of (3) from a potential system, which may be due, for example, to the complexity of the coefficients in diffusion and nonlinear terms in (3). Typical statical structures described within (3), when $\varepsilon = 0$, have the form

$$\psi^{(0)} = \Phi^{(0)}(\rho)\, e^{im\phi}, \quad \rho^2 = x^2 + y^2, \quad \phi = \text{arc tg}\,\frac{y}{x}, \quad m = 1,2... \tag{4}$$

and may be called the "cores" of a spiral wave. It is known that usually simple spiral waves with m = 1 are the solutions which are stable in time when $\varepsilon = 0$ /7/. Figure 2 presents the distribution $\Phi^{02}(\rho)$ with /8/:

$$f(|\psi|^2) = -1 + \beta|\psi|^2 - |\psi|^4 \ . \tag{5}$$

The family of solutions (4) depends on three arbitrary parameters ($x_0$, $y_0$ and $\phi_0$) which characterize the position of the spiral core and the direction of its arms. Below we shall need an explicit asymptotic form of the field of an individual spiral that can be readily obtained from the linearized equations (3):

$$\Phi^{(0)}(\rho) \underset{\rho\to\infty}{\to} \frac{ke^{-\rho}}{\sqrt{\rho}}, \quad k = \text{const} \ . \tag{6}$$

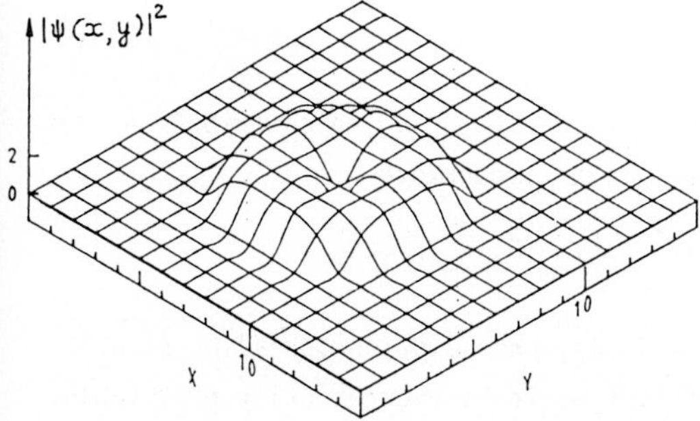


Figure 2. Intensity distribution $|a(x,y)|^2$ of the metastable spiral wave for model (3) with dunction f in the form (5), $\beta = 2.5$.

The spirals described by (4) with function (5) corresponding to the medium with hard exitation are unstable. However, these spirals may be long-lived. This may be demonstrated by simple considerations.

In the first place, stationary localized structures exist only when $\beta \geqq 2$. This follows from the fact that if the function $|\psi|^2$ has a maximum at point $x^*, y^*$, then $\Delta|\psi_{max}|^2 \leqq 0$. The function satisfies the equation

$$-|\psi|^2 + \beta|\psi|^4 - |\psi|^6 + \frac{1}{2}\Delta|\psi|^2 - |\nabla\psi|^2 = 0 . \tag{7}$$

In virtue of

$$\left.\frac{1}{2}\Delta|\psi|^2 - |\nabla\psi|^2\right|_{\substack{x=x^* \\ y=y^*}} \leqq 0 ,$$

we have $-|\psi_{max}|^2 + \beta|\psi_{max}|^4 - |\psi_{max}|^6 \leqq 0$, which is impossible when $|\psi_{max}|^2 \geqq 0$ and $\beta < 2$. Consequently, $|\psi|^2 = 0$ is the only steady-state solution for $\beta < 2$.

Localized structures are possible when $\beta > 2$ /8/ and their stability can be investigated by means of a free energy functional

$$F = \int\{|\psi|^2 - \frac{\beta|\psi|^4}{2} + \frac{|\psi|^6}{3} + |\nabla\psi|^2\}dx\, dy , \tag{8}$$

which can be represented in the form

$$F = \int\{|\psi|^2(1 - \frac{|\psi|^2}{\sqrt{3}})^2 + (\frac{2}{\sqrt{3}} - \frac{\beta}{2})|\psi|^4 + |\nabla\psi|^2\}dx\, dy . \tag{9}$$

The functional is positive when $2 \leqq \beta \leqq 4/\sqrt{3} = 2.3094$ and reaches its maximum in a trivial solution $\psi = 0$. Therefore it is energetically profitable for an arbitrary localized initial perturbation to collapse for $\beta \leqq 4/\sqrt{3}$.

On the other hand, the analysis shows that for $\beta > \beta_c = 4/\sqrt{3}$ localized perturbations generally benefit by spreading. In this case, a travelling transfer front is observed. Indeed, let us consider a solution in the form of a cylindrical front:

$$|\psi|^2 = \begin{cases} |\psi_m|^2 = \beta/2 + \sqrt{\beta^2/4 - 1} & , \quad r \leqq r_0, \quad r_0 \gg 1 , \\ \sim|\psi_m|^2\, e^{2(r-r_0)} \end{cases} \tag{10}$$

where $|\psi_m|^2$ is a maximum possible steady-state value of $|\psi|^2$. The contribution from the exponential "tail" can be neglected when $r_0 \gg 1$, therefore

$$F \approx \pi r_0^2\{|\psi_m|^2 - \frac{1}{2}\beta|\psi|^4 + \frac{1}{3}|\psi_m|^6\} = \pi r_0^2\{\frac{1}{2} - \frac{1}{6}|\psi_m|^4\}|\psi_m|^2 .$$

For $|\psi_m|^2 > \sqrt{3}$ (which corresponds to $\beta > \beta_c = 4/\sqrt{3}$), the solution spreading (i.e. the increase of $r_0$) leads to the decrease of F, i.e. it is energetically profitable. This situation describes the propagation of a cylindrical front. The same considerations are valid for a three-dimensional case.

Thus, localized spirals in media with hard excitation and long-wave instability are metastable structures. Their lifetime, however, may be arbitrary long when $\beta \to \beta_c$ because the spreading velocity is proportional to $|\beta - \beta_c|$.

3.2. The second example is the Ginzburg-Landau equation describing the media with hard excitation and short-wave instability:

$$\psi_t = -\psi + \beta|\psi|^2\psi - |\psi|^4\psi - (k_0^2+\Delta)^2\psi. \tag{11}$$

This equation describes the situation that is implemented with thermoconvection in a binary liquid when instability develops with a finite wave number and a finite frequency /9,10/. Equation (11) can be considered as the generalized Swift-Hohenberg equation for an oscillatory instability /11/.

Equation (11) has spiral solutions in the form (4) that are asymptotically stable (Fig.3) which was proved by numerical experiment. It is difficult to prove the stability analytically, but some considerations can be inferred from the free energy functional

$$F = \int\{|\psi|^2 - \frac{1}{2}\beta|\psi|^4 + \frac{1}{3}|\psi|^6 + |(k_0^2+\Delta)\psi|^2\}dx\, dy . \tag{12}$$

Apparently, the front propagation is not profitable in the considered model because of the structure of the differential operator. On the contrary, strongly inhomogeneous distributions, like the ones shown in Fig.3, must minimize the functional.

In contrast to the spirals described by Eq.(3), the spirals considered here have an oscillating "tail", the asymptotic form of which is given by the expression

$$|\psi(\rho,0)| = \frac{A_0}{\sqrt{\rho}}\, e^{-\alpha\rho} \cos(k\rho+\xi_0) , \tag{13}$$

where

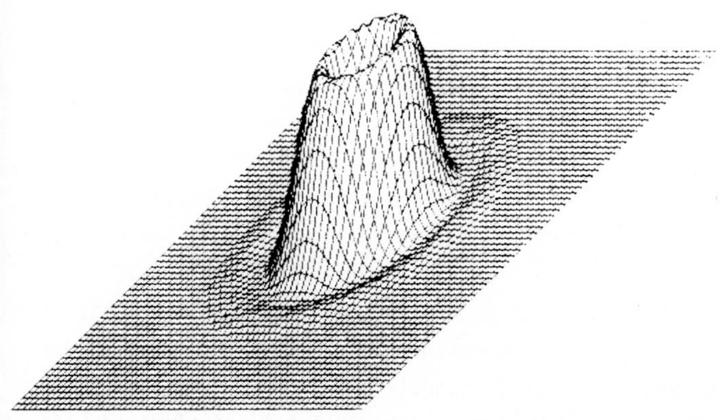

Figure 3. Distribution of the $|a(x,y)|^2$ field for an asymptotically stable spiral in the Ginzburg-Landau model with hard excitation and short-wave instability: $\beta$ = 2.5, $k_0$ = 1.

$$\alpha = \left(\frac{1}{2}\left(\sqrt{k_0^2+1} - k_0^2\right)\right)^{1/2}$$

$$k = \left(\frac{1}{2}\left(\sqrt{k_0^2+1} + 1\right)\right)^{1/2}$$

and $A_0$, $\xi_0$ are some constants.

3.3. The third example /12/ is related to a model described by a coupled system of the scalar Klein-Gordon equation

$$\frac{\partial^2 u}{\partial t^2} - c^2 \Delta u + g(u) = r(u,v) \tag{14}$$

and the equation taking into account spatial dispersion:

$$\frac{\partial^2 v}{\partial t^2} + (1+\Delta)^2 + f(u) = s(u,v) , \tag{15}$$

where functions r and s describe the coupling between fields u and v. It is essential to take into account spatial dispersion in (15) since it may determine the characteristic spatial scale of the sought localized formations. The system (14)-(15) was investigated for the nonlinear functions g and f corresponding to the double-pit potentials $g = u(1-\gamma u^2)$, $f(v) = v(\alpha-\beta v+v^2)$ and $r = v$, $s = \xi uv$. With such r, s, system (14)-(15) no longer belongs to the class of potential systems but retains its conservatism, which is apparent in a spatially discrete approximation. The resulting infinitely-dimensional dynamic system retains in time its phase space volume (the vector field divergence in the phase space of this system is equal to zero).

The static solutions of system (14)-(15) (when $\partial^2/\partial t^2 = 0$) were described using the dissipative analogy /13/:

$$\frac{\partial u}{\partial t} - c^2 \Delta u - u(1-\gamma u^2) = -v , \tag{16}$$

$$\frac{\partial v}{\partial t} + (1+\Delta)^2 v - v(\alpha-\beta v+v^2) = \xi uv \tag{17}$$

with the boundary conditions

$$u\Big|_s = 0 , \quad \frac{\partial v}{\partial n}\Big|_s = 0 . \tag{18}$$

Note that system (16)-(17) is in itself of interest since it can be considered as a generalization of the known Swift-Hohenberg model (see (17)) describing the formation of ensembles of structures and the defect motion in two-dimensional models of three-dimensional convection /3/. Variables u and v in this system can be interpreted, for example, as nonequilibrium density (v) and temperature (u) in a medium where an exothermic reaction proceeds. The systems (14)-(15) and (16)-(17) have identical equilibrium states, therefore stable static localized (particle-like) solutions of (16)-(17) may be expected in the initial system. Computer experiments showed that the model (14)-(15), indeed, describes the self-generation and stable existence of three-

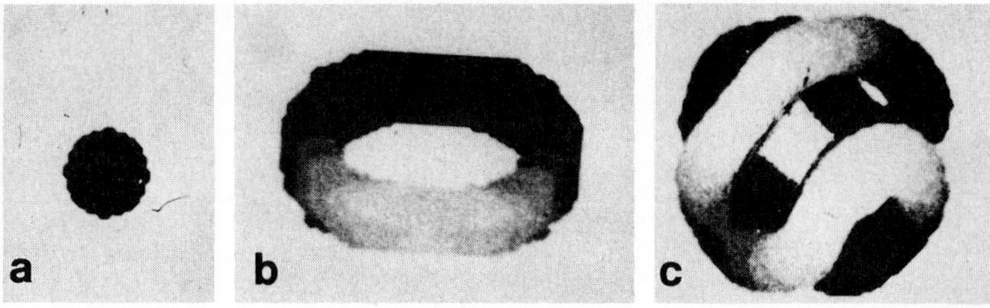


Figure 4. The "elementary particle" described by (16), (17): a) a ball; b) a torus; c) a baseball. The parameter values are: $\alpha = 0.3$, $\beta = 1$, $\gamma = 4$ and $c^2 = 0.1$ (here and in Figs.5 and 6 the spatial surface distribution is $v$ = const = 0.4; the parameter values are constant).

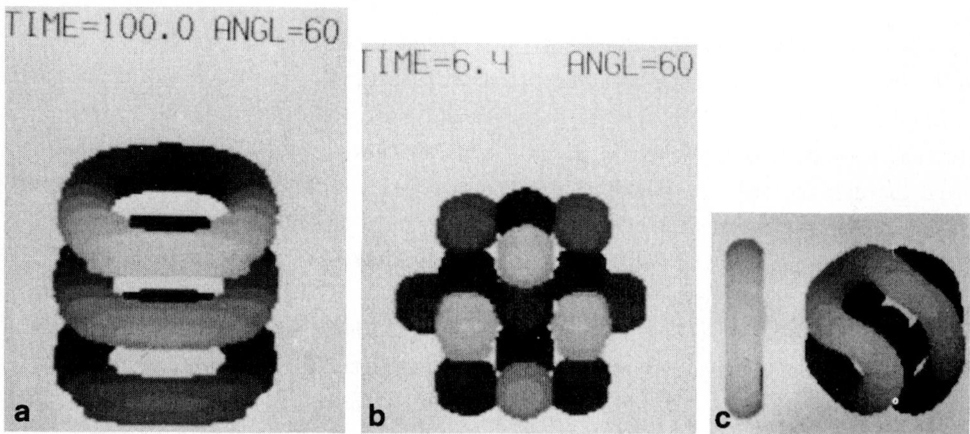


Figure 5. Stable bound states for system (16)-(17): a) three tori; b) a cluster of balls; c) a torus and a baseball with a ball inside.

dimensional particle-like solutions of different topologies. Three types of "elementary" particles were revealed: a ball, a torus and a baseball (Fig.4) with a characteristic size ~1. Under appropriate initial conditions stable solutions were realized in the form of bound states of elementary particles, identical and different (see Fig.5). The orientation of elementary particles in space is determined only by initial conditions and their topology and size are universal. Note that various structures that are not directly bound states of elementary particles such as, for example, a metastable helix in Fig.6, can be formed within the frames of the model (14)-(15). However, such structures are not attractors (in this case trivial attractors or equilibrium states) of the system of interest and transit to bound states as $t \to \infty$.

Direct computer experiments confirm stable existence of elementary particles in terms of the initial conservative field equations (14)-(15). In particular, a ball and bound states belonging to this type of elementary particles demonstrate stable existence without pronounced distortions or deformations.

Figure 6. Metastable helix of system (16)-(17).

Since system (14)-(17) is autonomous (i.e. there is no explicit dependence on coordinates and time in these equations) and isotropic, all localized solutions must be invariant with respect to spatial shears and turns, i.e. the families of these solutions depend, generally speaking, on six parameters ($\vec{r}_0 = \{x_0, y_0, z_0\}$ are the centre coordinates and $\phi_0$, $\theta_0$ and $\eta_0$ are the angles characterizing their orientation in space). The six parameters, however, determine only the elementary solution in the form of a baseball, the families of other elementary particles contain, in view of apparent structural symmetries, a smaller number of parameters (a torus contains five, $\vec{r}_0$, $\phi_0$, $\theta_0$ and a ball, three ($\vec{r}_0$) parameters). There are no direct computer indications on the presence of other (internal) parameters. (This problem is quite complicated and delicate and needs further investigation).

When seeking the asymptotic forms of localized solutions, it should be borne in mind that all elementary particles exist against the background of a stable homogeneous state of the system ($u = 0$, $v = -\gamma^{-1/2}$). Therefore the asymptotic forms of the localized solutions, $u_\infty^{(0)}$ and $v_\infty^{(0)}$ satisfy the equations

$$\begin{aligned} &\left[(1+\Delta)^2 - (\alpha+\gamma^{-1/2})\right] v_\infty^{(0)} = 0 , \\ &\left[c^2\Delta + 2\right] u_\infty^{(0)} = \xi v_\infty^{(0)} . \end{aligned} \tag{19}$$

Since the first equation in (19) can be represented in the form $\hat{A}_+\hat{A}_- v_\infty^{(0)} = 0$, where $\hat{A}_\pm = \Delta + 1 \pm i(\alpha+\gamma^{-1/2})^{1/2}$, we obtain

$$\begin{Bmatrix} u^{(0)}(r) \\ v^{(0)}(r) \end{Bmatrix} \xrightarrow[r\to\infty]{} \begin{Bmatrix} (c^2k^2 + 1) \\ 1 \end{Bmatrix} \frac{e^{ikr}}{r} \tag{20}$$

where $\mathrm{Re}k = \mathrm{Re}\sqrt{1+i(\alpha+\gamma^{-1/2})^{1/2}}$ and $I_m k = I_m\sqrt{1+i(\alpha+\gamma^{-1/2})^{1/2}}$. Note that the asymptotic solutions which are oscillatory and decreasing exponentially against r are inherent in all types of elementary particles (only the multipliers defining angular dependences may be different) and, besides, in two-and one-dimensional solutions, if any.

Concluding this short survey on localized structures we would like to note that the enumerated properties of such patterns are sufficient for the analytic description of various aspects of their nonlinear dynamics.

## 4. WEAK INTERACTION. ADIABATIC THEORY

4.1. The general concept of the adiabatic description of the dynamics of particle-like solutions is quite clear and close to that in the investigation of the interaction and evolution under the action of various perturbations of nonlinear solitary waves, i.e. solitons /14,15/. In this respect the approach presented below can be considered as the development on the basic idea set forth in those papers concerning the localized patterns of arbitrary physical origin.

With the perturbations (inhomogeneous or nonstationary medium, exposure to external forces, the effect of weak fields on other particle-like solutions, etc.) taken into account, the solution of interest near the given localized structure can be sought in the form of series

$$\psi(r,t) = \psi^{(0)}(\vec{r}) + \sum\varepsilon^n\psi^{(n)}(\vec{r},t) \ , \tag{21}$$

where the dominant term of the series, $\psi^{(0)}(r)$, is a known elementary localized solution and $\varepsilon$ is a small parameter of the problem. When we deal with the interaction of localized patterns, this parameter is equal to the ratio of the fields of external localized structures at the site of the given localized structure to the maximal value of the field of this structure. Equations (1) and (21) yield in a standard fashion the following successive approaximations for $\psi^{(n)}$:

$$\hat{L}\psi^{(n)} = H^{(n)} \tag{22}$$

Here $H^{(n)}$ contain only the functions of preceding approximations and the specified weak fields of foreign localized structures. The solution of the linear evolution problem (22) can be constructed by the expansion of $\psi^{(n)}(\vec{r},t)$ in a complete set of eigenfunctions of the operator $\hat{L}$. A typical peculiarity of the solutions obtained in this fashion is a secular divergence (in time) of the part of the series related to the localized eigenfunctions belonging to the zero eigenvalue of the discrete spectrum of operator L /14/. These localized eigenfunctions always include the ones obatined by the variation of the generating solution $\psi^{(0)}(\vec{r},\vec{C})$ over the parameters $C(\vec{C} = \{C_1,C_2...C_m\})$ characterizing the family of such solutions. This suggests that introducing into $\psi^0$ and, consequently, into $H^{(1)}$ an arbitrariness in the form of an unknown dependence of the parameters $\vec{C}_\ell$ on slow time ($\tau=\varepsilon t$), the secular divergence caused by these eigenfunctions can be suppressed, provided that the modified right-hand sides of $\tilde{H}$ in (21) ($\tilde{H} = \vec{C}_t\nabla_{\vec{C}}\psi + H$) are orthogonal to the corresponding eigenfunctions $\psi^+$ of the operator conjugate to $\hat{L}$:

$$\int\psi^+\tilde{H}^{(n)}d\vec{r} = 0 \ . \tag{23}$$

If a subsystem of the localized functions corresponding to a zero eigenvalue consists only of $\partial_{C_\ell}\psi^{(0)}$ ($\ell=1,2...$), then it may be argued, at least for the solutions to a first approximation (n=1), that with the use of the orthogonality conditions (23) the primary divergence is eliminated and the initial distributed system can be described approximately by a reduced finite-dimensional system:

$$\sum_{\ell=1}^{m} \int \psi_{C_k}^{+} \psi_C \frac{dC_\ell}{dt} d\vec{r} = \int \psi_{C_k}^{+} H^{(1)} d\vec{r} . \tag{24}$$

The most essential problem that remains uncertain when using (24) is the determination of the number of parameters characterizing the families of the localized structures of interest. This problem, in its turn, is closely connected with the symmetry of initial field equations and with the structure of localized solutions. Thus, the number of the "external" parameters describing the position and the orientation of static localized structures in space is maximal (six) in the case of an isotropic and a homogeneous model (1): three angular parameters ($\phi_0$, $\theta_0$, $\eta_0$) and centre coordinates ($\vec{r}_0 = \{x_0,y_0,z_0\}$). In concervative models (1), besides these six "external" parameters, four more may appear: three velocity components ($v_x,v_y,v_z$) and energy which are, evidently, due to the inertial transformation invariance of the coordinate system and to the time shift invariance. In the general case not all these parameters are independent. For example, for the solutions depending on $\vec{r} - \vec{v}t$ the time shifts are equivalent to the space shifts as a result of which energy is not an independent parameter. The presence of "internal" parameters is related to the symmetry of the field equations. In particular, the invariance of the unperturbed Ginzburg-Landau equation (3) as well as of equation (11) to the phase shifts of the complex-valued field function ($\psi \to \psi e^{i\phi_0}$) is apparent. A complete solution of the problem on the number of "internal" parameters is still in question. Below we shall assume that a family of localized solutions of Eqs.(3) and (11) contains such a parameter while the solutions of (14)-(17) have no parameters of this type. Restricting ourselves to such a set of "external" and "internal" parameters, we shall briefly describe the derivation of (24) for the problem on the interaction of the elementary structures described by the models presented in the previous section.

4.2. Interaction of Elementary Ball-Type Structures. First let us consider the problem in terms of the nonconservative model (16)-(17). In this case the linear operator L has a form

$$\hat{L} = \begin{pmatrix} \partial_t - c^2\Delta - (1-3\gamma u^{2(0)}); & -1 \\ -\xi u^{(0)}; & \partial_t + (1+\Delta)^2 + \alpha - 2pv^{(0)2} - \xi u^{(0)} \end{pmatrix} \tag{25}$$

and is a non-self-conjugate one. Because the variable coefficients of operator $\hat{L}$ (as well as of $\hat{L}^+$) depend only on $r = |\vec{r}|$, all their eigenfunctions can be sought by the separation of variables $\{u^+(\vec{r}),v^+(\vec{r})\} = y_p^{(q)}(\theta,\phi)\{u^+(r),v^+(r)$ where $y_p^{(q)}$ are spherical functions. The eigenfunctions needed in (25) correspond to the translational mode and are proportional to $y_1$, therefore they can be written as

$$\dot{\vec{r}}\{u^+(r),v^+(r)\} = \nabla\{\tilde{u}^+(r),\tilde{v}^+(r)\} . \tag{26}$$

The gradient structure of the eigenfunctions and the spherical geometry of $\tilde{v}^+(r)$ and $\tilde{u}^+(r)$ contribute to an exquisite form of Eq.(24). Thus, owing to the symmetry of $u^{(0)}$, $v^{(0)}$ and $\tilde{u}^+$, $\tilde{v}^+$, and to the consequent explicit identities of the form $\int d\vec{r}\; u_x \tilde{u}_y{}^+$, system (24) takes on a normal form relative to the velocity of the ball centres:

$$M \frac{dr_{12}}{dt} = \int\{\tilde{u}^+,\tilde{v}^+\}H_{1,2}{}^{(1)}\, d\vec{r} , \tag{27}$$

where

$$M = \frac{8\pi}{3}\int_0^\infty r\; u^{(0)}(r)\tilde{u}^+(r) + v^{(0)}(r)\tilde{v}^+(r)\; dr .$$

The gradient structure of the eigenfunctions of L specifies the gradient form of the right-hand sides of (27). Indeed, the right-hand side of (22) written, e.g. for the first particle, in the first approximation is equal to

$$H_1{}^{(1)} = \left\{ \begin{array}{l} 3\gamma\; u_1{}^{2(0)} u_2{}^{(0)} \\ 3v_1{}^{2(0)} v_2{}^{(0)} - 2\beta v_1{}^{(0)} v_2{}^{(0)} - \zeta u_1{}^{(0)} v_2{}^{(0)} - \xi u_2{}^{(0)} v_1{}^{(0)} \end{array} \right\} \tag{28}$$

where $u_1{}^{(0)}$ and $v_1{}^{(0)}$ depend only on r while $u_2{}^{(0)}$ and $v_2{}^{(0)}$ depend on $|\vec{r}-\vec{R}|$ in the coordinate system originating from the first particle; and $|\vec{R}| = |\vec{r}_1-\vec{r}_2|$ is the distance between the particles. Substituting (28) into (27) and integrating by parts, (27) can take on a form

$$\int\{\tilde{u}_1{}^+,\tilde{v}_1{}^+\}H_1{}^{(1)}d\vec{r} - \int\{U(r)\nabla u_2{}^{(0)}(|\vec{r}-\vec{R}|) + v(r)\nabla v_2{}^{(0)}(|\vec{r}-\vec{R}|)\}d\vec{r} , \tag{29}$$

where

$$U(r) = \int^r dr'(\xi v_1{}^0 \tilde{v}_r{}^+ - 3\gamma u^{(0)^2} \tilde{u}_r{}^+) ,$$

$$v(r) = \int^r dr'(2\beta v_1{}^{(0)} + \xi u_1{}^0 - 3v_1{}^{(0)^2})\tilde{v}_r{}^+ .$$

Since $\nabla_r = -\nabla_{r_1}$ in (29), operations $\nabla$ and $\int dr$ may exchange places and Eq.(27) will take on a gradient form. Replacing $u_2{}^{(0)}$ and $v^{(0)}$ by their asymptotic forms (20) and using the expansion

$$\mathrm{Re} = \frac{e^{ik|\vec{r}-\vec{R}|}}{|\vec{r}-\vec{R}|} \simeq R^{-1}e^{-\mathrm{Im}k(R-r\cos\theta)} \cdot \cos \mathrm{Re}\, K(R-\cos\theta r)$$

where $\theta$ is the angle between vectors $\vec{r}$ and $\vec{R}$, we obtain (27) in an explicit form

$$M \frac{d\vec{r}_1}{dt} = + I\nabla_{\vec{r}_1} \frac{\mathrm{Re}\, e^{ikR}}{R} , \tag{30}$$

where $I = (A^2+B^2)^{1/2}\exp \mathrm{Im}\, kR_0$; $R_0 = (\mathrm{Re}\, k)^{-1}\phi_0$, $\mathrm{tg}\phi_0 = B/A$,

$$A = \int d\vec{r}\{U(r)(D \cos \eta + E \sin \eta) + V(r) \cos \eta\} e^{\mathrm{Im}\, kr \cos \theta},$$

$$B = \int d\vec{r}\{U(r)(D \sin \eta + E \cos \eta) + V(r) \sin \eta\} e^{\mathrm{Im}\, kr \cos \theta},$$

$$D = \mathrm{Re}(c^2k^2+1), \quad E = \mathrm{Im}(c^2k^2+1), \quad \eta = (\mathrm{Re}\, k)r \cos \theta.$$

The equation for $\vec{r}_2$ can be obtained in a similar fashion. Combining these equations (the right-hand sides depend only on $\vec{R}$), we finally obtain an equation for $\vec{R}$:

$$\frac{d\vec{R}}{d\tilde{t}} = + \nabla_{\vec{R}} \frac{\mathrm{Re}\, e^{ikR}}{R} \quad . \tag{31}$$

Here the numerical coefficients M and I are eliminated by an apparent change of the scale $\tilde{t} \to It/M$. System (31) has two integrals $xy = c_1$ and $xz = c_2$ indicating that the two particles move along the line connecting their centres. The time variation of the distance between the particles is determined by the equation

$$R = \frac{1}{R}\left(\mathrm{Re}\, \frac{e^{ikR}}{R}\right)_R \, .$$

In this case there is a countable (infinite) number of the equilibrium points which correspond to the bound states of the elementary structures like the spheres; the motion starting with an arbitrary initial distance R(t=0) between the particles ceases at the nearest stable equilibrium state. Equation (30) is easily generalized to the case of the interaction of an arbitrary number of localized structures by adding the terms $\mathrm{Re}\left[(e^{ik}R_{1j})/R_{1j}\right]$ and $R_{1j} = |\vec{r}_1 - \vec{r}_j|$ to the right-hand side of (30). It is already difficult to enumerate the configurations of the bound states of an arbitrary number of particles (regular and irregular polyhedrons and polygons, finite and infinite lattices, etc.). All these bound states may be multiplied by a similar extension of the figures until they reach the next stable equilibrium state (this, of course, affects the stability factor of the bound state).

To conclude this section, we shall briefly consider the problem of the interaction of the localized structures of the same type but in terms of the conservative model (14)-(15). In this case the family of statical solutions is a part of the points of the common family that is described by three more parameters (velocity $\vec{v}$) and the derivation of the equation of interaction needs, generally speaking, to know the structural dependence of the generating solution of $\vec{v}$. However, the derivation of the sought equations in the limit $\vec{v} \to 0$ is analogous to the one described above. The only difference is that all transformations are performed in the second approximation (the first approaximation gives an apparent relation , cf. /14/) $d_{\vec{r}_{12}}/d\tilde{t} = \vec{v}_{12}$:

$$\frac{d\vec{v}_{12}}{d\tilde{t}} = \nabla_{\vec{r}_{12}} \frac{\mathrm{Re}\, e^{ikR}}{R} \quad . \tag{32}$$

Equation (32) is an example of a classical two-body problem. It is well known that this problem may be reduced to squaring. Note that all statical bound states of systems (32) and (30) coincide completely. The essential difference is that in this case nonstatical bound states which correspond to the mutual rotation of the particles around the common centre of gravity are also possible.

4.3. Spiral Wave Interaction. A particular form of the equation of spiral motion can be obtained when the structure of the eigenfunctions belonging to the zero eigenvalue of the discrete spectrum and the structure of $H^{(1)}$ are known. The operator of the linearized problem (3) has the form

$$\hat{L} = \begin{pmatrix} \partial_t-(\Delta+f(\Phi^{(0)2}) + 2u^{2(0)}f'(\Phi^{(0)2}); & -2u^{(0)}v^{(0)}f'(\Phi^{(0)2}) \\ -2u^{(0)}v^{(0)}f'(\Phi^{(0)2}); & \partial_t + (\Delta+f(\Phi^{(0)2})) + 2v^{(0)2}f'(\Phi^{(0)2}) \end{pmatrix} . \quad (33)$$

It is easily seen that $\hat{L}$ is a self-conjugate operator and the eigenfunctions obtained by the variation of the generating solution over the parameters $x_0$, $y_0$ and $z_0$ can be taken as $\psi^+$, i.e. $\psi_{x_0}^{(0)}$, $\psi_{y_0}^{(0)}$, $\psi_{z_0}^{(0)}$. Let us consider the interaction of widely spaced spirals. Then a two-spiral solution can approximately be represented as a superposition of individual spirals:

$$\psi = \Phi^{(0)}(\rho_1)e^{i(\theta_1-\phi_1)} + \Phi^{(0)}(\rho_2)e^{i(m\theta_2-\phi_2)} \quad (34)$$

where

$$\rho_{1,2} = \sqrt{(x-x_{1,2})^2 + (y-y_{1,2})^2}, \quad \theta_{1,2} = \mathrm{arctg}\,\frac{y-y_{1,2}}{x-x_{1,2}} .$$

The like spirals (i.e. the ones rotating in one direction) correspond to m = 1, while the unlike spirals (rotating in opposite directions) correspond to m = -1. Correction $H_1^{(1)}$ to the first spiral is written in the form

$$H_1^{(1)} = -\frac{\partial\psi_1^{(0)}}{\partial t} + (\beta|\psi_1^{(0)} + \psi_2^{(0)}|^2) - |\psi_1^{(0)} + \psi_2^{(0)}|^4)(\psi_1^{(0)} + \psi_2^{(0)}) -$$

$$-\beta(|\psi_1^{(0)}|^2\psi_1 + |\psi_2^{(0)}|^2\psi_2) + |\psi_1^{(0)}|^4\psi_1 + |\psi_2^{(0)}|^4\psi_2 . \quad (35)$$

Commuting the subscripts 1 and 2 in (35) yields an expression for $H_2^{(1)}$. The transformation of (24) gives the equations of spiral motion in a more exquisite form. The orhogonality conditions may, apparently, be represented in a vector form

$$(\hat{m}\vec{V}_1) = -\mathrm{Re}\int\nabla_1\psi^{(0)*}\tilde{\Phi}dx\,dy , \quad (36)$$

where

$$\nabla_1 = (\frac{\partial}{\partial x_1}, \frac{\partial}{\partial y_1}, \frac{\partial}{\partial\phi_1}) , \quad \hat{m} = \begin{pmatrix} m_x & 0 & 0 \\ 0 & m_y & 0 \\ 0 & 0 & m_\phi \end{pmatrix}$$

$\vec{V} = (\dot{x}_1, \dot{y}_1, \dot{\phi}_1)$ is the velocity vector, $\hat{m}$ is the mass tensor where

$$m_x = m_y = \frac{1}{2}\int|\nabla\psi^{(0)}|^2dx\,dy \text{ and } \quad m_\phi = \int|\psi^{(0)}|^2dx\,dy,$$

and $\hat{\Phi}$ are the perturbations generated in the interaction. The terms in (35) which do not contain the product of $\psi_1^{(0)}$ and $\psi_2^{(0)}$, can be omitted because they either do not contribute to the orthogonality condition or have the next order of smallness. Besides, it is convenient to rewrite the orthogonality conditions in a more symmetric form. Taking into account that $\nabla_1\psi_2^{(0)} = 0$, we obtain

$$\mathrm{Re}\int \nabla_1\psi_1^{(0)*}\tilde{\Phi}_1 dx\, dy = \mathrm{Re}\int \nabla_1(\psi_1^{(0)}+\psi_2^{(0)})^*\tilde{\Phi}_1 dx\, dy =$$

$$= \mathrm{Re}\int \nabla_1(\psi_1^{(0)}+\psi_2^{(0)})^*\ \beta|\psi_1^{(0)}+\psi_2^{(0)}|^2(\psi_1^{(0)}+\psi_2^{(0)}) - |\psi_1^{(0)}+\psi_2^{(0)}|^4(\psi_1^{(0)}+\psi_2^{(0)})\ \cdot$$

$$\cdot\ dx\, dy = \nabla_1 \frac{1}{2}\int\{\beta\frac{|\psi_1^{(0)}+\psi_2^{(0)}|^4}{2} - \frac{|\psi_1^{(0)}+\psi_2^{(0)}|^6}{3}\} dx\, dy = \nabla_1 u(r_{12},\phi_{12}) .$$

The value

$$u = \frac{1}{2}\int\{\beta\ \frac{1}{2}(|\psi_1^{(0)}+\psi_2^{(0)}|^4) - \frac{1}{3}(|\psi_1^{(0)}+\psi_2^{(0)}|^6)\ dx\, dy ,$$

which coincides with a nonsquare part of the free energy functional can be considered as the potential of pair interaction. Then, the equations of spiral motion can be written in the gradient form

$$\begin{aligned} m_x\, \dot{x}_{1,2} &= -\frac{\partial}{\partial x_{1,2}}\, u \\ m_y\, \dot{y}_{1,2} &= -\frac{\partial}{\partial y_{1,2}}\, u \\ m_\phi\, \dot{\phi}_{1,2} &= -\frac{\partial}{\partial \phi_{1,2}}\, u . \end{aligned} \tag{37}$$

The expression for the potential u can be calculated taking into account that the distance between the spirals $R = \left[(x_1-x_2)^2+(y_1-y_2)^2\right]^{1/2}$ is large and the field of the second spiral can be replaced by an asymptotic expression for $\rho \to \infty$. Then

$$u = \frac{1}{2}\int(\beta|\psi_1^{(0)}|^2-|\psi_1^{(0)}|^4)(\psi_1^{(0)}\psi_2^{(0)*}+\psi_1^{(0)*}\psi_2^{(0)})dx\, dy =$$

$$= \mathrm{Re}\int(\beta|\psi_1^{(0)}|^2-|\psi_1^{(0)}|^4)\psi_1^{(0)}\psi_2^{(0)*}\ dx\, dy . \tag{38}$$

In the framework of the Ginzburg-Landau equation with long-wave instability (3), the asymptotic expression for the field of the second spiral can be written in the form

$$\psi_2^{(0)}(\rho) \approx A_0\rho_2^{-1/2}\ e^{-\rho_2+i(m\theta-\phi_2)} \approx A_0R^{-1/2}e^{-R+i(m\psi_{12}-\phi_2)}\ \cdot\ e^{+r\cos(\theta_1-\psi_{12})} \tag{39}$$

where $x = x_2-x_1$, $y = y_2-y_1$, $R = (x^2+y^2)^{1/2}$ are the different coordinates and $\psi_{12} =$ $=$ arctg $y/x$ is the viewing angle of the second spiral from the site of the first spiral, $r = \left[(x-x_1)^2+(y-y_1)^2\right]^{1/2}$.

Substituting (39) into (38) for the like spirals (m = 1) yields

$$u = cR^{-1/2}e^{-R}\cos(\phi_1 - \phi_2) , \tag{40}$$

where $c = A_0\mathrm{Re}\int r(\beta\Phi^{(0)3}(r) - \Phi^{(0)5}(r))e^{-r\cos\xi-i\xi}d\xi\, dr \equiv \mathrm{const}$ .

As a result, we obtain an equation for difference coordinates:

$$\dot{X} = -2c_1 \cos\phi \frac{\partial}{\partial X}(e^{-R}/\sqrt{R})$$
$$\dot{Y} = -2c_1 \cos\phi \frac{\partial}{\partial Y}(e^{-R}/\sqrt{R}) \tag{41}$$
$$\dot{\phi} = 2c_2 \sin\phi \, e^{-R}/\sqrt{R}$$

where $c_{1,2} \equiv$ const and $\phi = \phi_2 - \phi_1$. This system is readily integrated because the value $X/Y$ = const is retained. The spirals either collapse or vanish to the infinity without forming a bound state depending on the sign of the constant $c_2 = c/m_\phi$.

For $m = -1$ (the unlike spirals) we obtain the following expression for the potential:

$$u = \frac{ce^{-R}}{R} \cos(\phi + 2\psi_{12}) \ . \tag{42}$$

Simple verification of the equations generated by this potential shows that it has no fixed points. Hence, in view of the gradient form of the system, unlike spirals do not form bound states either and tend to scatter one another. The situation is essentially different in media with short-wave instability. The asymptotic expressions for the spiral tail have the following form

$$\psi_2^{(0)} \approx A_0 R^{-1/2} e^{-\alpha R + i(m\psi_{12} - \phi_2)} \cos(k_0 R + \xi_0) e^{-\alpha r \cos(\theta - \psi_{12})} + \mathcal{O}(r/R) \ .$$

Then, for the like spirals we have

$$u \sim cR^{-1/2} e^{-\alpha R} \cos\phi \cos(k_0 R + \xi_0)$$

and for unlike spirals we shall have

$$u \sim cR^{-1/2} e^{-\alpha R} \cos(\phi + 2\psi_{12}) \cos(k_0 R + \xi_0) \ .$$

In both cases these potentials have a countable number of stable fixed points corresponding to bound states: spiral dipoles (when $m = -1$) and double spirals (when $m = +1$). These states were observed in a direct numerical experiment with Eq.(11). The results are shown in Fig.7.

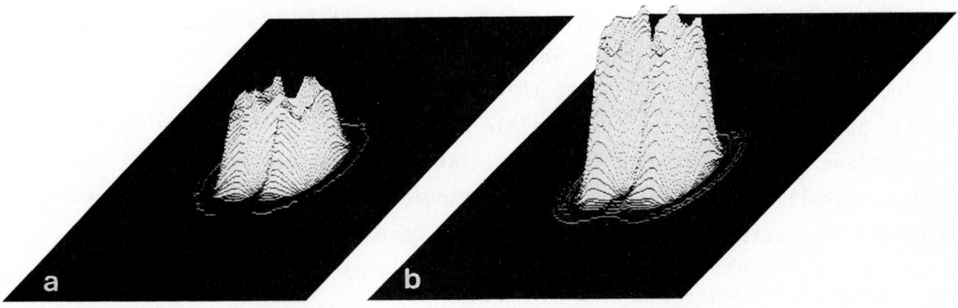


Figure 7. Distribution $|\psi|^2$ for Eq.(11) for $\beta = 2.5$: a) a spiral dipole ($m = -1$), b) a double spiral ($m = 1$).

4.4. Chaotic Drift of Localized Spirals. In this section we shall consider a case demonstrating nontrivial dynamics of localized structures in an external field taking as an example a weakly perturbed Ginzburg-Landau equation

$$\frac{\partial\psi}{\partial t} = -\psi+\beta|\psi|^2\psi-\psi|\psi|^4-(k_0^2+\Delta)^2\psi+i\beta'\psi|\psi|^2+\varepsilon\chi(|\psi|^2)f(\varepsilon x,\varepsilon y) \ , \tag{43}$$

where $\chi(|\psi|^2)$ describes the nonlinear response of the medium to the external effect the spatial distribution of which is characterized by function $f(\varepsilon x,\varepsilon y)$ that is smooth in comparison with the size of the spiral. Function $\chi(|\psi|^2)$ can be represented as a series $\chi(|\psi|^2) = A_0 + A_1|\psi|^2 + \dots$ . Getting ahead we can say that the term does not make a contribution in the first approximation; therefore, we shall assume that $A_0 = 0$ (coefficient $A_1$ can be taken to be equal to unity). Equations (36) describing the evolution of the spiral parameters ($x_0$, $y_0$ and $\phi_0$) are obtained using the same eigenfunctions as in the problem of interaction described in Section 4.1. Substituting $H^{(1)} = i\beta(|\psi^{(0)}|^2\psi^{(0)}+\varepsilon|\psi^{(0)}|^2 f(\varepsilon x,\varepsilon y)-\psi_t^{(0)}$ into (36), we obtain

$$\begin{aligned}
\dot{x}_0 &= B\ \mathrm{Re}\{fe^{i\phi_0}\}\Big|_{\substack{x=x_0\\ y=y_0}} \\
\dot{y}_0 &= B\ \mathrm{Im}\{fe^{i\phi_0}\}\Big|_{\substack{x=x_0\\ y=y_0}} \\
\phi_0 &= \Omega
\end{aligned} \tag{44}$$

where $B = \frac{1}{3}\varepsilon\int_{\infty}^{\infty}\rho\Phi^{(0)3}d\rho/my$ and $\Omega = \beta'\int^{\infty}\rho\Phi^{(0)4}d\rho/m\phi$. Let us represent function $f(\varepsilon x,\varepsilon y)$ in the form $f = \cos ky + i \sin kx$ and analyze, first, the situation when the spiral does not rotate ($\Omega = 0$). Then Eq. (44) can be represented in the form of a Hamiltonian system with the Hamiltonian

$$\begin{aligned}
\dot{x}_0 &= B \cos ky_0 = \partial H/\partial y_0 \\
\dot{y}_0 &= B \sin kx_0 = -\ \partial H/\partial x_0 \ .
\end{aligned} \tag{45}$$

The physe portrait of system (45) is shown in Fig.8a. It is seen that system (45) has a separatrix network covering all the $x_0,y_0$-plane. It is apparent that the separatrix network will be disintegrated when the nonstationary perturbations are weak and a stochastic spider-web similar to that considered in /16/ will appear. The stochastic spider-web indicates the possible stochastic drift of a spiral wave along the $x_0,y_0$-plane at arbitrary long distances. Taking into account the spiral rotation ($\beta' \neq 0$), we shall get nonstationary (time periodic) perturbations. In this case we shall obtain the equations for the spiral centre coordinates:

$$\begin{aligned}
\dot{x}_0 &= B(\cos \Omega t \cos ky_0 - \sin \Omega t \sin kx_0) \\
\dot{y}_0 &= B(\cos \Omega t \sin kx_0 + \sin \Omega t \cos ky_0)
\end{aligned} \tag{46}$$

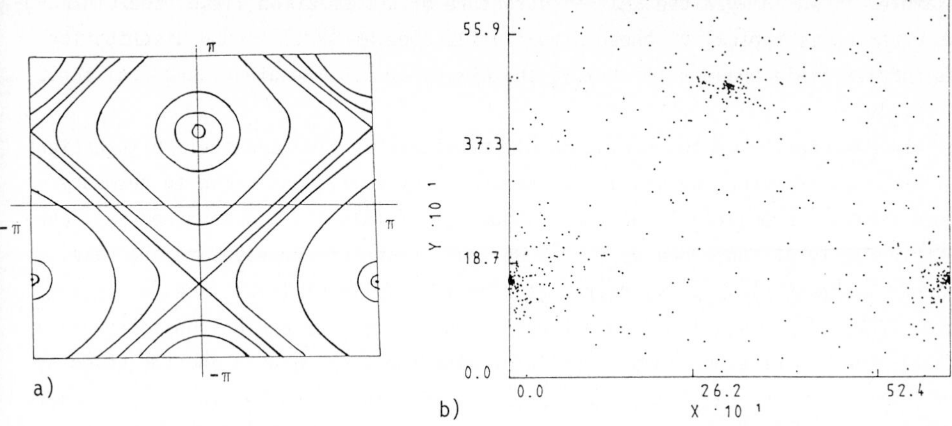


Figure 8. a) Structure of the phase plane of (45) for H = B/k(sin ky + cos kx); b) Poincaré section in period $2\pi/\Omega$ in (46) for $\tilde{\Omega}$ = k = 1, B = 3.

which depend explicitly on time, thus guaranteeing the destruction of the separatrix network in the unperturbed system (46) (see Fig.8b).

Since the spiral core motion is described by quite general third-order equations for which chaotic solutions are rather typical, the phenomenon of the spiral stochastic drift in external fields must be sufficiently widely spread.

## 5. STRONG INTERACTION. EXAMPLES OF STRUCTURAL BIFURCATIONS

It has already been said that the interactions resulting in qualitative deformation of structures, including their conversion into one another, are qualified as strong interactions of structures. Strong interaction of structures can be classified into two groups. The first class is the interaction resulting in the formation of bound states. The shape of individual patterns can be distorted significantly in this case but no qualitative changes in topology are observed. Such interactions are typical of the solitons of a sine-Gordon equation in the formation of a breather, of the spiral pairs of the Ginzburg-Landau equation with short-wave instability (11), etc. In contrast to weak interactions, they cannot be described asymptotically using an adiabatic approximation.

The other class includes the interactions causing qualitative changes in the topology of structures, i.e. bifurcations. Apparently, the asymptotic methods considered above are not valid in this case. The investigations into the mutual conversion of structures need a different approach based on the analysis of the topology of the spatial picture of a frozen field (i.e. its snapshot) /17/.

Assume that a spatial image, i,e, a shapshot of the structures observed in a two-dimensional nonequilibrium medium, is described by function $W_{t_0}(x,y)$ which is expressed through the relation of two rather smooth functions P(x,y) and Q(x,y) at a moment $t_0$. This function may have different physical meaning in different models.

For example, if we analyze the spatial structure of a modulation field (the Ginzburg-Landau equation is typical of these situations), then $W_{t_0}(x,y)$ is the distribution of the complex-field phase while $P(x,y)$ and $Q(x,y)$ are its imaginary and real parts, respectively.

For the description of the evolution of structures in time and their bifurcations with a varying parameter, we shall use the theory of foliations known in geometry. Assign different structures to different topology of foliations determined by a one-parametric family of functions $W_{t_0}(x,y)$. We shall restrict ourselves to the case when the foliations of $W_{t_0}(x,y)$ represent simply a family of level lines $W_{t_0}(x,y) =$ = const. If $W_{t_0}(x,y)$ is a function without singularities, it can be considered as a potential function of a nonlinear oscillator with the x-coordinate and the velocity y. The foliation of $W_{t_0}(x,y)$ is, then, a family of integral curves in the phase surface of a Hamiltonian system where only centres or saddle points are possible (Fig.9). In the general case, function $W_{t_0}(x,y)$ has various singularities and its foliation topology is more diverse than the phase portraits generated by the second-order Hamiltonian system.

Figure 9. Nonsingular structure of the foliation of $W_{t_0}(x,y)$.

Complex structures are the result of the successive evolution of spatial excitations that may have the same order of intensity. However, if we take interest in bifurcations, i.e. in the disappearance and birth of different structures, then only the interaction of different-amplitude excitations should be analyzed. Indeed, let the evolution of initial instability in the medium lead to the onset of a structure described by a regular sequence of centres and saddle points. As the medium evolves far from equilibrium, these initial excitations give birth to secondary excitations that change the topology of the snapshot when their amplitude reaches a critical value; this amplitude may be much higher than the amplitude of the initial excitation. Thus, for the analysis of the bifurcations of structures it is necessary to find the excitations appearing on the background of the existing ones and determine the accompanying change in the foliation topology.

Let us consider possible bifurcations of structures when the excitations are described by the popular nonlinear Ginzburg-Landau equation that can be written in the form

$$\frac{\partial \psi}{\partial t} = \psi - (1+i\beta)|\psi|^2\psi + e(1-ic)\Delta\psi \tag{47}$$

where $\beta$, c, e > 0.

The transformation of spatial structures corresponds to the transition from leading centres to spiral vortices. It follows from /17,18/ that with decreasing e the leading centres, i.e. the extreme points of $W_{t_0}(x,y)$, converge and then merge at some critical e ($W_{t_0}(x,y)$ acquires singularities), after which the singular points diverge in the transversal direction forming a spiral pair (Fig.10).

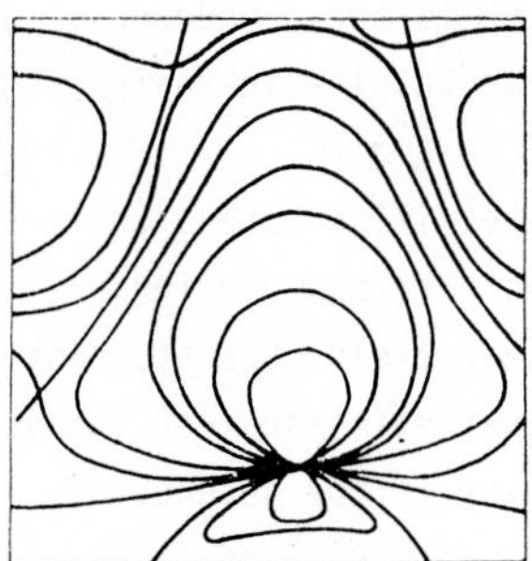

Figure 10. Singular structure of $W_{t_0}(x,y)$. Appearance of a spiral pair.

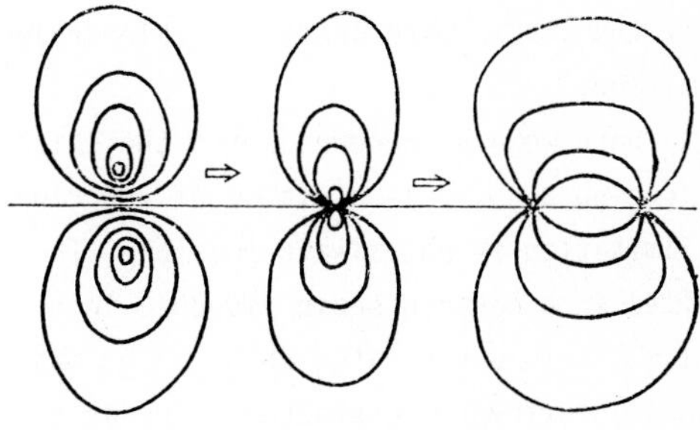

Figure 11. Scenario of spiral pair formation.

The bifurcations of spatial structures described above were revealed in a numerical experiment on model (47). In particular, the complex dynamics of spiral waves was observed with the interactions taking place exactly according to the scenario shown in Fig.11. This gives grounds to conclude that the mechanism of pair formation or disappearance of spiral (field singularities) is quite general. There are many signs and examples indicating the universality of pattern conversion. These processes are determined not so much by a particular form of equations as by more general properties, such as the types of symmetry and the peculiarities of the topological structure of configuration space.

## 6. THE CHAOTIZATION OF LOCALIZED STRUCTURES IN EXTERNAL FIELDS. ANALOGY WITH THE MOTION OF A CLASSICAL PARTICLE

Localized structures allow a qualitatively new approach to the analysis of the complex dynamics of a nonlinear field. Since the use of asymptotic methods for the "mass centre" of localized structures ("particles") yields equations in total derivatives which are similar to the equations of Newton's dynamics, there naturally occurs an urge to represent the turbulent behaviour of a nonlinear field like chaotic dynamics of the "gas" of such particles. Such an approach is, however, fraught with unexpected consequences and should be used with ample understanding. The evaluation of the disturbances in the adiabaticity of the particle behaviour is one of "thin places" in such a description. This problem will be considered below. First we shall elucidate the nonlinear mechanism of the organization of the spatio-temporal chaos of the fields.

The chaotic behaviour of deterministic systems is naturally related to nonlinearity However, while the mechanisms responsible for the stochastic motion are quite clear in pointed systems, the relation between nonlinearity and chaos in the fields is far from being a simple problem. Indeed, in a rough approximation we can say that the effect of nonlinearity in pointed systems reduces to the limitation of an accessible phase space inside of which individual motions are unstable. It is the boundedness of the phase space where unstable trajectories are located that results in their mixing.

When we are concerned with the chaotic behaviour of the wave fields such a challenging and simple division of functions between linear instability and nonlinear limitation is not always possible. The point is that even linear fields are stochastized in a certain sense /19,20/. In this case, however, we observe a transient chaos, i.e. it is degenerated when $\varepsilon \to 0$; but the field stochastization in the absence of nonlinearity is considered to be an established fact /21/. The mechanism of linear field stochastization is related to the random walk of narrow wave trains in an inhomogeneous and/or nonstationary medium. At short enough times, such a train behaves as a classical particle and moves chaotically even in an one-dimensional situation with the appropriate potential profile which regularly varies in space and time /22, 23/. The train broadens with time due to dispersion (or diffraction) and, finally, spreads having encompassed all admissible region (that is bounded, eg. by the resonator walls). The chaos is degenerated /20-22/ and is replaced by a regular (nearly periodic in time) motion of the field.

The above said suggest that nonlinearity may play a peculiar role in the organization of random motion of deterministic wave fields; it may have stabilizing functions protecting the wave train from spreading rather than mixing functions. If this is possible, then a narrow wave train will become steady-state and will exist at any $t \to \infty$, i.e. the particle-like chaos will no longer be observed as arbitrary long.

In this section we shall verify the concepts suggested above by means of a comparative analysis of the behaviour: a) of a particle in a periodic nonstationary potential, b) of a narrow wave train in a linear field, and c) of a nonlinear wave train (soliton) having the same potential profile.

We investigated the dynamics of a wave train in the framework of a Schrödinger equation

$$\psi_t = i(\psi_{xx} + n(x,t)\psi) \tag{48}$$

where the dynamics of a classical particles is described by

$$m\ddot{x} = -\frac{\partial}{\partial x} n(x,t) \tag{49}$$

and the dynamics of a soliton in the framework of a nonlinear Schrödinger equation

$$\psi_t = i(\psi_{xx} + \psi|\psi|^2 + n(x,t)\psi) . \tag{50}$$

The boundary conditions in Eqs.(48) and (50) are assumed to be periodic ones, m is taken to be an effective mass of the particle while a given smooth function of x and t is used as the potential n(x,t).

The analogy between the motion of classical particles and narrow trains is well known /22/. Indeed, representing the solution of (48) in the form

$$(x,t) = A(x,t)e^{i\phi(x,t)} \tag{51}$$

in the approximation of geometrical optics for function $\phi$ having a sense of an eikonal, we obtain the equation

$$-\phi_t + (\phi_x)^2 - n(x,t) = 0 . \tag{52}$$

Considering (52) like the Hamilton-Jacobi equation, we obtain the following equation

$$2\ddot{x} = - \frac{\partial n(x,t)}{\partial x} \tag{53}$$

for the mass centre of the wave train. It is apparent that Eq.(53) coincides with (45) if the effective mass is taken to be m = 2. It is also rather easy to estimate the times at which our approximations are valid. Assuming the potential n(x,t) to be constant (the potential nonstationarity and inhomogeneity may only decrease the applicability time because of appearing acoustics), the solution of (48) will be specified by the integral

$$\psi(x,t) = \int e^{-i((k^2-n)t-kx)}S_0(k)dk \tag{54}$$

where $S_0(k)$ is the spectrum of the initial (when t = 0) disturbance. The approximation of geometrical optics is valid (the wave train motion is, correspondingly, described by (48)) until the train width equals the characteristic scale of the variation of potential n(x,t) which is assumed to be of the order of $1/\varepsilon$ ($\varepsilon \ll 1$). Taking for simplicity $S_0(k) = e^{-k^2}$, i.e. the initial wave train width $\Delta x_0$ is of order of unity, after rather simple transformations in (54) we obtain that in time t the wave train width will be $(\Delta x)^2 \sim (\Delta x_0)^2 t$. Consequently, the time of applicability for our approximate description must satisfy the inequality $t \ll 1/\varepsilon^2$. On the other hand, the characteristic variation of the wave train velocity occurs in time $1/\varepsilon$, i.e. the transient chaos exists in the time interval $1/\varepsilon \ll t \ll 1/\varepsilon^2$. If $\varepsilon$ is chosen to be small enough, a sufficient number of wave train oscillations in potential n(x,t) may be described rather well by Eq.(53).

The "soliton - classical particle" analogy is well known for the nonlinear Schrödinger equation (50) (see, e.g. /15,24/). The soliton solution of (50) for a constant potential n(x,t) has a form

$$\psi^{(0)}(x,t) = \Phi^{(0)}(\xi)e^{i(\frac{v}{2}\xi+\phi)} \quad \psi^{(0)}(\xi) = \sqrt{2}\,\lambda \mathrm{ch}^{-1}\lambda\xi \tag{55}$$

where $\lambda = \omega + n - v^2/4$, $\xi = x - vt$, $\phi = \omega t$, v is the soliton velocity, and $\omega$ is the background frequency. If the potential h(x,t) is not constant, then the adiabatic theory of localized structures (see Sect.4.1) can be used for the description of soliton evolution. In this case Eqs.(24) for soliton parameters take on the form

$$d\lambda/dt = 0 \tag{56a}$$

$$dv/dt = d^2x/dt^2 = -2\,\frac{\partial n(x,t)}{\partial x} \tag{56b}$$

The first equation is a consequence of the validity of the integral of (50), N = $= \int|\psi|^2dx$ = const that has a sense of a "total number of quanta". Equation (56a) also means that the soliton amplitude motion is retained, i.e. $\lambda$ is an adiabatic invariant. Equation (56b) coincides with (49) having the effective mass m = 1/2. Thus, the soliton mass is one fourth of the wave train mass.

The problem of the applicability interval of the approximate description for a soliton is not solved yet. This is explained, in particular, by the fact that the method itself has no rigorous mathematical substantiation. The minimal applicability time is estimated to be $1/\varepsilon$, i.e. of the order of the time of velocity variation. This value, however, is strongly underestimated. We understand intuitively that because a soliton (in contrast to the wave train) does not spread in the absence of disturbances, i.e. it lives infinitely long, smooth disturbances are not likely to destruct the soliton quickly enough. In any case, the Hamiltonian perturbations of interest must retain the soliton at times of order $1/\varepsilon^2$ (we hope it will be proved rigorously).

A more detailed analysis of the soliton motion using the perturbation method shows that besides the effect of velocity modulation and soliton frequency there appear in subsequent approximations nonlocalized distortions which are interpreted like the wave radiation by a soliton (see, eg. /25/). However, numerous estimates of the radiation energy show that this effect is exponentially small $e^{-1/\varepsilon}$. Thus, the soliton spreading due to the wave generation is possible only at very large times, those which exceed $1/\varepsilon$ significantly. Moreover, an opposite effect - soliton absorption of the generated waves - is possible under periodic boundary conditions. Therefore the soliton lifetime in a smoothly varying potential can, in principle, be expected to be arbitrary long.

In order to obtain the proof or at least convincing arguments for these concepts, we carried out a direct numerical integration of the initial equations (48) and (50).

The integration was performed under periodic boundary conditions in the system $0 \leq x \leq L$, L = 100 with a time step $\tau = 0.04$. Altogether 512 harmonics of a fast Fourier transform were used. The numerical scheme was as follows:

$$\psi(t+\tau,x) = \exp\{i\left[\frac{\tau}{2}\,|\psi(t+\tau,x)|^2 + n(t+\tau,x)\right]\}\cdot\exp(i\tau\,\frac{\partial^2}{\partial x^2})\,\cdot$$

$$\cdot\,\exp\{i\,\frac{\tau}{2}\left[|\psi(t,x)|^2 + n(t,x)\right]\}\,\cdot\,\psi(t,x)\,.$$

In contrast to traditional codes /26/, our scheme had the order of three and was a conservative one. Although the scheme seems to be implicit, it does not need iterations because the value is found explicitly. Parameter $\varepsilon$ was about 0.03. The integration time was much higher than $1/\varepsilon^3$ and amounted to 4000. The potential was taken in the form $n(t,x) = c_0 \sin\omega_0 t \cos k_0 x$, with values $c_0$, $\omega_0$ and $k_0$ and the initial conditions chosen so that Eq.(56b) contained chaos due to the overlapping of the main resonances /27/. The initial (when $t = 0$) and the final ($t = 3952$) field states in (50) are presented in Fig.12. It is seen from Fig.12 that the behaviour of the system at such large times is described perfectly well only by one soliton. The radiation level was also extremely low (less than $10^{-7}$ of the soliton amplitude) at all times. The variations in the soliton amplitude during the motion did not exceed 3% and did not tend to systematic drift in either side.

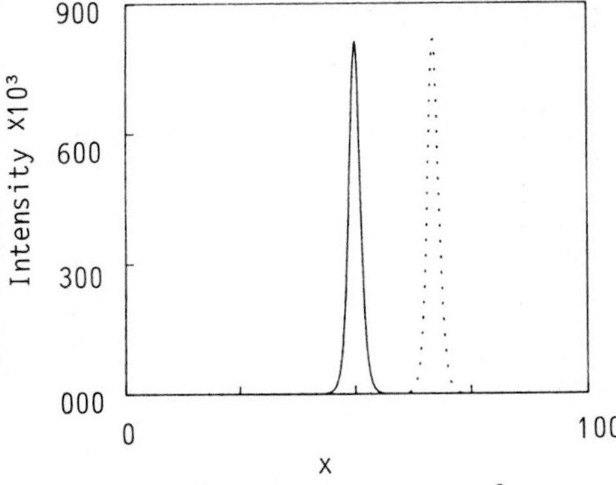


Figure 12. Distribution $|\psi(x,t)|^2$ in Eq,(50). The solid curve is for the initial condition ($t = 0$), the dotted curve is for the final state ($t = 3952$). The initial condition was chosen to meet (55) with $\lambda = 0.6375$, $n(x,t) = 0.67 \sin(\pi t/L)\cos(2\pi x/L)$. $L = 100$, $v = 1/2$.

A spatio-temporal diagram for Eq.(50) is shown in Fig.13. It is seen from this figure that the soliton trajectory behaves rather regularly: the soliton moves like a particle in a periodic potential. To verify this supposition we compared the soliton trajectory determined by the maximum $|\psi|^2$ in Fig.13 and the trajectory of Eq.(56b) having the same parameters and initial conditions. The results are presented in stroboscopic planes (Figs.14a and 14b). In spite of a rather approximate description, the figures demonstrate essential similarity. This suggests that model (56b) is valid at times much larger than $1/\varepsilon^2$.

The investigation of the linear Schrödinger equation (48) with the same parameters showed that the lifetime of the initial soliton perturbation modes do not

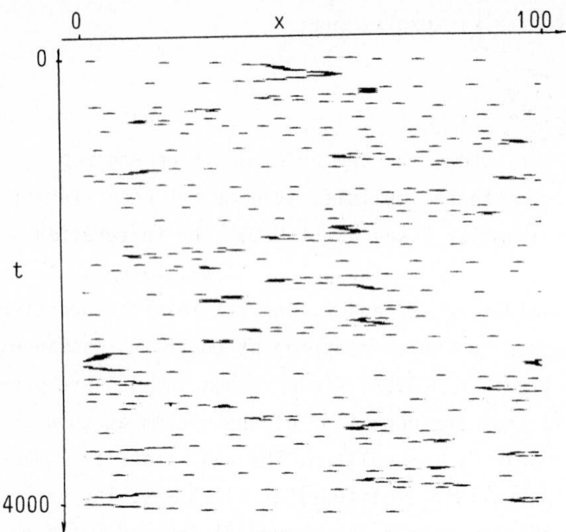


Figure 13. Spatio-temporal diagram for Eq.(50). The dark sections correspond to higher-intensity regions. The initial condition are the same as in Fig.12.

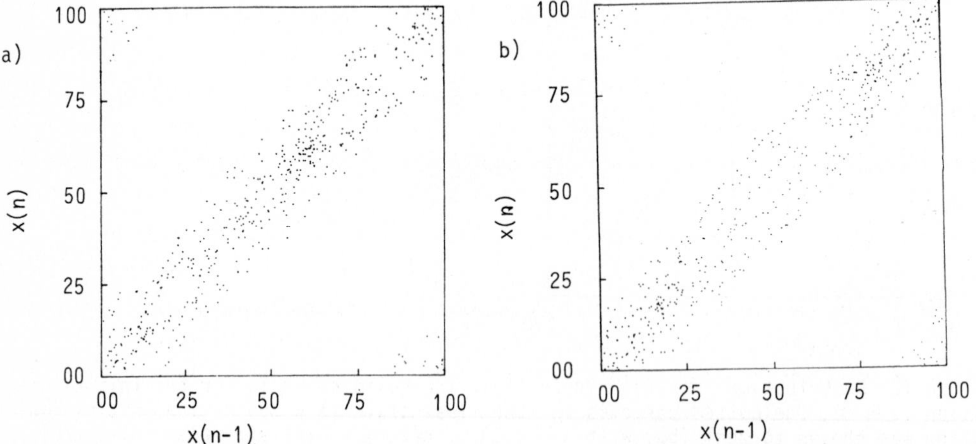


Figure 14. Stroboscopic planes ($x_n = x(t)$, $x_{n-1} = x(t-8)$): a) for the trajectory of a soliton described by (50), b) for the trajectory described by (49).

exceed 80 time units, which also corresponds to the theoretical concepts. During this time the wave train moves a distance of order of 100 units (see Fig.15). At larger times the wave train spreads completely and the system behaves nearly periodically relative to t and x.

Thus, we have proved our theoretical assumptions on the non-trivial role of non-linearity in the stochastization of wave trains.

Having made sure that the chaotization of wave field with the localizing effect of nonlinearity is retained at very large ($\sim 1/\varepsilon^2$) and, possibly, at arbitrary large times, we can come back to the problem on the conditions needed for this mechanism

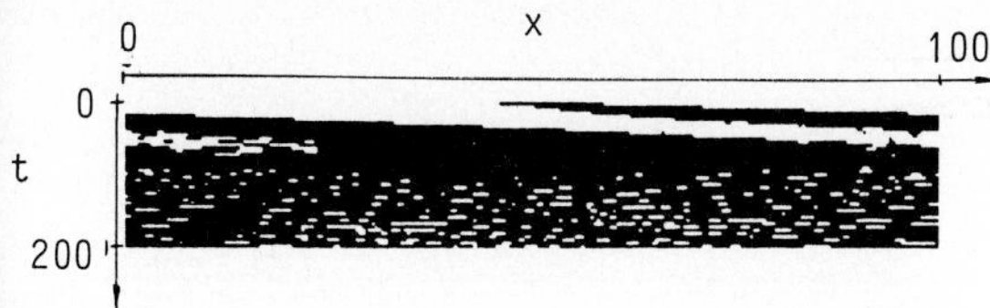


Figure 15. Spatio-temporal diagram for the linear Schrödinger equation (48) with soliton initial conditions $v = 1$, $\lambda = 0.8$, n is same as in Fig.12.

to operate. The nontriviality of this problem is related, in particular, to the fact that solitons were opposed to chaos even in the solution of the classical Fermi-Pasta-Ulam paradox /28/. It was revealed that in the parameter region where the soliton is unstable, elementary disturbances exchange energy irregularly in time and mix. In the region of a stable soliton either a periodic regime or mutual conversion of cnoidal waves into quasi-harmonic ones (recurrency) or a soliton regime with weak modulation were observed under most initial conditions /28/.The main distinction between that situation and the one observed in this paper is a different nature of the instability of initial disturbances. In the case of interest, stochastic instability is caused by the presence of an external oscillating field that destructs most integrals of motion of the unperturbed nonlinearity of the system where solitons (or their bound states) are elementary excitations (nonlinear modes). The soliton motion in an inhomogeneous potential in the presence of an oscillating field is not quasi-periodic, like the motions of a parametrically excited pendulum are not quasi-periodic either /27/. If there is no nonlinearity, the narrow wave train that is first stochastized will become, in a rather short time, just a set of independent quasi-sinusoidal modes, i.e. independent of their number, large as it may be, we shall observe a quasi-periodic behaviour of the field (possibly with a weak exponential growth due to parametric amplification). This situation is completely analogous to the transient chaos in quantum systems /29/. Naturally, the revealed mechanism of chaos ceases to operate in the case of nonadiabatic perturbations, in particular, if the soliton is placed in an oscillating rectangular potential well, it may be destructed due to the transformations as the result of reflections drom the walls and a nonsoliton behaviour will be observed again (Fig.16).

To conclude this Section, we can say that similarly to the traditional problems, nonlinearity in the mechanism studied above results in the limitation of the phase space inside of which a system moves. In the case of interest within this limitation an infinite number of the degrees of freedom of a distributed system are no longer independent (they form a stable pattern - soliton), which results in a finite-dimensional dynamics in a purely Hamiltonian system.

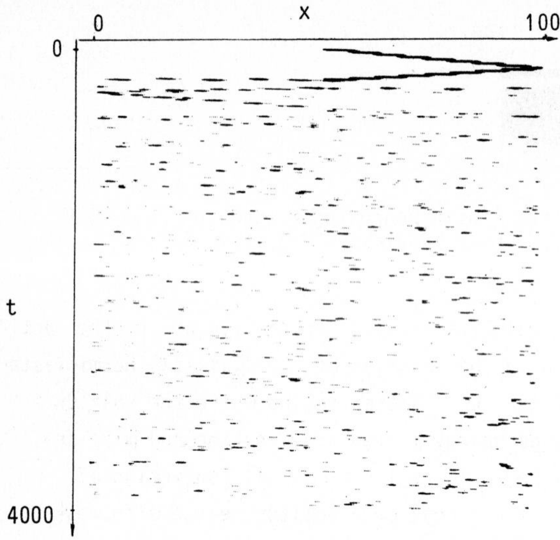


Figure 16. Spatio-temporal diagram of (50) for "nonadiabatic" potential $n(x,t) = 0.37 \sin(\pi t/L) f(x-L/2)$, $f(y) = \begin{cases} 1, & y<0 \\ -1, & y \leq 0 \end{cases}$. All other parameters are the same as in Fig.12.

## 7. CONCLUSION

The behaviour of "particles" with intrinsic dynamics is one of the most fascinating and, in fact, unexplored problems related to the analysis of particle-like solutions of multidimensional fields. The "dynamic particles", in particular, the oscillating or planetary ones are, naturally, most characteristic of Hamiltonian fields. However, such structures (that are, evidently, not described by gradient equations) are quite typical of dissipative nonequilibrium media as well. Take as an example a two-dimensional nonequilibrium medium such as a monitor screen in the "monitor - video camera - feedback" system (Fig.17 /30/). For a sufficiently high supercriticality (e.g. the amplification coefficient of the monitor), a localized pattern with a shape varying periodically in time is established in such a medium. It is apparent from the

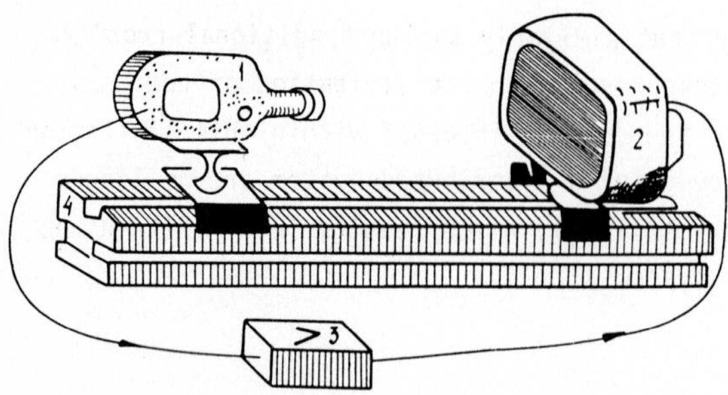


Figure 17. Scheme of "TV - video camera - feedback" system.

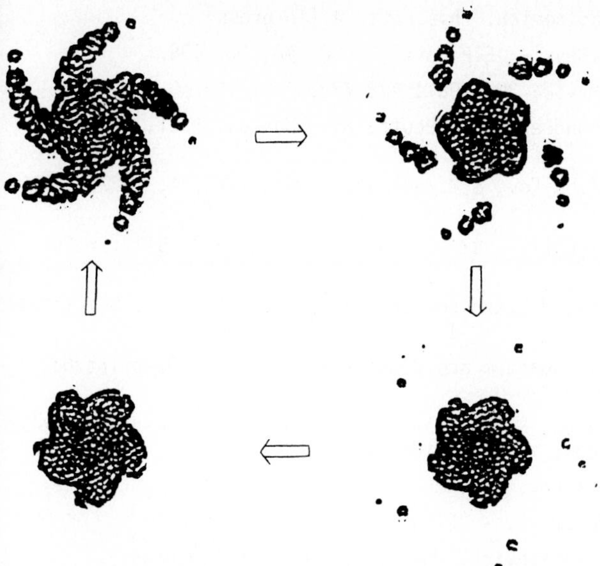

Figure 18. Periodic sequence of localized structures on the screen "TV - video camera-feedback" system.

motion picture exposure in Fig.18. These pulses become chaotic with a further increase of supercriticality. The investigation of the interaction of particles having chaotic intrinsic dynamics seems to be most nontrivial.

## REFERENCES


1. V.G.Makhan'kov. Fiz.Elementarn.Chastits i Yadra, 1983, 14, 123 (in Russian).
2. H.Hagen. Synergetics. Hierarchy of Instabilities in Self-Organized Systems and Devices. Mir, Moscow, 1985, 485 p.
3. Cellular Structures in Instabilities. Ed. by J.E.Wesfreid and S.Zaleski. Lect. Notes in Physics, 1984, 210, 390.
4. H.Linde. Topological Similarities in Dissipative Structures in Self-Organization Autowaves and Structures Far from Equilibrium. Ed. by V.I.Krinski, Springer, Berlin, 1984, 157-157.
5. Y.Kuramoto. Chemical Oscillations. Waves and Turbulence. Springer, Berlin, 1984, 365.
6. A.V.Gaponov-Grekhov, A.S.Lomov, G.V.Osipov, M.I.Rabinovich. In: Nonlinear Waves I (Dynamics and Evolution), Springer, Berlin, 1989.
7. P.S.Hagan. SIAM J.Appl.Math., 1982, 42, 762-786.
8. I.S.Aranson, M.I.Rabinovich. J.Physics A (in press).
9. E.Moses, V.Steinberg. Phys.Rev. A, 1986, 34, 1, 693-696.
10. I.Rehberg, S.Rasenat, J.Fineberg, M.Juarer, V.Steinberg. Phys.Rev.Lett., 1988, 61, 21, 2443-2552.
11. B.A.Malomed. Sov.Phys.-Dokl., 1986, 291, 2, 327-332.

12. K.A.Gorshkov, A.S.Lomov, M.I.Rabinovich. Phys.Lett. A (in press).
13. A.S.Lomov, M.I.Rabinovich. Sov.Phys.- JETP Lett., 1988, 48, 11, 598.
14. K.A.Gorshkov, L.A.Ostrovsky. Physica 3D, 1981, 428-438.
15. Solitons in Action. Ed. by K.Lonngren and E.Scott. Mir, Moscow, 1981, 312 p. (Russian edition).
16. A.A.Chernikov, R.Z.Sagdeev, D.A.Usikov, G.M.Zaslavsky. Phys.Lett. A, 1987, 125, 101-106.
17. I.S.Aranson, M.I.Rabinovich. Izvestiya VUZ'ov, 1986, 29, 12, 1514-1517 (in Russian).
18. I.S.Aranson, A.V.Gaponov-Grekhov, M.I.Rabinovich. Izvestiya USSR Acad. Sci., 1987, 51, 6, 1133-1150 (in Russian).
19. B.V.Chirikov. Transient Chaos in Quantum and Classical Mechanics. Preprint No. 85-55, Novosibirsk, USSR, 1985.
20. G.M.Zaslavsky. In: Nonlinear Waves, Self-Organization. Ed. by. A.V.Gaponov-Grekhov and M.I.Rabinovich, Nauka, Moscow, 1983, 96-106 (in Russian).
21. S.W.McDonald, A.N.Kauffman. Phys.Rev. A, 1986, 37, 3067-3086.
22. S.S.Abdulaev, G.M.Zaslavsky. Sov.Phys. - JETP, 1983, 85, 1573-1583.
23. I.S.Aranson, K.A.Gorshkov, M.I.Rabinovich. Phys.Lett. A, 1989 (in press).
24. I.S.Aranson, K.A.Gorshkov, M.I.Rabinovich. Sov.Phys. - JETP, 1984, 86, 3, 929-936.
25. B.A.Malomed. Phys.Scripta, 1987, 38, 66-74.
26. T.R.Tana, M.J.Ablowitz. J.Comp.Phys., 1984, 55, 109-230.
27. D.F.Escande, F.Doveil. J.Stat.Phys., 1981, 26, 237-239.
28. B.V.Chirikov, F.M.Izrailev. Sov.Phys. - Dokl., 1966, 166, 57-59.
29. B.V.Chirikov, F.M.Izrailev, D.L.Shepelyansky. Sov.Sci.Rev., 1981, 21, 209-267.
30. V.N.Golubev, M.I.Rabinovich, V.V.Talanov, V.V.Schclover, V.G.Yakhno. Sov.Phys.-JETP Lett., 1985, 42, 84-87.

# The Turbulent Alfvén Layer

*V.Yu. Trakhtengerts*

Institute of Applied Physics, USSR Academy of Sciences,
46 Ulyanov Str., 603600 Gorky, USSR

The interaction between ionized and neutral envelopes in stellar and planetary atmospheres results in the development of Alfven mode instability when the velocity of neutral component with respect to ionized one exceeds some threshold value. At the final stage the turbulent boundary layer (TBL) is formed which plays an important role in the atmospheric dynamics. On the other hand, TBL is also of interest as a new object for nonlinear wave dynamics investigation.

1. In the atmosphere of active stars and planets with a magnetic field, a peculiar transient layer is formed which separates the plasma from the frozen-in magnetic field and the neutral gas. Such an interface is defined by the violation of the condition for the existence of frozen-in magnetic field and is situated at the atmospheric heights where the following equality is satisfied (see Fig.1):

$$\Omega_B \sim \nu_{in} \tag{1}$$

where $\Omega_B$ is the ion gyrofrequency, $\nu_{in}$ is the ion-neutral collision frequency. Because of the sharp decrease of atmospheric density with height, this layer is thin as compared to the characteristic variation scale of plasma parameters. With relative motions of plasma and neutral gas (surface activity), such a boundary layer can play a vital role in the conversion of the mechanical energy of the neutral medium motions into electromagnetic energy and plasma energy. Such a problem arises in the analysis of phenomena in the outer photospere and cromosphere on the Sun, as well as in the auroral region in the Earth's atmosphere. Taking the Earth's atmosphere as an example, we shall analyze a boundary flow caused by the interaction of magnetospheric convection with atmospheric substrate: the transition of the flow to turbulent regime, the energetics and dynamics of the turbulent boundary layer (TBL), the role of TBL in magnetospheric phenomena. Also, the TBL theory generates a number of interesting problems in the nonlinear wave dynamics.

2. In general, the structure of the boundary layer described above is rather complicated and includes the interrelated dynamics of the neutral medium and the magnetized plasma. Bearing in mind also the inhomogeneity of the parameters of the medium, it seems problematic to obtain a conceivable solution. Nevertheless, if the ionization rate in the transition layer region ($\Omega_B \sim \nu_{in}$) is small, then the motion of neutrals in the first approximation can be assumed as given. The second essential

Research Reports in Physics **Nonlinear Waves 3**
Editors: A.V. Gaponov-Grekhov · M.I. Rabinovich · J. Engelbrecht

simplification of the problem can be obtained by taking into account the small thickness of the boundary layer. In this approximation, the interaction of plasma flow with neutral gas can be described using the so-called impedance boundary condition (see below). Under real conditions, a significant role is played by the plasma structure above the transition layer. As a rule, the plasma density has a maximum at some height above the boundary layer (see Fig.1). Essentially, such a plasma layer represents a resonator for Alfven waves; this makes them most important among other types of wave perurbations and sharply lowers the threshold velocity at which the transition to turbulent flow occurs. Alfven waves have one more remarkable property. In a relatively cold plasma, their scale across the magnetic field is actually arbitrary; therefore, Alfven perturbations "adapt" well to any type of neutral gas motion such as a homogeneous flow, convective cells or cyclonic vortices.

3. Bearing in mind the above mentioned simplifications (given motion of neutral gas and also impedance boundary condition), we now turn to a qualitative consideration using the simplest model of stratified flow across the magnetic field (Fig.1). Let us first consider some results of the linear theory. Initial here are the wave equation of Alfven waves and two boundary conditions: the wave escaping condition at infinity and the impedance boundary condition on the atmospheric "substrate". The Fourier transformation of this equation ($J \sim j(z)\exp\{-i\omega t + i\vec{\kappa}\vec{\rho}\}$) yields the following system (in a coordinate system moving together with plasma):

$$\frac{d^2 j}{dz^2} + \frac{\omega^2}{V_A^2(z)} \cdot j = 0, \tag{2}$$

$$z = 0 : \left(\Sigma_p j + \frac{c^2}{4\pi i\omega}\frac{\partial j}{\partial z}\right) = 0, \tag{3}$$

$$z \to \infty : j \to \exp\{-i\omega t + i\int^{z}(\omega/V_A(z'))dz' \;, \tag{4}$$

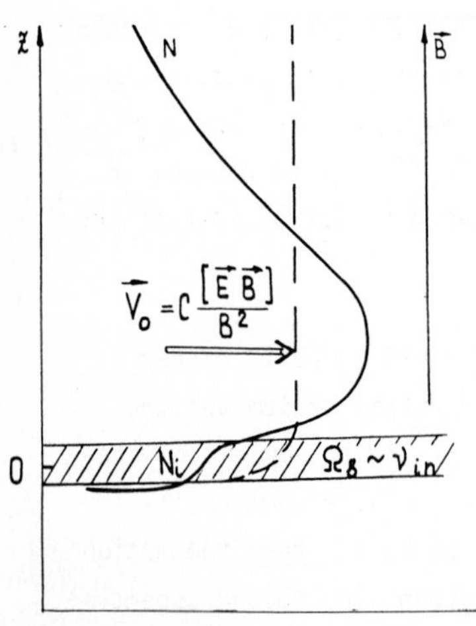


Figure 1. Structure of the boundary layer between plasma flow and neutral gas.

where $j(z)$ is the longitudinal current in the Alfven wave; the integral Pedersen conductivity $\Sigma_p$ of the transition layer has the form:

$$\Sigma_p = \int \frac{\omega_{pi}{}^2 \nu_{in} \omega dz}{4\pi(\nu_{in}{}^2 + \Omega_B{}^2 \omega_D}, \quad \omega_D = \omega - \frac{\nu_{in} V_0}{\Omega_B{}^2 + \nu_{in}{}^2} (\kappa_j \nu_{in} + \kappa_x \Omega_B), \tag{5}$$

where $\vec{\kappa}$ is the transverse wave vector, $\vec{V}_0$ is the flow velocity of neutral gas, and $\omega_{pi}$ is the plasma frequency of ions in the transition layer. Equation (4) includes a shear flow which leads to an instability. Unstable are the eigenmodes of the Alfven resonator (AR) through which the feedback is realized. The AR eigenmode spectrum is defined by dependence $V_A(z)$ and by the value of $\Sigma_p$. There is an energy leak through the upper wall of the resonator, which, however, is not significant for low eigenmodes in the case of a large difference in plasma density at the center and at the periphery ($z \to \infty$) of the AR. Referring to a detailed derivation to original papers /1-3/, we confine ourselves to the expression for the instability growth rate of the fundamental eigenmode, which follows from the analysis (1)-(4):

$$\gamma = \frac{V_{A0}}{\ell} \cdot \frac{\delta}{1+\delta^2} (1 - V_t/V_0), \tag{6}$$

where $V_{A0}$ is the Alfven velocity in the region of the plasma density maximum N in the AR, $\ell$ is a characteristic scale of N decrease above the maximum, $\delta = \Sigma_{p0}/3\Sigma_w$, $\Sigma_{p0} = \Sigma_p(V_0=0)$, $\Sigma_w = c^2/4\pi V_{A0}$ is the wave conductivity and $V_t$ is the threshold velocity determined by the linear energy losses of Alfven waves in the AR and in the atmospheric substrate. The optimal transverse scale is defined, within the order of magnitude, from the double resonance condition:

$$\omega_0 = \frac{4V_{A0}}{\ell} \approx \kappa_{opt} \cdot V_0, \quad \kappa_{opt} = \frac{4V_{A0}}{V_0 \ell}, \tag{7}$$

where $\omega_0$ is the frequency of the fundamental mode of the AR. Thus, at $V_0 \ll V_{A0}$, strongly anisotropic Alfven vortices are excited with a transverse scale much less than the longitudinal one ($\kappa_{opt}^{-1} \ll \ell$). In expression (5), of interest is the dependence on the coefficient $\delta$, which is proportional to the ion density in the transition layer and characterizes the electrodynamic coupling between AR and substrate. Under real conditions (e.g. in the nighttime ionosphere) $\delta \ll 1$. If with an increase in intensity of Alfven vortices, the acceleration of electrons is capable of generating additional ionization in the transition layer, then the instability increase due to the growth of $\delta$ can become explosive. There are real mechanisms of particle acceleration with increasing amplitude of Alfven waves. One of such mechanisms is the switching on of anomalous resistance due to the excess of the longitudinal current amplitude in the Alfven wave over a critical value corresponding to current instability threshold. A more detailed analysis of this mechanism is given in /4/. Anomalous resistance leads, on the one hand, to turbulent plasma heating and, on the other hand, to the appearance of "runaway" accelerated electrons.

4. Based on the physical concept described above, we can construct a semiphenomenological nonlinear theory of TBL. In this theory, we shall include the turbulent plasma heating and the electron acceleration caused by the anomalous dissipation of Alfven vortices as well as the additional ionization into the ionospheric sublayer. Corresponding equations in the case of homogeneous TBL across the magnetic field have the form (for more details see /4/):

$$\frac{dW}{dt} = \gamma W - Q_1 - Q_2, \quad Q_1 = a_1 j^3, \quad Q_2 = a_2 j^p, \quad p > 3, \tag{8}$$

$$\frac{dN_i}{dt} = I_0 + \frac{Q_2}{w_i} - \alpha N_i^2 . \tag{9}$$

Equation (8) represents the energy conservation law for Alfven vortices. Here the Alfven wave energy density $W = (2\pi j^2/\kappa_{opt}^2 c^2)$, $\kappa_{opt} \sim V_0^{-1}$, $Q_1$ characterizes the turbulent heating and $Q_2$ is due to the appearance of accelerated electrons, $\gamma \sim$ $\sim N_i(1+\nu^2 N_i^2)^{-1}(1-V_t/V_0)$, $N_i$ is the ion density in the transition layer, $I_0$ is the amplitude of the external ionization source, $w_i$ is the ionization potential, and is the recombination coefficient. The solution of system (8)-(9) can easily be investigated in the phase plane $(W,N_i)$. First we shall consider some peculiarities of the stationary state of TBL ($dW/dt = dN_i/dt = 0$). Figure 2 shows the curve of stationary values of current amplitude j in the Alfven vortices as a function of V. This curve is of a hysteresial character. When critical flow velocity $V_*$ is achieved, there occurs an explosive amplification of TBL (the transition from state 1 to state 2 in Fig.2), with $Q_2 \sim V_0^2$ in the upper state.

We now turn to some specific features of the spatio-temporal dynamics of TBL. With smooth evolution of flow parameters the critical velocity $V_*$ under real conditions is achieved at one point in the first place. What is the scenario of the spatio-temporal

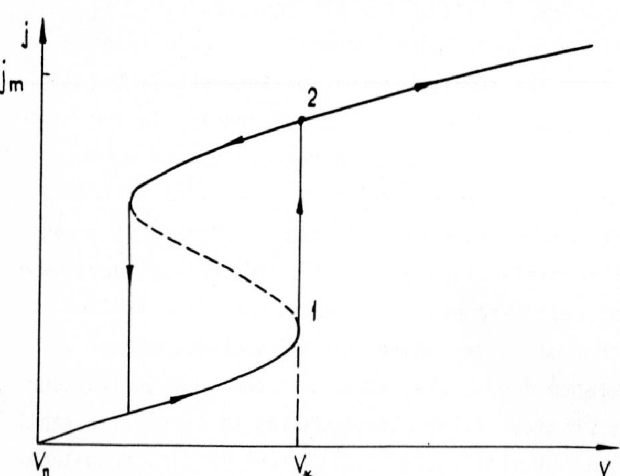


Figure 2. The curve of the TBL stationary state. The arrows indicate the motion of the system with flow velocity variation.

development of TBL? Let us try to answer this question in the case where the plasma motion with respect to neutral gas is due to the electric drift across the magnetic field like the process in the Earth's magnetosphere. As the TBL is locally switched on, the flow is reconstructed self-consistently because of the evolution of the "substrate" parameters (increase of $N_i$). A corresponding set of equations can be written as:

$$\frac{\partial N_i}{\partial t} + (\vec{V}_0 \nabla) N_i = I_0 + Q_2/w_i - \alpha N_i^2 \,, \tag{10}$$

$$\mathrm{Div}_{z=0}(\hat{\Sigma} + \Sigma_w)\vec{E} = 0, \quad \hat{\Sigma} = \hat{\Sigma}_0 (N_i/N_0), \tag{11}$$

where the source $Q_2$ responsible for TBL switching on can accordingly (Fig.2) be put into the form:

$$Q_2 = Q_0 (E/E_0)^2 \cdot 1(E - E_*). \tag{12}$$

Velocity $V_* = cE_*/B$, the substrate conductivity tensor

$$\hat{\Sigma} = \begin{pmatrix} \Sigma_p & \Sigma_H \\ -\Sigma_H & \Sigma_p \end{pmatrix}$$

where $\Sigma_H$ is the Hall conductivity. Equation (11) describes the reconstruction of the electric field (and, therefore, the flow velocity variation) because of the change of $N_i$ in the transition layer (Div is two-dimensional divergence in the horizontal plane). It is easy to see that Eqs. (10)-(11) are similar to the set of equations describing the electric discharge in gas with an ionization source $Q_2$. By this analogy, the spatio-temporal development of TBL will occur in the form of the so-called streamer - an external electric field $\vec{E}_0$ aligned to the region of increased ionization $N_i$, which moves at a velocity dependent on the source $Q_2$ and the background conductivity $\Sigma_{p0}$. In front of the streamer the electric field increases sharply and exceeds the critical field $E_*$ thereby leading to a further motion of the streamer. Quantitative estimates of the streamer characteristics can be found in /5/.

5. The proposed TBL theory is rather effective in explaining the phenomena arising in the polar ionosphere during the magnetic substorms. These periods are characterized, as a rule, by an amplification of the global electric field determining the plasma flow at ionospheric heights and accompanied by the intensification of electron precipitation and the appearance of fast-moving regions of increased ionospheric ionization. These regions resemble a streamer with a corresponding system of electric currents and electric field amplification near the streamer head. Using the TBL theory it is possible to explain quantitatively the characteristics of the phenomena mentioned above. Thus, the proposed concept of turbulent Alfven boundary layer defines, essentially, a new energy source in the Earth's magnetosphere, which

takes energy directly from the magnetospheric convection and can contribute significantly to the energy budget of a magnetospheric substorm.

The idea of Alfven TBL can be fruitful in explaining the physical processes in the solar atmosphere. In particular, some structures in the solar chromosphere, such as spikules, can result from the excitation of the Alfven vortices by photospheric convective cells. Of interest is the TBL contribution to the energy budget of solar flares. These problems seem to be new interesting objects of investigation in the field of nonlinear wave dynamics.

## REFERENCES


1. V.Yu.Trakhtengerts, A.Ya.Feldstein. On stratification of magnetospheric convection. Fizika Plazmy, 1982, 8, 140 (in Russian).
2. V.Yu.Trakhtengerts, A.Ya.Feldstein. Quiet auroral arcs: ionospheric effect. Planet. Space.Sci., 1984, 32, 127.
3. V.Yu.Trakhtengerts, A.Ya.Feldstein. On excitation of small-scale electromagnetic perturbations in the ionospheric Alfven resonator. Geomagnetizm i Aeronomiya, 1987, 27, 315 (in Russian).
4. V.Yu.Trakhtengerts, A.Ya.Feldstein. The turbulent regime of magnetospheric convection. Geomagnetizm i Aeronomiya, 1987, 27, 258 (in Russian).
5. V.Yu.Trakhtengerts, A.Ya.Feldstein. The explosive phase of substorm as a consequence of the turbulent regime of magnetospheric convection. Geomagnetizm i Aeronomiya, 1988, 28, 743 (in Russian).

# Frequency Selfconversion and Reflectionless Propagation of a Powerful Electromagnetic Pulse in an Ionized Medium

*V.B. Gil'denburg, A.V. Kim, V.A. Krupnov, and V.E. Semyonov*

Institute of Applied Physics, USSR Academy of Sciences,
46 Ulyanov Str., 603600 Gorky, USSR

Frequency self-conversion of a powerful electromagnetic pulse in plasma is investigated. The possibilities of strong frequency increase and the resulting total prevention of a signal cut-off during a breakdown are analyzed in some simple models.

## 1. INTRODUCTION

A high-frequency electromagnetic pulse (a wave packet) is considered to propagate without reflection in a homogeneous or a smoothly inhomogeneous gaseous medium only when its power (at the assigned duration) does not exceed a threshold value which is determined by the conditions of the breakdown emergence. The condition when the electron concentration N during a breakdown (for the time which is less than the pulse duration) reaches the known critical value $N_{co} = m\omega_0^2/4\pi e^2$, is usually taken as a sufficient condition at low frequencies of electron collisions $\nu \ll \omega_0$ for locking (i.e. cut-off) of a powerful pulse introduced into the gas at frequency $\omega_0$. Thus the value $N_{co}$ is considered as an upper limit of N in the propagating pulse.

However, this criterion of the signal cut-off ($N = N_{co}$) does not take into account a possibility of a parametric transformation (increase) of the initial signal frequency $\omega_0$ which is caused by nonstationary breakdown plasma. Though a conversion effect has been studied earlier both for the case of assigned variations of the medium parameters /1-2/ and for some problems of the self-interaction of a wave packet in nonlinear media /3-4/, it has been practically neglected in all investigations considering dynamics of a high-frequency breakdown and only a possible role of a medium breakdown in widening of the spectrum of powerful laser pulses has been noted /5/. Our brief report does not aim at a thorough analysis of frequency conversion in an ionizing medium but we want to illustrate its role in breakdown dynamics on some examples which make clear some new important possibilities, such as: a) a strong increase in the signal frequency (in the final state $\omega \gg \omega_0$); b) reaching of $N \gg N_{co}$ in a reflectionless regime; c) total preventing of a cut-off during a breakdown.

These effects are clearly observed in a breakdown in the so-called fields of "superhigh intensity" /6/ where the energy of electron oscillations is great enough if compared to the molecule ionization potential and where the variation of the electron momentum due to collisions takes place during time $1/\nu$ which is much greater than the characteristic time of the electron concentration growth $1/\nu_i$. Otherwise,

Research Reports in Physics Nonlinear Waves 3
Editors: A.V. Gaponov-Grekhov · M.I. Rabinovich · J. Engelbrecht

when $\nu_i \ll \nu$, the wave is strongly damped earlier than its frequency varies. Assuming further that the condition

$$\omega \gg \nu_i \gg \nu \tag{1}$$

is fulfilled and making a supposition on a zero value (or on isotropic distribution) of the velocity of electrons at the moment of their birth, let us write down an equation for the wave electric field E(x,t) in the form

$$c^2 \frac{\partial^2 E}{\partial x^2} - \frac{\partial^2 E}{\partial t^2} = \omega_p{}^2 E, \qquad \omega_p{}^2 = \frac{4\pi e^2}{m} N . \tag{2}$$

Consider two different mechanisms of gas ionization in an electromagnetic wave field: 1) ionization of molecules by an electron impact; 2) direct ionization by a wave field (multiphoton ionization, tunnel electron detachment).

## 2. IONIZATION BY AN ELECTRON IMPACT

A balance equation for the electron concentration (or for $\omega_p{}^2$ which is proportional to the concentration) has in this case the form

$$\frac{\partial \omega_p{}^2}{\partial t} = \nu_i \omega_p{}^2 \tag{3}$$

(the frequency of electron losses $\nu_a$ is considered neglible if compared to the ionization frequency $\nu_i$).

In a locally quasiharmonic field $E(x,t) = A(x,t)\exp i\phi(x,t)$ ($\omega = \partial\phi/\partial t \gg \nu_i$, $k = -\partial\phi/\partial x \gg L^{-1}$, $\omega\tau \gg 1$; L and $\tau$ are space and time pulse duration, respectively) the ionization frequency $\nu_i$ does not depend on the oscillation phase and can be considered as the assigned function of the relation $A/\omega = U$; $\nu_i = \nu_i(U)$ /6/ that enables us to treat Eq.(2) in the approximation of geometrical optics and to write down equations for the transfer of the local frequency $\omega$ and the local amplitude A as

$$\frac{\partial \omega^2}{\partial t} + V \frac{\partial \omega^2}{\partial x} = \frac{\partial \omega_p{}^2}{\partial t}, \tag{4}$$

$$\frac{\partial}{\partial t} (A^2\omega) + \frac{\partial}{\partial x} (VA^2\omega) = 0 , \tag{5}$$

where $V = c(1-\omega_p{}^2/\omega^2)^{1/2} = c(1-N/N_c)^{1/2}$ is a local group velocity of the wave and $N_c = m\omega^2/4\pi e^2$. A local wave number here is $k = (\omega/c)(1-\omega_p{}^2/\omega^2)^{1/2}$. Equation (5) can be also written down as

$$\frac{\partial A^2}{\partial t} + \frac{\partial (A^2 V)}{\partial x} = - \frac{1}{2} \nu_i A^2 \frac{\omega_p{}^2}{\omega^2} . \tag{6}$$

Note that though Eqs. (2,4-6) do not contain a term containing collision frequency, and do not describe the wave absorption caused by electron collisions, they take into account another source of dissipation which is much stronger at $\nu_i \gg \nu$, i.e. the transfer of energy to newly born electrons which acquire in the field both the ordered oscillatory velocity component (adiabatically varying with the field amplitude) and the constant component (which depends on the field phase at the moment of birth and remains after the pulse).

It follows from (4) that a change in the difference $\omega^2 - \omega_p^2$ along the group trajectory (i.e. characteristics) $x = x(t)$, $dx/dt = V$ is described by the equation

$$\frac{d}{dt}(\omega^2-\omega_p^2) = -V\frac{\partial\omega^2_p}{\partial x} . \qquad (7)$$

It means that plasma will always remain transparent ($\omega > \omega_p$) during the breakdown, if $\partial\omega_p^2/\partial x \leq 0$. Below we give some of the solutions of system (3)-(5) which meet this condition and which have been obtained for cases permitting various simplifications.

i) A homogeneous discharge created by a homogeneous plane wave

Supposing A, $\omega$ and $\omega_p$ to be independent of x, we get from (4), (5) that

$$\omega^2-\omega_p^2 = \omega_0^2-\omega_{p0}^2 = k^2c^2 = \text{const}, \quad A^2\omega = A_0^2\omega_0 = \text{const}, \quad V = V_0\frac{\omega_0}{\omega} , \qquad (8)$$

where $A_0$, $\omega_0$, $\omega_{p0}$, $V_0$ are the assigned initial values of A, $\omega$, $\omega_p$, V at t = 0. As a result the problem is reduced to integration of the first-order equation (3) with the given dependence $\nu_i[U(\omega_p)]$, $U = A/\omega = A_0\omega_0^{1/2} \cdot (\omega_p^2+k^2c^2)^{-3/4}$. It is evident that the frequencies $\omega$ and $\omega_p$ increase with the growth of t while amplitude A decreases. This process continues until U, which determines the oscillatory electron energy falling to some limit $U_c = A_c/\omega_0$ beyond which the field ceases to be "superhigh". Then the inequality $\nu_i >> \nu$ breaks and we observe some strong additional wave damping due to ordinary collisions. Corresponding limiting values of $\omega$, $\omega_p$ and V at $A_0 >> A_c$ are

$$\omega_{p\,max} \cong \omega_{max} = \omega_0(A_0/A_c)^{2/3} >> \omega_0; \quad V_{min} \cong V_0(A_c/A_0)^{2/3} << V_0 . \qquad (9)$$

ii) An initial problem for a semi-limited pulse at $\nu_i(U>U_c)$ = const

Let $\omega = \omega_0$, $\omega_p = \omega_{p0} << \omega_0$, $A(x<0) = A_0$, $A(x>0) = 0$ be assigned at t = 0. (Here a generalization for a pulse of finite duration is evident). In the region x < 0 Eqs.(8) and (9) are valid and $\omega_p^2 = \omega_{p0}^2 \exp(\nu_i t)$. Since at x > 0 the leading wave edge moves with the velocity close to c so that at 0 < x < ct, $A/\omega > U_c$, we have $\omega_p^2 = \omega_{p0}^2 \exp[\nu_i(t-x/c)]$, $\partial\omega_p^2/\partial x < 0$ and, consequently, the difference $\omega^2 - \omega_p^2$ increases while the group fronts move along x(t). It can be shown that at x = x(t) > > 0

$$dV/dt = (c\nu_i/2)(1-V/c)^2(1+V/c), \qquad (10)$$

$$\omega(1-V/c) = \text{const}, \quad \omega_p^2(1-V/c)(1+V/c)^{-1} = \text{const}; \qquad (11)$$

$$\omega A^2(Vc-V_1^2) = \text{const}, \qquad (12)$$

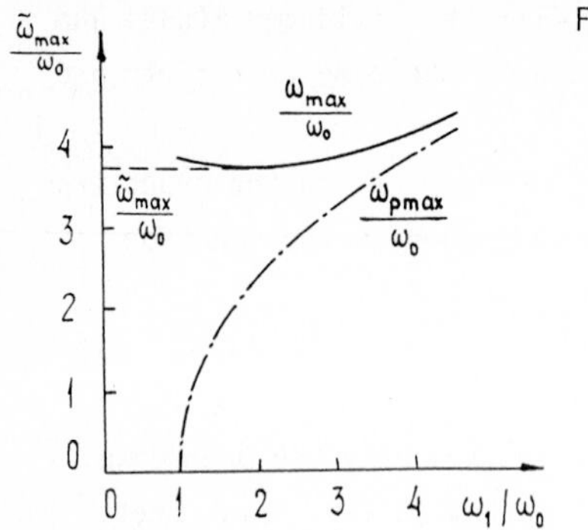


Figure 1.

where $A_1$, $\omega_1$, $V_1$ are initial values of respective magnitudes which are given for each group front at the moment $t_1$ when the front passes through point $x = 0$ based on the solution for $x < 0$. Maximum values of $\omega$ and $\omega_p$ that are reached for $x > 0$ on each characteristic are shown qualitatively for $A_0/A_c = 10$ in Fig.1 as functions of $\omega_1$. It is seen that the maximum of $\omega$ does not drop below $\omega_{max} \cong \omega_0(A_0/A_c)$ which corresponds to the pulse element undergoing greater elongation due to inhomogeneity V.

iii) Pulse entrance into the ionized medium. Approximation V = const

If a wave with given parameters $A_0$ and $\omega_0$ is incident to a medium with a sharp boundary, then frequency $\omega$ in the boundary region remains all the time equal to the initial value $\omega_0$ and the time of the signal locking at the boundary (i.e. the threshold duration $\tau_c$) is determined in a usual way, that is, $\tau_c = \nu_{i0}^{-1}\ell n(\omega_0^2/\omega_{p0}^2)$ where $\nu_{i0} = \nu_i(A_0/\ _0)$. With receding from the boundary, the frequency of the transmitted pulse (with the duration $\tau < \tau_c$) continuously increases up to the limit determined by the ratio $A_0/A_c >> 1$. In particular, at $\tau << \tau_c$ the approximation $V = const = c$ is correct, frequency $\omega$ reaches its maximum (9) at the distance $\ell \cong (c\omega_0^2/\nu_i\omega_{p0}^2) \cdot$ $\cdot \exp(-\nu_{i0}\tau) >> c/\nu_{i0}$ where $A = A_c$. In the case of a descending dependence $\nu_i(U) \sim$ $\sim 1/U^\gamma(\gamma\sim 1)$ which is typical for "superhigh" fields, the concentration grows as well and reaches values $N >> N_{c0}$: $\omega^2_{p\ max} \cong \omega^2_{max}c/(\nu_{i0}\ell)$, $\omega_{max} >> \omega_{p\ max} >> \omega_0$. An ionizing pulse with the overthreshold duration ($\tau >> \tau_c$) can be introduced into the medium without a cut-off if the boundary of the medium is diffused enough. If, in particular, the scale of the transition region is $L = \nu_i/(\partial\nu_i(U,x)/\partial x) >> c\tau$, then it follows from the estimation of $\partial\omega_p^2/\partial x$ that $\omega_p << \omega$, $V \cong c$ and the maxima of $\omega_p$ and $\omega$ are much greater than $\omega_0$.

Note that a possibility to fulfill the condition $\partial\omega_p^2/\partial x < 0$ providing $\omega^2 - \omega_p^2$ to be positive, is quite evident in a short pulse due to the finite velocity of the leading ionizing edge not only for the case of the inhomogeneous gas ($\nu_i = \nu_i(U,x)$) but also when the concentration $\omega_{p0}^2 = \omega_{p0}^2(x)$ is initially distributed inhomogeneously. Thus it follows that the ionization of the medium is not only an obligatory reason for the pulse locking (at $\omega_p > \omega_0$) but thanks to the frequency autoconversion it can result in the transparency of the opaque plasma created by an external source: under condition $\nu_i L > c$ the pulse with duration $\tau < L/c$ and the field amplitude $A_0 \ll \ll A_c$ passes through the region where $\omega_{p0}(x)$ increases up to the value exceeding $\omega_0$ by several times without reflection (though it looses some portion of the stored energy).

## 3. DIRECT IONIZATION BY THE ELECTRO-MAGNETIC WAVE FIELD

Such an ionization mechanism can be predominating during a breakdown of a rarefield gas by a powerful laser pulse. If the oscillation energy of a free electron in the field $w_\sim = e^2A^2/2m\omega^2$ is not large compared to the potential of the molecule ionization I, then the probability of the multiphoton ionization is a function of the field amplitude and the field frequency and the equation for $\omega_p^2$ has the form

$$\frac{\partial\omega_p^2}{\partial t} W(A,\omega) . \tag{13}$$

Unlike Eq.(3) this equation does not contain $\omega_p^2$ in its right hand side, i.e. the growth of N and $\omega_p^2$ is not avalanche-like. Though in this case a concrete form of the functions $\omega_p^2(x,t)$, $A(x,t)$, $\omega(x,t)$ varies if compared to the cases above, the solutions do not change qualitatively and new effects are not observed.

For the opposite limiting case, when $w_\sim \gg I$, the ionization velocity W (the right hand side of the balance equation (13)) is determined by a probability of the tunelling electron detachment /7/ and depends on temporal (real) value of the electric field $|\mathrm{Re}E(t)|$

$$W = K\beta \exp(-E_0/|\mathrm{Re}E|) , \tag{14}$$

where $K = (4\pi e^2/m)N_m$, $E_0 = (\frac{3}{4}\hbar e)^{-1}\sqrt{2m}\, I^{3/2}$, $\beta \sim I/\hbar$ and $N_m$ is the molecule concentration. The value of W in the field with linear polarization ($\mathrm{Re}E = A\cos\phi$) is a function of both the amplitude and the field phase and, consequently, $\omega_p^2$ has not only a slowly increasing part (in the scale $1/\omega$) determined by a mean value of W but also rapidly oscillating harmonics

$$\frac{\partial\omega_p^2}{\partial t} = W = W_0 + W_2 \cos 2\phi + \dots , \tag{15}$$

$$W_0 = \frac{2}{\pi}\int_0^{\pi/2} W(|A\cos\phi|)d\phi , \tag{16}$$

$$W_2 = \frac{4}{\pi}\int_0^{\pi/2} W(|A\cos\phi|)\cos 2\phi d\phi . \tag{17}$$

The zero and the second harmonics $\omega_p^2(x,t)$ taken into account in the right hand side of Eq.(2) and playing the main role in the above quasiharmonic approximation ($\partial\omega_p^2/\partial t \ll \omega^3$) lead to a new transfer equation for the amplitude of the main (first) field harmonic

$$\frac{\partial A^2}{\partial t} + \frac{\partial(A^2 V)}{\partial x} = F(A,\omega) \; , \tag{18}$$

$$F(A,\omega) = - \frac{A^2}{\omega^2} \langle \sin^2\phi W(A,\phi)\rangle = - \frac{A^2}{2\omega^2}\left(W_0 - \frac{1}{2} W_2\right) \tag{19}$$

(angular brackets mean the averaging over $\phi$). It is evident that the variation of the field amplitude A along the characteristics depends in this case on the character of the ionization velocity distribution over the field phases $\phi$. In particular, if the ionization takes place mainly at those values of the field which are close to the maximum, i.e. at $\cos\phi = \pm 1$, then the velocity of the field decrease (-F) strongly diminishes. It is caused by a decrease of the constant component of the velocity possessed by an electron at the moment of its birth. For a limiting case $W \sim \delta(|\mathrm{Re}E| - A) \sim \delta(\sin\phi)$ it follows from (19) that $F = 0$, i.e. the amplitude during the frequency conversion remains constant. From (14) it is evident that a similar situation is realized at $A \ll E_0$.

## REFERENCES


1. N.S.Stepanov. Izvestiya VUZ'ov, Radiofizika, 1976, 19, 960 (in Russian).
2. Yu.A.Kravtsov, Yu.I.Orlov. Geometrical Optics of Inhomogeneous Media. Moscow, Nauka, 1980 (in Russian).
3. L.A.Ostrovsky. Zh.Eksp.Teor.Fiz., 1968, 54, 1235 (in Russian).
4. D.Mestdagh, M.Haelterman. Opt.Comm., 1987, 61, 291.
5. N.Blombergen. Opt.Comm., 1973, 8, 285.
6. S.G.Arutynyan, A.A.Rukhadze. Fizika Plazmy, 1979, 5, 702 (in Russian).
7. Yu.P.Raizer. A Laser Spark and Discharge Propagation. Moscow, Nauka, 1974 (in Russian).

# Nonlinear Stability of an Ideal Plasma

*V.A. Gordin and V.I. Petviashvili*

Kurchatov Institute of Nuclear Energy, 123182 Moscow, USSR

The functional series for the first integrals of helicity are used for the derivation of the sufficient stability conditions of ideal plasma in the sense of Lyapunov. The Euler equations for the Lyapunov functional imply the Grad-Shafranov equation with pressure profile determined by the toroidal field. The calculations were carried out both for cylindrical and toroidal configurations. In the considered example the magnetic field is mostly created by currents in plasma. The safety factor is less than one and decreases in the direction of the plasma boundary. The toroidal field behaves similarly: decreasing but not turning to zero. The pressure and density of helicity and their gradients, however, turn to zero at the boundary.

1. The role of stability in the problem of plasma confinement by constant magnetic field in the approximation of ideal magnetohydrodynamics (MHD) is rather important. The definition of stability of a stationary point in a system of ordinary differential equations was proposed by LYAPUNOV /1/. He also formulated the sufficient conditions of such a stability based on the properties of extremal points of a certain function called the Lyapunov function, which is defined over the phase space of the system.

The ideas of Lyapunov's theory in application to the evolution of partial differential equations, i.e. in the case of infinite dimensional phase space, started to develop in the fifties but became spread with respect to the hydrodynamical equations only in the second half of the sixties /2-4, see also the references therein/. To this time the linear MHD theory providing the necessary conditions of stability has been well developed /5,6/. In the case of ideal MHD the linear theory may give the subset of the equilibrium states for which all frequencies of linear oscillations are real or damped. As it is known, this is only a necessary, but not a sufficient condition for stability of the initial nonlinear system /1/. For example, the possibility of threshold-free instability caused by interaction of three and more linear waves in inhomogeneous medium is not taken into account. The main point of the Lyapunov's theory is the finding of the corresponding Lyapunov function (functional). For this a series of the first integrals connected with the helicity /7,8/ turns out to be useful. The solutions of the MHD equations as equilibrium states which can be tested for the stability in the sense of Lyapunov, have to be the extremals, i.e. to be the solutions of the Euler equations for the corresponding Lyapunov functional. Therefore, the class of equilibrium solutions is restricted, because the functional exists not for all solutions of the Grad-Shafranov equation /6/.

Research Reports in Physics **Nonlinear Waves 3**
Editors: A.V. Gaponov-Grekhov · M.I. Rabinovich · J. Engelbrecht

The next step is to test the strong positive definiteness of the second variation of the Lyapunov functional in the neighbourhood of the extremal. Here the difficulties arise due to the infinite dimension of the phase space, i.e. the possibility of the continuous spectrum for the Jacobi equation with coefficients having their poles on the resonance surfaces, and due to the difference in the notion of positive definiteness and strong positive definiteness.

## 2. THE EQUATIONS OF IDEAL MHD

$$\begin{aligned}
&\partial_t \underline{B} = \mathrm{curl}(\underline{v}\mathrm{x}\underline{B}), \qquad \mathrm{div}\ \underline{B} = 0\ , \\
&d_t p + \gamma p\ \mathrm{div}\ \underline{v} = 0\ , \\
&\partial_t \rho + \mathrm{div}\ \rho\ \underline{v} = 0\ , \\
&\rho d_t \underline{v} = -\nabla p + \mathrm{curl}\ \underline{B}\mathrm{x}\underline{B}/4\pi\ ,
\end{aligned} \tag{1}$$

where $\gamma$ is the adiabatic index and all other notations are common. System (1) has the following first integrals:

$$\begin{aligned}
&\int \rho v d^3x && \text{(the components of momentum)}, \\
&W = \int (B^2/8\pi + p/(\gamma-1) + \rho v^2/2) d^3x && \text{(energy)}, \\
&\int (\underline{A}\cdot\underline{B}) d^3x\ , \quad \int \rho(\underline{v},\underline{B}) d^3x && \text{(helicity and cross helicity)}, \\
&\int_D \rho f(p/\rho^\gamma) d^3x && \text{("entropy series")},
\end{aligned} \tag{2}$$

where f is an arbitrary function, D is the arbitrary volume of fluid, it is supposed that when the domain of integration is not given explicitly, then the integration is carried out over the whole volume V inside the conducting wall.

As to the boundary $\partial V$, the conditions of nonpercolation and of infinite conductivity are assumed to be satisfied, i.e.

$$(\underline{n}\cdot\underline{v})\Big|_{\partial V} = 0\ , \quad (\underline{n}\cdot\underline{B})\Big|_{\partial V} = 0\ , \tag{3}$$

where unit vector n is normal to the boundary.

From the first equation of set (1) follows the equation for vector-potential

$$\partial_t \underline{A} = \underline{v}\mathrm{x}\underline{B} - \nabla\phi(\underline{x},t)\ , \tag{4}$$

where $\phi$ is an arbitrary function. Let $\phi = \underline{A}\cdot\underline{v}$. Then the helicity density $h = \underline{A}\cdot\underline{B}$ satisfies the continuity equation

$$\partial_t h + \mathrm{div}\ h\underline{v} = 0\ . \tag{5}$$

It follows that the quantity $\mu = p^{1/\gamma}/h$ is conserved in the particles: $\partial_t \mu + \underline{v}\cdot\nabla\mu = 0$.

For the expanded set of equations (1,4) there exists the new functional series of the first integrals:

$$K_{F,D} = \int_D hF(\rho/h,\mu)d^3x \ , \qquad (6)$$

where F is an arbitrary function of two variables. D is the liquid volume of plasma.

3. Let us choose the first integral as a sum of the energy and helicity to be the Lyapunov functional:

$$L = W + K_F \ . \qquad (7)$$

The minimization should be carried out both over the potential and the vorticil part of $\underline{A}$. Function F can be chosen as $F = g(\mu) + C_1p^2/h^2 - C_2p/h$, where $C_1, C_2$ = const, g will be determined below. For the sake of simplicity, let us suppose that the vacuum domain is missing, D = V. The boundary conditions (3) are also assumed to be satisfied for the variations. Minimizing (7) over $\underline{v}$, we find that the equilibrium takes place only when the motion is absent: $\underline{v} = 0$. Minimizing further the functional over variable p, we obtain $p = (C_2/2C_1)h$. The positive definiteness of (7) is satisfied over variables p and $\underline{v}$, when $C_1h > 0$. Hence the density of helicity h for the extremal of (7) cannot change its sign.

Minimizing the functional over pressure p, one gets the Euler equation:

$$\gamma p + (\gamma-1)g'(\mu) = 0 \ , \qquad (8)$$

where dash means the derivative in respect to the argument. It follows that in the state of equilibrium the pressure depends only on helicity.

At last, taking variation in respect to $\underline{A}$, one gets the Euler equation:

$$\underline{j} + 2G\underline{B} + \nabla G \times \underline{A} = 0 \ , \quad \underline{j} = \mathrm{curl}\ \underline{B}/4\pi \ , \quad G = g - \mu g' - C_2^2/4C_1 \ . \qquad (9)$$

Acting on equation (8) with operator grad , and on the equation (9) with operator div , one gets:

$$\nabla p = h\nabla G \quad (\underline{B},\nabla p) = 0 \ . \qquad (10)$$

Taking the vector product of (9) and $\underline{A}$, and taking into account (10), the condition of equilibrium is obtained:

$$\underline{j} \times \underline{B} = \nabla p \ , \qquad (11)$$

which can be deduced straightforwardly from (1). However, the set of Euler equations (8), (9) together with condition (11), need an additional restriction on current:

$$(\underline{A},\underline{j}) + 2Gh = 0 \ , \qquad (12)$$

which can be derived if one takes scalar product of (9) and $\underline{A}$. Therefore, not every equilibrium plasma configuration is an extremal of some L.

Turning to the investigation of the second variation of L, we find that from the expression

$$\delta^2 L = \int\{(\delta \underline{B})^2/8\pi + (\delta p)^2/2\gamma p + \rho(\delta \underline{v})^2/2 + Gh_2 - G'p^{1/\gamma}(\delta p/\gamma p - h_1/h)^2/2 +$$

$$+ C_1\Big[(\sigma\rho)^2/h - \rho\sigma\rho h_1/h^2 - \rho^2 h_2/h^2 + \rho^2 h_1{}^2/h^3\Big]\}d^3x \ ,$$

where $h_1 = \underline{A}\cdot\delta\underline{B} + \delta\underline{A}\cdot\underline{B}$, $h_2 = \delta\underline{A}\cdot\delta\underline{B}$, the variations of density and velocity can be excluded algebraically during the minimization. The pressure variations can be excluded similarly but with some restrictions on the extremal. Besides, let us note that potential variations $\delta\underline{A}$ are possible, at which $\underline{B} = 0$, $h_2 = 0$, but $h_1 \neq 0$. The positive definition of $\delta^2 L$ at variations of this kind leads to some additional restrictions. Summing up all the restrictions /7,8/, we obtain:

$$G' \geqq 0 \iff \mu g'' \geqq 0 \ ; \quad 0 \leq hd_h p \leq \gamma p \ . \tag{13}$$

The careful introduction of the norm is necessary for proving strong positive definition of second variation

$$\delta^2 L\Big[\delta X\Big] > C_0||\delta X||^2 \ .$$

If inequality in (13) is fulfilled in an open region, then the second variation is only nonnegative and the Lyapunov stability conditions are violated. In examples considered below, all inequalities are strong, except at the plasma boundary $\partial V$. It follows from (8) that $\mu = \mu(h)$, so we can introduce function $U = U(h)$ defined as: $U'(h) = 4\pi G(\mu(h))$. Then the expression for pressure follows from (8,10):

$$4\pi p(h) = \ hU''dh \ . \tag{14}$$

Further, $\delta^2 L$ can be reduced to the dependence from $\delta\underline{A}$ only. This makes it possible to introduce reduced functional:

$$Y = \int\Big[B^2/2 + U(h)\Big]d^3x \ . \tag{15}$$

It is easy to show that the condition of strong positivity of $\delta^2 Y$ is satisfied at the same conditions as that of $\delta^2 L$. Further, function U is selected in a way to make $\delta^2 Y$ strongly positive with possible maximal confined pressure in accordance with condition (13) and at the given profile of boundary. The first variations of Y by $\underline{A}$ under bounda condition $\underline{n}\cdot\underline{B}\big|_{\partial V} = 0$ leads to the Euler equation equivalent to (9)

$$\mathrm{curl}\ \underline{B} + 2\cdot U'\cdot\underline{B} + \Big[\nabla U' x \underline{A}\Big] = 0 \ . \tag{16}$$

Density $\rho$ as before is proportional to h in extremum, and pressure is given by (14). So the extremum and the corresponding plasma configuration is defined by function U and the profile of the wall. In addition, the condition of the thermal insulation is to be met: the pressure vanishes at the boundary:

$$P\Big|_{\partial V} = 0 \ . \tag{17}$$

From this, together with (13), (14), the equivalent conditions follow:

$$h\Big|_{\partial V} = \nabla h\Big|_{\partial V} = \nabla p\Big|_{\partial V} = 0 \ ; \qquad 4\pi p = \int_0^h hu''dh \ . \tag{18}$$

In the problem of strong positive definition of

$$\delta^2 Y = \int(\delta B^2/2 + U'h_2 + U''h_1{}^2/2)d^3x \ ,$$

the definition of the norm in the space of perturbation is necessary. Here the norm is selected carefully depending on the extremal near which the positiveness is investigated in the form $\delta^2 Y \geq C||\delta A||^2$, where C is a positive constant. So we take:

$$||\delta A||^2 = ||\delta B||^2_{L^2} + ||h_1||^2_{L^2} \ . \tag{19}$$

The Jacobi equation has the form:

$$\mathrm{curl}\ \delta\underline{B} + 2U'\delta\underline{B} + \left[\nabla U' x \delta\underline{A}\right] + U''h_1\underline{B} + \mathrm{curl}(U''h_1\underline{A}) = 0 \ . \tag{20}$$

The solution of (20) and complete investigation of positiveness of $\delta^2 Y$ is possible if the wall and extremum $\underline{A}$ have some symmetry. Here the cylindrical and axial symmetry cases will be considered, /8/.

4. The cylindrical equilibrium where extremal depends only on r in cylindrical system of coordinates $(r,\phi,z)$, will be considered first. We introduce a more convenient function $\theta(r) = U'(h)$ and assume h(r) to be monotonic. In extremal $A_r = 0$. At the axis $r = 0$ we assume: $A_\phi = 0$, $d_r A_z = 0$. The Euler equations (16) take the form

$$\begin{aligned} &d_r B_z = 2\theta B_\phi - A_z d_r \theta, \quad d_r A_z = -B_\phi \ , \\ &r^{-1} d_r (rB_\phi) = -2\theta B_z - A_\theta d_r \ , r^{-1} d_r (rA_\phi) = B_z \ . \end{aligned} \tag{21}$$

The variation $\delta\underline{A}$, due to cylindrical symmetry of background, can be expanded in double Fourier series with coefficients depending on r:

$$\delta\underline{A} = \sum_{k,m} \underline{a}_{k,m}(r)\exp(i(m\phi+kz)) \ .$$

Then (20) can be reduced to only one equation relative to $b(r,k,m) = (\delta B)_r$:

$$d_r{}^2 b + \frac{1}{r}\Gamma_1 d_r b = (\Gamma_2/r^2 + \Gamma_3/\sigma + \Gamma_4/\sigma^2)b \ , \tag{22}$$

where $\Gamma_i$ are smooth functions inside plasma cylinder $0 < r \leq r_b$ with boundary at $r=r_b$, which are given in /8/; $\sigma = (m-kLq)B_\phi/r$, $q = rB_z/B_\phi L$, q is safety factor, L is the length of the cylinder.

Equation (22) has singular points at $\sigma = 0$ inside the plasma. The components of $\underline{A}$ can be expressed uniquely through b. The local stability criterion near the singular points $\sigma = 0$ in (22) should be fulfilled. This coincides with the Suidam criterion for stability of cylindrical plasma in linear approximation /6/.

In the case of axial symmetry, equation (21) was solved and toroidal equilibrium was obtained numerically with $\underline{A}$ unique and smooth everywhere. The Mercier criterion for stability was used during the calculations /8/. The conditions on pressure distribution (13) restrict the class of equilibrium only to systems with ultra-low q: q is a monotonical, decreasing function from centre of plasma and $0 < q < 1$ everywhere. The toroidal component of the magnetic field is decreasing too but does not change its sign /8/. This structure has a close similarity with the installation Repute-1 from the Tokyo university. Due to the resistivity, the safety factor is not decreasing up to the boundary, but begins to grow in the periphery region /9/. This probably leads to MHD instabilities observed in the Repute. The current drive is needed to compensate the effect of finite conductivity and to make q decreasing up to the boundary.

## REFERENCES


1. A.M.Lyapunov. Collection of Works, v.2. Gostekhizdat, Leningrad, 1956 (in Russian).
2. V.I.Arnold. Mathematical Methods of Classical Mechanics. Nauka, Moscow, 1974 (in Russian).
3. L.A.Diky. Hydrodynamical Stability and Atmospheric Dynamics. Hydrometeoizdat, Leningrad, 1976 (in Russian).
4. V.A.Gordin. Mathematical Problems of the Hydrodynamical Weather Forecast, v.1,2. Hydrometeoizdat, Leningrad, 1987 (in Russian).
5. B.B.Kadomtsev. In: Reviews of Plasma Physics, v.2. Consultants Bureau, 1966.
6. G.Bateman. MHD Instabilities. The MIT Press, 1978.
7. V.A.Gordin, V.I.Petviashvili. Sov.Phys.-JETP. Lett., 1987, 45, 267.
8. V.A.Gordin, V.I.Petviashvili. Sov.Phys.-JETP, 1989, 97, 1711.
9. Y.Murakami, Z.Yoshida, N.Ynoue. Nuclear Fusion, 1988, 28, 449.

# Dynamics of a High-Frequency Streamer

*V.B. Gil'denburg, I.S. Gushin, S.A. Dvinin, and A.V. Kim*

Institute of Applied Physics, USSR Academy of Sciences,
46 Ulyanov Str., 603600 Gorky, USSR

The space-time evolution of small-size high-frequency plasmoids created by a high-frequency gas break-down around a solitary primary ionization centre is investigated. It is found analytically and by some computer simulation that a quasi-spherical plasmoid being formed at the first stage of the avalanche-like process in a course of time stretches in the direction of the external electric field and turns into a rapidly growing "high-frequency streamer". This effect is due to an increase of the electric field amplitude at "polar" regions of the plasmoid.

1. One of the key problems in the theory of high-frequency discharge in gases is to study the space-time evolution of small-size isolated plasma formations (plasmoids) which appear as a result of independent (nonoverlapping) electron avalanches around discrete primary ionization centres /1/. Single rarely-spaced electrons, liquid and solid aerosols, small-scale inhomogeneities formed due to some discharge instabilities, sources of artificial ionization etc. can play a role of such centres.

The aim of this paper is to investigate (i) theoretically and (ii) numerically by computer simulation the dynamics of a gas-discharge plasmoid with dimensions less than the wave-length and the skin-layer depth in a superbreakdown field with linear polarization. The main physical factors determining the character of the discharge propagation from the region of an initial breakdown in the case of interest are the diffusion (both free and ambipolar), the electron impact ionization of molecules and the effect of formed plasma on the applied high-frequency field. A corresponding initial system of equations includes the equation of a quasistatic (vortexless) approximation for a complex amplitude of a quasiharmonic electric field $\vec{E}(\vec{r},t)\exp(i\omega t) = -\nabla\psi(\vec{r},t)\exp(i\omega t)$ and the equation of ionization balance for the electron concentration N:

$$\mathrm{div}(\varepsilon\nabla\psi) = 0, \quad \varepsilon = 1 - \frac{N}{N_c}\left(1 + i\frac{\nu}{\omega}\right), \tag{1}$$

$$\frac{\partial N}{\partial t} = D\nabla^2 N + (\nu_i - \nu_a)N - \alpha N^2 . \tag{2}$$

Here $\varepsilon$ is the complex plasma permittivity, $N_c = m(\omega^2+\nu^2)/4\pi e^2$, $\omega$ is the angular field frequency, $\nu$ is the effective collision frequency of electrons, D and $\alpha$ are the diffusion and recombination coefficients, respectively, $\nu_i$ is the frequency of ionizing collisions, and $\nu_a$ is the attachement frequency of electrons to molecules.

Consider the difference of the frequencies of ionization and attachment as a given rapidly increasing function of the field amplitude $\nu_i - \nu_a = f(|\vec{E}|)$; for a large domain of conditions we can use a power approximation:

$$\nu_i - \nu_a = \nu_a\left[\left(\frac{|\vec{E}|}{E_c}\right)^\beta - 1\right] . \tag{3}$$

Here $E_c$ is a so-called breakdown field. A supposed local dependence $\nu_i(|\vec{E}|)$ (which allows us to avoid the temperature dependence) occurs when the conditions of local and inertialess heating of electrons are fulfilled: $\delta_T\nu\tau >> 1$, $\Lambda >> \ell\sqrt{\delta_T}$ (where $\delta_T$ is a portion of energy lost by an electron during its collision with a molecule, $\tau$ and $\Lambda$ are characteristic time and space scales of the amplitude variations, respectively, and $\ell$ is the length of the electron mean free path).

2. Let us analyze the solution of the system (1),(2) describing the propagation of ionization from a certain primary ionization centre in the external homogeneous (in the absence of plasma) field $\vec{E}_0$ with the linear polarization. Let the functions $N(\vec{r},t)$, $\psi(\vec{r},t)$ satisfy the conditions

$$N(r,0) = N_\Sigma\delta(r) , \tag{4}$$

$$\Delta\psi(\infty,t) = -\vec{x}_0E_0 . \tag{5}$$

Here $r = |\vec{r}|$ is a distance from the origin, $\vec{x}_0$ is a unit vector parallel to the unperturbed external field, $\delta(r)$ is a "three-dimensional" delta function. If the conditions $N_0 \equiv N_\Sigma(\nu_i-\nu_a)/D^{3/2} << N_c$, $\alpha N_0 << \nu_i-\nu_a$ are satisfied (and that is usually so during the formation of discrete avalanches at isolated primary electrons), then an initial stage of the considered electron evolution (until $\alpha N << \nu_i-\nu_a$, $|\varepsilon-1| << 1$) is determined by the known spherically symmetrical solution for the linear equation (2) with $\alpha = 0$, $\nu_i-\nu_a = \text{const}$, $|\nabla\psi| \equiv E_0$:

$$N(r,t) = \frac{N_\Sigma}{8(\pi Dt)^{3/2}} \exp - \frac{r^2}{4Dt} + (\nu_i-\nu_a)t . \tag{6}$$

Neglecting the small initial time interval $t \lesssim (\nu_i-\nu_a)^{-1}$ where function $N(0,t)$ drops to $N \cong N_0$ due to diffusion, this solution describes a rapid increase of the electron concentration (which at $r < 2\sqrt{D(\nu_i-\nu_a)}t$ is close to the concentration in the avalanche).

The character of the further evolution of discharge strongly depends on the dominant mechanism of nonlinearity. If concentration N can reach its recombination limit $N_\alpha = (\nu_i-\nu_a)/\alpha$ without perturbing the field (i.e. $|\varepsilon(N_\alpha) - 1| << 1$), then the dynamics of a forming plasmoid is determined by a spherically symmetrical solution of Eq.(2) with a constant coefficient ($|\nabla\psi| = E_0 = \text{const}$). At large t the discharge is a quasi-homogeneous sphere with the concentration $N \cong N_\alpha$ expanding with the known diffusion velocity /2/

$$v_D = 2\sqrt{D(\nu_i-\nu_a)} . \tag{7}$$

For a more interesting and important case $|\varepsilon(N_\alpha) - 1| >> 1$ when the main mechanism of nonlinearity mechanism in the considered system is not the recombination but the variations of field amplitude and the ionization frequency $\nu_i$ in forming plasma, the plasmoid cannot remain spherically symmetrical and its configuration and structure strongly depend on $\nu/\omega$. Further we pay our attention mainly to the case of $\nu/\omega >> 1$ which is observed in high-pressure discharges.

At $\nu/\omega >> 1$, the field perturbations become noticeable when the concentration N in the centre of the plasmoid reaches $N \gtrsim \omega N_c/\nu \cong \nu N_{c0}/\omega$ ($N_{c0} = m^2/4 e^2$). Such a plasma is, actually, a conductor ($\varepsilon - 1 = -i\omega N/\nu N_{c0}$) where the field amplitude decreases with the growth of N thus resulting in the decrease of the avalanche velocity $\nu_i - \nu_a$ in the central region and also in gradual flattening of the concentration distribution N(r) (similar to the recombination case). The avalanche velocity in the periphery, actually determining the velocity of the discharge propagation into a non-ionized region depends on local amplitudes which for large $\nu/\omega$ (i.e. in case of conducting plasma) are enlarged in "polar" plasmoid regions while in "equatorial" ones they are diminished (here the polar axis OX is parallel to the external field $\vec{E}_0$). Thus the velocity of the discharge propagation depends on the direction and the plasmoid stretches along the external field.(Papers /1,3/ deal also with the possible role of this effect in experiments which resulted in stretched plasmoids at single electron breakdowns /1/ and in a branching semi-self-maintained discharge in underbreakdown fields /4/). Analogous to similar phenomenon during a discharge in a static field, such an extending plasmoid will be called a high-frequency (HF) streamer.

3. Let us analyze the dynamics of a HF streamer using a qualitative model with the following simplifying assumptions:

1) A plasmoid is a homogeneous ellipsoid with a sharp boundary (the width of the transition surface layer is much less than the radii of the boundary curvature).

2) The elongation velocities of the large (x - parallel) and the small semi-axis of the ellipsoid coincide with the corresponding local propagation velocities of a one-dimensional discharge with a plain boundary.

According to the results given in /2/, these velocities can be calculated using expression (7) where the value of $\nu_i$ at the poles is determined by the amplitude of a normal field component outside plasma, while on the equator it is determined by a continuous amplitude of the tangent component.

A system of equations describing the plasmoid evolution for the assumptions above (and for $\alpha = 0$), can be written as

$$E_i = x_0 \frac{E_0}{1+(\varepsilon-1)n_x} , \quad \varepsilon = 1 - \frac{N}{N_c}\left(1 + i\,\frac{\nu}{\omega}\right) , \tag{8}$$

$$\frac{\partial N}{\partial t} = \left[\nu_i(E_i) - \nu_a\right]N , \tag{9}$$

$$\frac{\partial a}{\partial t} = 2\sqrt{D\left[\nu_i(E_n) - \nu_a\right]} , \tag{10}$$

$$\frac{\partial b}{\partial t} = 2\sqrt{D[\nu_i(E_i) - \nu_a]}, \tag{11}$$

$$E_n = |\varepsilon| E_i, \quad E_i = |\vec{E}_i| . \tag{12}$$

Here $\vec{E}_i$ is an electric field vector inside the ellipsoid, $E_n$ is the amplitude of the external field at the ellipsoid tops (poles) which is $|\varepsilon|$-times larger than the amplitude $E_i$ in the equatorial region, a and b are the large and the small semi-axes of the ellipsoid, respectively, $n_x$ is the a/b-dependent depolarization coefficient in the x direction (see, e.g. /5/); in particular at $a \gg b$

$$n_x = \left(\frac{b}{a}\right)^2 \ell n \frac{2a}{b} . \tag{13}$$

It is evident from (8) and (12) that the behaviour of functions $E_i(t)$)and $E_n(t)$ and, subsequently, the whole character of the considered evolution is determined by the competition of two factors: the growth of $|\varepsilon|$ and the decrease of depolarization coefficient $n_x$ due to the increase of the relation a/b. It is easy to demonstrate that at rapid growth of the function $\nu_i(|\vec{E}|)$ (that usually occurs in the region $|\vec{E}| \gtrsim E_c$) the asymptotic form of the solution of (8)-(12) at large times (when $|\varepsilon| \gg 1$, $a \gg b$) is such that

$$|\varepsilon| \cdot n_x \to 0 , \quad E_i \to E_0 . \tag{14}$$

In particular, for the power approximation (3) with the index $\beta > 1$ (for air $\beta \cong 2 \div 5$ at $5E_c > |E| > E_c$), the asymptotic solution ($t \to \infty$) has the form

$$E_i = E_0 , \quad N = N_0 e^{\gamma t} , \tag{15}$$

$$v_{||} = \frac{da}{dt} = v_a e^{\beta\gamma t/2} , \tag{16}$$

$$v_{\perp} = \frac{db}{dt} = v_b , \tag{17}$$

$$\frac{a}{b} \sim \frac{1}{\gamma t} e^{\beta\gamma t/2} , \quad |\varepsilon| n_x \sim e^{-\gamma(\beta-1)t} , \tag{18}$$

where $v_a = 2\sqrt{D\nu_i(E_0)}$, $v_b = 2\sqrt{D\gamma}$, $\gamma = \nu_i(E_0) - \nu_a$ . It is clear that the propagation of a HF streamer in the $\vec{E}_0$-field direction goes on with the exponentially growing velocity $v_{||}$ (determined by the field $E_n$ at the top). The accompanying rapid decrease of the depolarization coefficient $n_x$ sustains the field $E_i$ in plasma at the level of the unperturbed field $E_0$. Thus the avalanche inside the plasmoid continues with a nondecreasing velocity. So this process (as well as the growth of the velocity $v_{||}$) can be stopped by some effects which have not been considered here; such as recombination, a decrease of $\beta$ to values less than a unity at large $E_n$ or field skinning in plasma. Note that the main condition for the existence of the above solution (15)-(18) is a rather strong dependence of velocity $v_{||}$ on the field amplitude ($v_{||}^2 \sim E_n^{\beta}$, $\beta > 1$). Thus the character of the solution remains the same for other (nondiffusive) mechanisms of the discharge propagation, for example, for a regime where the streamer top is a source of the ultraviolet ionizing radiation /1-6/. This regime is realized in strong fields and provides for higher velocities $v_{||}$.

The discharge evolution in case of low collision frequencies ($\nu \ll \omega$) is more complex and its analytical description (even using qualitative models) presents another separate problem. Here we shall note only three new aspects of such a case which are of particular importance: (i) an increase of the field (to be more exact, of its component parallel to $\nabla N$) near the plasma resonance surface ($N = N_c$); (ii) a possibility of the dipole (or multipole) plasmoid resonance (For a homogeneous sphere the condition of the m-th multipole resonance is $\vec{\varepsilon}m + m + 1 = 0$ ($m = 1,2,3,...$); in inhomogeneous objects resonances are strongly suppressed due to losses near the plasma resonance surface /7/); (iii) the transfer of the external field maximum (in some range of mean values of $N \sim N_c$) from polar to equatorial regions of a plasmoid. The last phenomenon can result in the tendency of the plasmoid expansion (on some intermediate stage of the evolution) in the direction perpendicular to field $\vec{E}_0$ (with the formation of an oblate ellipsoid). However, on subsequent stages at $N \gg N_c$) the field maxima have to shift to the poles and the discharge evolution seems to go on according to the above mentioned type (at $\nu/\omega \gg 1$).

4. Numerical simulation of the dynamics of a HF streamer has been carried out for a two-dimensional case $\psi = \psi(x,y)$, $N = N(x,y)$ for the following initial and boundary conditions

$$N(x,y,0) = \begin{cases} N_1 , & |x|,|y| < \ell , \\ 0 , & |x|,|y| > \ell , \end{cases} \tag{19}$$

$$N(\pm L_1, \pm L_2, t) = 0 , \tag{20}$$

$$\nabla\psi(\pm L_1, \pm L_2, t) = -\vec{x}_0 E_0 .$$

The boundaries $x = \pm L_1$, $y = \pm L_2$ were far enough from the ionization region and the solution in the whole calculated interval of time was in good agreement with the asymptotic condition (5). The system of equations (1),(2) was solved using expression (3) by the Newton iterative method in its finite-difference form in a rectangular net with the nonuniform integration step. The results of the numerical calculations for the following parameters $\beta = 4$, $\alpha = 0$, $N_1/N_c = 10^{-2}$, $\ell\sqrt{\nu_i/D} = 1$, $\nu/\omega = 10$, $\mathcal{E}_0 = E_0/E_c = 1.3$ are shown in Figs.1-3 as functions of the concentration $n = N/N_c$ and the field amplitude $\mathcal{E} = |E|/E_c$ depending on the longitudinal (x) and transverse (y) coordinates for different t. The attachment time $\nu_a^{-1}$ and the diffusive attachment length $L_a = \sqrt{D/\nu_a}$ are taken as units of scale of time and length, respectively (dimensionless variables $t \to \nu_a t$, $x \to x/L_a$, $y \to y/L_a$ are introduced).

The discharge evolution defined by the numerical solution is observed to be close to the one predicted using the above described qualitative model. At the initial stage, when $|\varepsilon - 1| \ll 1$, the numerical solution is close to the known (analogous to

Figure 1.

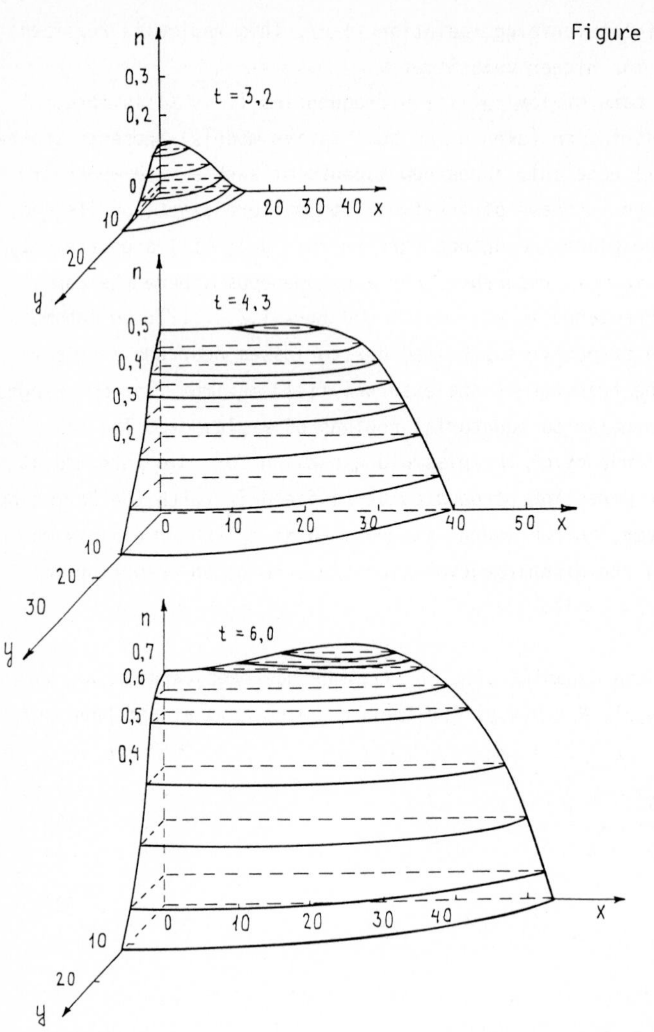


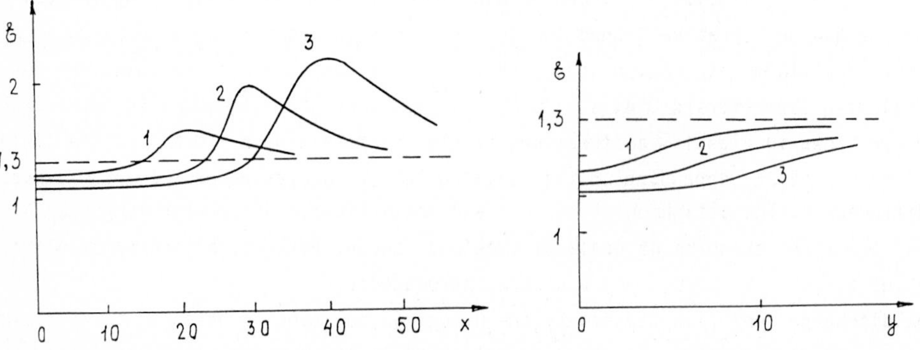


Figure 2. 1 - t=3.2; 2 - t=4.3; 3 - t=6.    Figure 3. 1 - t=3.2; 2 - t=4.3; 3 - t=6.

(6)) symmetrical solution of the initial problem for a linear two-dimensional diffusion equation. At $t \gtrsim 1$, when $|\varepsilon-1| \gtrsim 1$, the symmetry breaks up: the field in the polar regions greatly increases and the discharge stretches along the external field $\vec{E}_0$. In particular, at t = 6 the relation of longitudinal (a) and transverse (b) dimensions of the region where the concentration at the boundary is two times less than the maximum, is $a/b \cong 10$. The time dependences of longitudinal and transverse velocities ($da/dt = v_{||}$, $db/dt = v_{\perp}$) found as a result of the numerical solution are in good agreement with (7) where the ionization frequency $\nu_i$ stands for the maximum ionization frequency in a respective part of the plasma boundary region. In the claculated time interval ($t \leqq 6$) the asymptotic regime of (15)-(18) (which requires much computer time for its realization) seems not to have been reached and some deviations from the model of a homogeneous ellipsoid have been observed: 1) the plasmoid boundary remained to be rather diffuse and small maxima of concentration (10% greater than that in the centre) were formed near the plasmoid ends (y=0, x=25) at t = 6; 2) the maxima of the field amplitudes at the ends did not reach the values of $E_n = |\varepsilon| \cdot E_i$ determined in (12) and the field amplitude in the centre decreased in a course of time (though at $t \geqq 4$ this decrease practically ceased having reached not more than 10%).

As a conclusion, note that the considered mechanism of the HF streamer seems to have been realized experimentally /1,4,9/ and can serve as an alternative to the mechanism of thermal ionization instability of a homogeneous HF discharge /10,11/ when explaining a number of other experiments where a discharge comprising some filament-like plasmoids has been observed (see, e.g. /8,12/).

REFERENCES


1. A.L.Vikharev, V.B.Gil'denburg, O.A.Ivanov et al. Fizika Plazmy, 1986, 12, 1503 (in Russian).
2. A.N.Kholmogorov, I.G.Petrovsky, N.S.Piskunov. Voprosy Kibernetiki, Moscow, 1975, No.12, 3 (in Russian).
3. S.A.Dvinin. Vestnik MGU, ser.3, Fizika, Astronomiya, 1985, 26, 30 (in Russian).
4. G.M.Batanov, S.I.Gritsynin, I.A.Kossyi et al. Voprosy Fiziki Plazmy i Plazmennoi Elektroniki, Trudy FIAN, 160, Moscow, Nauka, 1985, 174 (in Russian).
5. L.D.Landau, E.M.Lifshits. Electrodynamics of Continuous Media, Moscow, Nauka, 1982 1982, 37 (in Russian).
6. V.I.Fisher. Zh.Eksp.Teor.Fiz., 1980, 79, 2142.
7. V.B.Gil'denburg. Zh.Eksp.Teor.Fiz., 1963, 45, 1978.
8. A.M.Hovatson. Introduction into the Theory of a Gas Discharge, Moscow, Atomizdat, 1980, 56, 84 (in Russian).
9. G.V.Bogomolov, Yu.D.Dubrovsky, A.A.Letunov, V.D.Peskov. Zh.Eksp.Teor.Fiz., 1978, 93, 519.
10. V.B.Gil'denburg, A.V.Kim. Fizika Plazmy, 1980, 6, 904 (in Russian).
11. A.V.Kim, G.M.Fraiman. Fizika Plazmy, 1983, 9, 601 (in Russian).
12. A.L.Vikharev, V.B.Gil'denburg, S.V.Golubev et al. Zh.Eksp.Teor.Fiz., 1988, 94, 136.

# The Propagation of the Front of Parametrically Excited Capillary Ripples

*P.A. Matusov and L.Sh. Tsimring*

Institute of Applied Physics, USSR Academy of Sciences, 46 Ulyanov Str., 603600 Gorky, USSR

The propagation of the capillary ripple front of the surface of a fluid in a periodically oscillating vessel is investigated theoretically and experimentally. The propagation velocity of the parametric instability front is found from the linear theory; it is demonstrated that the nonlinear front propagates with linear velocity. The front of a parameterically excited ripple is obtained in laboratory conditions and the dependence of the front velocity on supercriticality is found experimentally.

The autowave processes in the active nonlinear media are the subject of essential interest with respect to different problems of physics, chemistry, biology, etc. Nowadays the autowave processes in diffusive systems with a spatially distributed source of energy are the most attractive ones /1/. However, it should be noted that the dynamics of a number of physically interesting active systems cannot be described by simple diffusion equations due to a possibly important role of dispersion, inherent in the medium. Such a situation takes place in the case of parametric wave generation by a homogeneous oscillating field /2/.

The results presented here were obtained in a theoretical and experimental investigation of simple non-stationary and spatially inhomogeneous cases of parametric instability (PI) - the propagation of the Faraday ripple front as a result of the evolution of the localized initial surface disturbance of a liquid, placed in an oscillating gravity field.

The evolution of a small disturbance of a liquid surface in a vertically oscillating tank in its initial stage may be described by a linear equation

$$\eta_{tt}'' + \tilde{g}\eta_z' + \frac{\sigma}{\rho}\Delta_{x,y}\eta_z' = 0 \, , \qquad (1)$$

where $\eta$ is a vertical displacement of liquid particles, t is time, x and y are horizontal and z is the vertical coordinate, $\Delta_{x,y} = \partial^2/\partial x^2 + \partial^2/\partial y^2$, $\sigma$ is the surface tension rate, $\rho$ is the liquid density, $\tilde{g} = g(1+\alpha \cos \Omega t)$ is the gravity acceleration in the reference frame moving with the tank, and $\Omega$ is the oscillation frequency of the tank. Equation (1) yields the relation between frequency $\omega$ and wavenumber $\vec{k}$ of the free (i.e. with $\alpha = 0$) surface wave being of the $\exp(kz+i\vec{k}\vec{r}-i\omega t)$ type:

$$\omega^2 = g|\vec{k}| + \frac{\sigma}{\rho}|\vec{k}|^3 \, , \qquad (2)$$

which is a well-known dispersion relation for gravity-capillary waves. Further we will take the pump frequency $\Omega$ high enough to neglect the first right-hand term in equation (2) and study the waves as purely capillary.

Let us assume the evolution of one-dimensional initial disturbance defined by its spatial Fourier transform:

$$\eta_k(t) = \int_{-\infty}^{\infty} \eta(x,t)e^{-ikx}dx . \tag{3}$$

The substitution of (3) into (1) yields the Matieu equation for $\eta_k(t)$:

$$\eta_{ktt}'' + \frac{\sigma}{\rho}k^3\eta_k = -\alpha gk\eta_k \cos \Omega t . \tag{4}$$

With $\mu = \alpha g\rho/\sigma k^2 << 1$ the dependence $\eta_k(t)$ must be sought in the form

$$\eta_k(t) = A_k e^{-i\omega t} + \sum_{n=1}^{\infty} \{\mu^n B_k^{(n)} e^{-i(n\Omega-\omega)t} + \mu^n C_k^{(n)} e^{-i(n\Omega+\omega)t}\} + \text{c.c.} . \tag{5}$$

Substituting (5) into (4), one easily finds the dispersion equation $\omega(k)$ for the ripple in oscillating gravity field with the accuracy of the 2nd order in $\mu$:

$$\left[\omega^2 - \frac{\sigma}{\rho}k^3\right]\left[(\omega-\Omega)^2 - \frac{\sigma}{\rho}k^3\right]\left[(\omega+\Omega)^2 - \frac{\sigma}{\rho}k^3\right] = \left(\frac{\alpha gk}{2}\right)^2\left[(\Omega+\omega)^2 + (\Omega-\omega)^2 - \frac{2\sigma}{\rho}k^3\right] + O(\mu^4). \tag{6}$$

This equation takes into account only the main parametric resonance with frequencies $\omega = \pm\Omega/2$. With $|\omega-\Omega/2| << \Omega$ the branch of the dispersion curve of interest may be described by a formula

$$\omega = \pm \frac{\Omega}{2} \pm \left\{\left[\frac{\sigma}{\rho}k^3 - \frac{\Omega}{2}\right]^2 - \frac{\alpha^2 g^2 p^2}{16\sigma k}\right\}^{1/2} . \tag{6'}$$

As can be seen from (6') with $\left[(\sigma k^3/p)^{1/2} - \Omega/2\right] >> \alpha^2 g^2 \rho/16\sigma k$ , the ripples are close to free capillary waves. In the neighbourhood of resonance

$$|k-k_0| \le \Delta k_0 , \quad k_0 = (\Omega\rho/2\sigma)^{2/3}, \quad \Delta k_0 = \alpha g/3\sigma k_0 \tag{7}$$

parametric instability appears. The presence of viscous dissipation inherent in a real liquid (the frequency in the right hand side of (6') must be amended by a decrement $-2i\nu k^2$) causes PI threshold

$$\alpha_{thr} \cong 8\Omega\nu k_0/9 . \tag{8}$$

Here we are going to consider the problem of the evolution of localized disturbances in the system governed by equation (6').

In the general formulation for arbitrary active systems the problem was recently discussed in /5/. Here only some comments are made. At times greater than the characteristic inverse increment, the disturbances become exponentially large within a bounded region with the boundaries, spreading with a fixed velocity. So it is reasonable to introduce the term "the front of instability" in a linear nonequilibrium system. Actually, the front velocity coincides with that of the frame of reference with the absolute-to-convective(or vice versa) instability type translation (cf./6/).

The velocity of the instability front can be easily proved to be the solution of the following system of equations for a complex wavenumber k and V /5/:

$$d\omega/dk = V; \tag{9}$$

$$\mathrm{Im}(\omega - kV) = 0 , \tag{10}$$

which appears as a direct generalization of the Briggs criterion (cf./6/) for the determination of the instability type in a given frame of reference. With respect to the present problem described by the dispersion relation (6) amended by viscous damping, system (9)-(10) can be transformed to the form

$$\tilde{k}/\sqrt{\tilde{k}^2-1} = \tilde{V} ; \quad \mathrm{Im}(1/\sqrt{\tilde{k}^2-1}) = -D , \tag{11}$$

where the dimensionless variables are $\tilde{k} = 6\sigma k_0(k-k_0)/\alpha\rho g$, $D = 8\Omega\nu k_0/\alpha g$, $V = V/V_g$; $V_g = \frac{3}{2}(\sigma k_0/\rho)^{1/2}$ is the group velocity of free capillary waves with $k = k_0$, $\omega = \Omega/2$. The solution of the above system are

$$\tilde{k}_s = \pm\tilde{V}(\tilde{V}^2-1)^{-1/2} ; \tag{13}$$

$$\tilde{V} = (1-D^2)^{1/2} \tag{14}$$

It is seen from (14) that at the instability threshold (D=1) the front velocity turns to zero, but with $D \to 0$ (high supercriticality), V tends to the group velocity of free capillary waves. Signs "+" and "-" in the expression for the velocity (14) correspond to forward and back fronts, respectively: the middle of the PI region is immobile, while the boundaries of this region move in opposite directions with equal velocity magnitudes. Note that formulae (13) and (14) are derived from (6'), i.e. with the PI range assumed to be narrow ($\Delta k_0 \ll k_0$). Therefore high supercriticality must be regarded only as $\alpha_{thr} \ll \alpha \ll \Omega^2/gk_0$.

Inside the instability region the wavefield is a superposition of opposite waves, with the frequency $\omega = \Omega/2$ and the wavenumbers $k \cong \pm k_0$. The wavefield grows exponentially with the increment that reaches its maximum at the centre of the instability region and monotonously tends to zero at the front location. It is clear that far from the front the exponential growth of waves must be suppressed by nonlinearity. With small supercriticality, the nonlinear damping /7/ and the frequency shift of the nonlinear surface wave removing a pair of waves out of resonance with the pump /2/ are the main mechanisms of suppression. Further, the PI range will be assumed to be close, i.e. $\alpha \ll \Omega^2/gk_0$. So the surface displacement may be expressed in the form

$$\eta(x,t) = \frac{1}{2}\left[a_+(x,t)e^{ik_0x} - a_-(x,t)e^{-ik_0x}\right]e^{-(i\Omega/2)t} + \text{c.c.} , \tag{15}$$

where $a_\pm(x,t)$ are smooth (in comparison with $\exp(-i\frac{\Omega}{2}t \pm ik_0x)$ functions of coordinates and time. The equations for complex amplitudes $a_\pm$ are easily obtained from the original equation (1) and supplemented by the terms describing the damping and the frequency shift (cf/4/):

$$\frac{\partial a_\pm}{\partial t} \pm V_g \frac{\partial a_\pm}{\partial x} + \gamma a_\pm = iHa_\mp^* - R|a_\pm|^2 a - iT|a_\pm|^2 a - iS|a_\mp|^2 a_\pm \ . \tag{16}$$

Here $\gamma = 2\nu k_0^2$ is the linear decrement of ripple damping, $H = \alpha g k_0/2\Omega$ is the increment of PI, $R \cong 8.8\nu k_0^4$ is the coefficient of nonlinear damping of capillary waves /7/, $T = 0.0312\ \Omega k_0^2$ and $S = 0.312\ \Omega k_0^2$ are the coefficients of nonlinear frequency shift determined by self-interaction and interaction with the opposite wave, respectively /4/. Equation (16) does not account the capillary wave dispersion due to its weakness as compared to the dispersion caused by PI in the vicinity of $\omega = \Omega/2$ for $\alpha << \Omega^2/gk_0$.

Let us consider first the simple case of T = S = 0. Here system (16) may be reduced to an equation for $a_+$:

$$\left(\frac{\partial}{\partial t} + \gamma + R|a_+|^2\right)^2 a_+ - V_g^2 \frac{\partial^2 a_+}{\partial x^2} = H^2 a_+ \ , \tag{17}$$

which is equivalent to a pair of real first order equations one of which corresponds to the instability:

$$\left(\frac{\partial}{\partial t} + \gamma + R|a_+|^2\right)a_+ = HA_+ + \frac{1}{2}\frac{V_g^2}{H}\frac{\partial^2 a_+}{\partial x^2} - \frac{V_g^4}{8H^3}\frac{\partial^4 a_+}{\partial x^4} + \dots \ . \tag{18}$$

The infinite sum of derivatives in the right hand side of equation (18) corresponds to the expansion of the square root in equation (6'). With $E = (H-\gamma)/H << 1$ (small supercriticality), the instability range is narrow as compared to $k_0$, so we may restrict ourselves only to the first and second terms in the right-hand side of equation (18). Under these assumptions equation (18) turns to a diffusion equation with a nonlinear source which is well-known in the theory of autowave processes /1/:

$$\frac{\partial a_+}{\partial t} - \frac{1}{2}\frac{V_g^2}{H}\frac{\partial^2 a_+}{\partial x^2} - EHa_+ + R|a_+|^2 a_+ = 0 \ . \tag{19}$$

In /8/ within the scope of (19), it was proved rigorously that the localized initial disturbances produce (when $t \to \infty$) the stationary fronts (the so-called "phase autowaves") with velocity being independent of R and equal to

$$V_{fr} = \sqrt{2E}\ V_g \ . \tag{20}$$

In the framework of the PI problem assumed here we consider the stationary envelope front, while the surface displacement itself may be non-stationary (the carrier speed $\Omega/2k$, in general, does not coincide with $V_{fr}$).

It can be easily seen that expression (14) found earlier from the linear theory turns to (20) with $E = 1-D << 1$. So, with small supercriticality, a stationary nonlinear autowave propagates with a "linear" velocity of the instability front. The dependence of the phase front velocity on the form and rate of nonlinear damping was discovered in 1937 /8/. However, the fact of the coincidence of this velocity with that of the instability front propagation in the linear system corresponding to (19) was pointed out only in a recent paper /11/ (cf. also /7/). When the supercriticality may be regarded as not too small, the rigorous proof, similar to /8/, can hardly be made, but the statement made earlier about the nonlinear autowave velocity being equal to the corresponding "linear" velocity obtained within a scope of (18), may be ensured by some physical implications.

Actually, the nonlinear damping causes the stationary nonlinear solution of (18) to be everywhere less than the one obtained with $R = 0$ and under the same initial conditions. From this it follows at once that the velocity of the stationary front cannot be greater than that of the instability front. On the other hand, it is clear that the stationary front moving slower than with "linear" velocity is unstable - small disturbances at the edge of the front (where $a_\pm \ll (R/EH)^{1/2}$) will exponentially grow and penetrate into the region ahead of the front. So, the stationary front moving with the speed of linear stage of instability is the only possible one from the point of view of existence and stability. Note that such an argumentation is applicable to an arbitrary system with linear instability and obviously stabilizing nonlinearity. However, the applicability of this argumentation to the full system (16) taking into account both nonlinear damping and nonlinear frequency shift, is not so obvious, so the above system was integrated numerically using the Lax method /10/.

With $t = 0$ small initial disturbance was set $a = a_0 \exp(-x^2/\Delta x^2)$. Numerical simulations showed that the front speed is independent of the values of the T, S parameters and is governed by the ratio $H/\gamma$ according to formula (14). The profiles of $\mathrm{Re}\, a_+(x)+a_-(x)$ are plotted in Fig.1. As can be seen from Fig.1, with $T = S = 0$ the envelope is smooth, but with $T \neq 0$ and $S \neq 0$ the oscillations appear on the front. These oscillations are possibly of the same nature as those observed in numerical simulations /3/ of a spatially homogeneous case of the growth of a parametric spin wave in ferromagnetics.

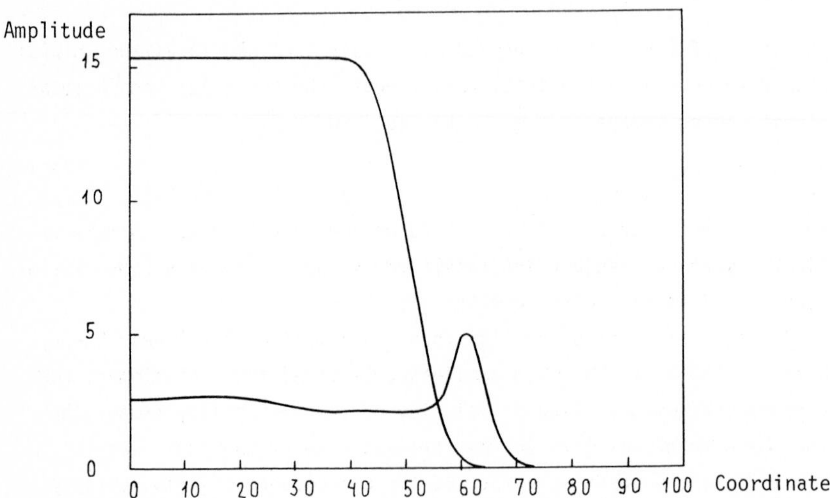


Figure 1. The envelope front structure for the capillary ripples (numerical simulations): a- H=3, $\gamma$=0.3, R=2.3, T=0, S=0; b- H=3, $\gamma$=0.3, R=2.3, T=6.25, S=0.625.

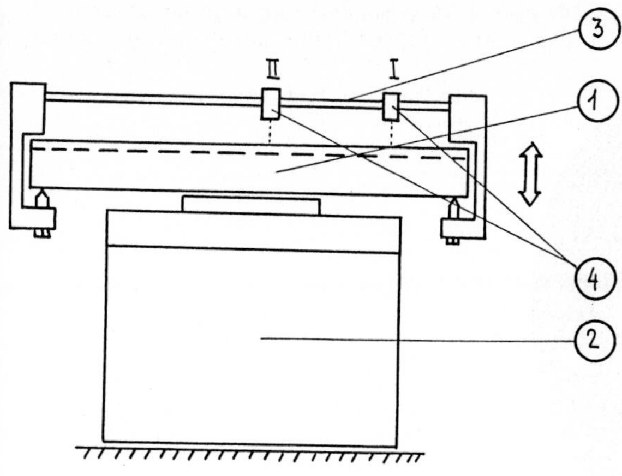


Figure 2. The experimental set-up: 1 - a silicon oil layer; 2 - a vibro-exciter; 3 - IR probe fastening; 4 - IR probes.

The propagation of the capillary ripple front was investigated experimentally at the surface of silicon oil ($\rho$ = 0.95 g/cm, $\sigma$ = 23.3 erg/cm, $\nu$ = 0.04 cm²/s) layer 7 mm thick in a circular tank with a diameter of 300 mm, fixed in a vibro-exciter (Fig.2). The waves in the oil surface were measured by two contactless infrared probes fastened on the tank walls at different distances from the tank center. The vibro-exciter was fed by sine current together with a simultaneously switched on D.C. governing the initial tank displacement. Thus, when the tank oscillations are switched on, a thin boundary layer appears at the tank walls which plays the role of a localized initial disturbance. Later, a region with capillary ripples appears within the tank walls which has a boundary moving with certain velocity towards the tank centre (see Fig.3). The ripple front speed was determined from the delay time between the origin of the IR probe output signals.

Three sets of experiments were performed with pump frequencies equal to 80, 60 and 40 Hz. Each set involved the measurements of the ripple front speed as a function of supercriticality within the range $0.1 < E < 2.6$. With $E$ exceeding 2.5, an intense drop break-off from the oil surface arouse inside the instability region.

Experimental results are plotted in Fig.4 as three sets of points. All these sets can be seen to be close to the theoretical law governed by expression (14). This expression, written for a plane wave case, was used because the wavelength- and front width-to-wavefront radius ratios at the probe locations were small. Note that with relatively high supercriticality the nonlinear ripple dynamics past the front is quite different from that discussed above. In particular, the evolution process involves a fast growth of a second wave pair with radial wavefronts (cf.Fig.3). It is interesting to note that a region with both wave pairs is also closed within a quite sharp boundary moving towards the tank centre, but the "second front" propagation was not investigated.

Figure 3. A momentary photograph of the capillary ripples front taken using a flash.

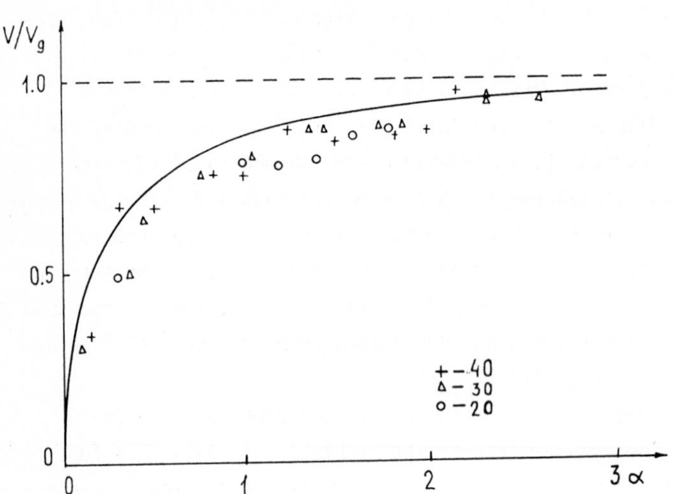


Figure 4. The front propagation speed vs supercriticality: 1 - theoretical law (14); 2 - experimental points for different wave frequencies.

The propagation of the parametrically excited capillary ripple front considered here is a typical graphic example of the evolution of a local disturbance in an unstable system. The speed of the instability front was predicted theoretically and obtained experimentally from the linear theory even at the late stage of instability evolution with essentially nonlinear disturbances past the front. That property seems to be general for a lot of linear-unstable systems.

Note, finally, that the presence of noise in the system results in a natural limitation for the observation length of the front propagation: this is the time of noise growth up to a stationary level. Here that time was about 1 s with supercriticality $E \approx 0.5$. The length of the instability front propagation within the tank was slightly less (≦0.4 s), so the front was easily determined.

## REFERENCES


1. W.A.Wasilyev, Yu.M.Romanovsky, W.G.Yakhno. Autowave Processes. Moscow, Nauka, 1987 (in Russian).
2. M.I.Rabinovitch, D.I.Trubetskov. Introduction to the Theory of Oscillations and Waves. Moscow, Nauka, 1984 (in Russian).
3. V.E.Zakharov, V.S.L'vov, S.L.Musher. Sov.Phys.-Solid State, 1972, 14, 2913 (in Russian).
4. A.B.Yezersky, M.I.Rabinovitch, B.P.Reutov, I.S.Starobinets. Sov.Phys.-Zh.Eksp. Teor.Fiz., 1986, 91, 6(12), 2070 (in Russian).
5. W.van Saarlos. Phys.Rev.A, 1988, 37, 1, 211.
6. L.D.Landau, Ye.M.Lifshits. The Electrodynamics of Continuous Media. Moscow, Nauka, 1981 (in Russian).
7. V.A.Krasil'nikov, V.I.Pavlov. Moscow State Univ.Trans., Physics. 1972, 1, 94 (in Russian).
8. A.N.Kolmogorov, I.G.Petrovsky, N.S.Piskunov. Moscow State Univ. Bulletin, Sect. A., 1937, 1, 6, 1 (in Russian).
9. G.Dee, J.S.Langer. Phys.Rev.Lett., 1983, 50, 383.
10. G.E.Forsythe, W.G.Vazow. Finite-Difference Methods for Partial Differential Equations. New York, Wiley, 1959.

# Nonlinear Surface and Internal Waves in Rotating Fluids

*L.A. Ostrovsky and Yu.A. Stepanyants*

Institute of Applied Physics, USSR Academy of Sciences,
46 Ulyanov Str., 603600 Gorky, USSR

A class of nonintegrable equations related to a wide range of physical problems including surface and internal waves in rotating ocean is considered. A characteristic feature of these equations is the presence of a broad "dispersionless" band in the frequency spectrum that separates the regions of low- and high-frequency dispersion. The structures of plane and two-dimensional steady-state solutions are studied analytically and numerically. Results of the numerical calculations of nonstationary perturbation dynamics under different initial conditions are presented.

## 1. INTRODUCTION

The 1970s were the years of impetuous progress in the "nonlinear mathematical physics" that was related to the development of exact methods for the solution of a definite class of integrable evolution equations. The mathematical body of nonintegrable equations is much less developed although they are typically more adequate to real physical models. There are, in fact, no universal methods of solution of such equations; an exception is the perturbation method that in some cases gives quite general results. Paraphrazing L.N.Tolstoy, we can say that all integrable models "are happy" (solvable) in a similar fashion while each nonintegrable model "is unhappy" in its own way. At the same time, "unhappy" nonintegrable equations often have an astonishing wealth of various solutions. Note that exact integrability is not a "rough" property, i.e. it may be disturbed even by slight changes in the structure of the equations (sometimes only with the change of numerical coefficients).

In our paper we shall consider a class of nonintegrable equations that are related to a rather wide range of physical problems and have been studied intensively in the recent decade. The title of this paper is, therefore, rather conventional since it represents only one type of physical systems belonging to the class of interest. However, the results presented below refer to any nonlinear systems the frequency spectrum of which has a broad "dispersionless" band separating the regions of low-frequency (LF) and high-frequency (HF) dispersion. Possible versions of dispersion dependences of such a type are shown in Fig.1. Examples of the waves having such a dispersion, besides the one mentioned in the title, are extraordinary electromagnetic and oblique magnetoacoustic waves in a magnetized plasma, the excitations in a chain of atoms described by the Frenkel-Kontorova



Research Reports in Physics Nonlinear Waves 3
Editors: A.V. Gaponov-Grekhov · M.I. Rabinovich · J. Engelbrecht

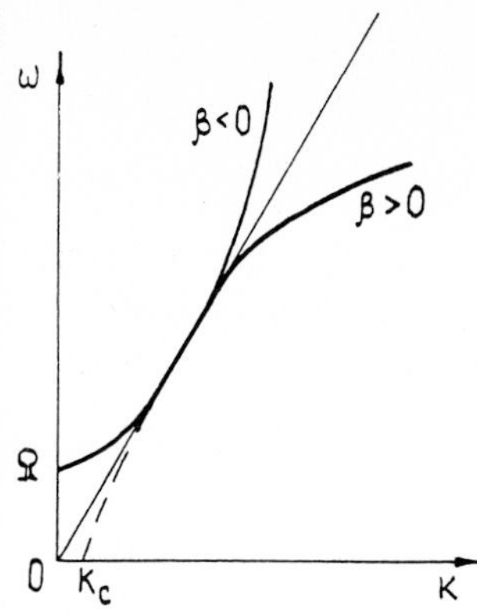


Figure 1. Dispersion curves with a broad dispersionless band (a linear section) in the frequency spectrum with adjoining regions of low- and high-frequency dispersion.

model with far interaction /1/, the waves in transmission lines of the band filter type (we shall consider them in Sect.3.2), etc. Below we shall describe the properties of this class of systems primarly based on an example of surface waves in a rotating fluid. First we shall briefly analyze the basic model equations, then investigate one-dimensional steady-state solutions of these equations, present results of numerical calculations of nonstationary perturbations and, finally, consider some properties of two-dimensional perturbations.

## 2. MODELS

Consider a homogeneous layer of a rotating fluid having the depth h that has a horizontal bottom and a free upper surface subject to capillary forces. The propagation of long (as compared to h) weakly sloping perturbations on the surface of such a layer is described by a two-dimensional system of Boussinesq equations /2/:

$$\vec{u}_t + (\vec{u}\nabla)\vec{u} + \left[\vec{\Omega}\vec{u}\right] + \nabla(g\eta+\beta\eta_{tt}/3h) = 0 \ ,$$
$$\eta_t + \operatorname{div}\left[(h+\eta)\vec{u}\right] = 0 \ . \qquad (1)$$

Here $\eta(x,y,t)$ is the perturbation of the free surface, $\vec{u}(x,y,t) = (u,v)$ is the depth-average horizontal particle velocity of the fluid, $\vec{\Omega}$ is the double frequency of fluid rotation around the axis normal to the x,y-plane ($\Omega$ is the Coriolis parameter), $\beta = h^3-3\sigma/\rho g$ is the parameter describing HF dispersion, $\sigma$ is the coefficient of surface tension, $\rho$ is the fluid density, and g is the acceleration due to gravity. Subscript t denotes the differentation with respect to the variable t (the same will further refer to x and y), $\nabla = (\partial/\partial x,\partial/\partial y)$. When $\Omega = \beta = 0$, system (1) transforms to a well-known set of "shallow water" equations /3/ which is similar to the set of equations of gas dynamics with the adiabatic index equal to two. In a one-dimensional case, when the perturbations depend only on one spatial coordinate x, system (1) reduces to one equation for the transverse velocity component v /4/:

$$v_{tt} - c_0^2 v_{xx} + \Omega^2 v - \beta v_{ttxx}/3 = \left(\frac{v_t v_x}{\Omega+v_x}\right)_t + \frac{\Omega}{2}\left[\left(\frac{v_t}{\Omega+v_x}\right)^2\right]_x \qquad (2)$$

where $c_0 = gh$. The disturbance of the free surface and the longitudinal velocity component are expressed here through v: $\eta = hv_x/\Omega$ and $u = -v_t/(\Omega+v_x)$. For moderate amplitude waves, when the right-hand-side of the equation may include only quadratic nonlinearity, Eq.(2) transforms to the generalized Boussinesq equation (known also as the nonlinear string equation) which contains an additional term taking into account the rotation of the fluid:

$$v_{tt} - c_0^2 v_{xx} + \Omega^2 v - \beta v_{ttxx}/3 = (v_x v_{tt} + 2v_t v_{xt})/\Omega \ . \qquad (3)$$

For infinitesimal low-amplitude perturbations proportional to exp $i(\omega t-kx)$ equations (2)-(3) yield the dispersion relation

$$\omega^2 = (\Omega^2 + c_0^2 k)(1+\beta k^2/3)^{-1} \ . \qquad (4)$$

The solid curve in Fig.1 is a qualitative form of $\omega(k)$ when $\beta > 0$ (gravity waves) and $\beta < 0$ (capillary waves).

For relatively short waves, where the term $\Omega^2$ is insignificant, it follows from (4) that $\omega \cong c_0 k(1-\beta k^2/6)$. This type of dispersion characterizes the waves described by the Korteweg-de Vries (KdV) equation /2/. Thus, Eqs.(2)-(3) have "two dispersions" in the long-wave (LF) and short-wave (HF) regions and a nearly dispersionless section at medium wavelengths. It is clear from (4) that LF dispersion in this case is due to rotation and vanishes when $\Omega \to 0$. Note, however, that in other physical systems LF dispersion may be related to different mechanisms rather than rotation (e.g. dispersion in transmission lines which will be considered below).

For the wave processes the spectrum of which belongs, essentially, to a dispersionless range and only slightly involves the regions of LF and HF dispersion, Eq.(4) can be written in the form

$$\omega \cong c_0 k - c_0 \beta k^3/6 + \Omega^2/2c_0 k \ . \qquad (5)$$

It is clear from (5) that the corresponding evolution equation should be related to the KdV equation with an additional term responsible for rotation. We omit here the derivation of this equation (for details see /5/), but analyze the low-amplitude perturbations in the reference system moving along x with a velocity $c_0$:

$$\left[v_t + 3c_0(v_\xi)^2/4h + c_0\beta v_{\xi\xi\xi}/6\right]_\xi = \Omega^2 v/2c_0 \ ,$$

where $\xi = x - c_0 t$. This equation can also be written for free surface perturbations $\eta = hv_\xi/\Omega$:

$$(\eta_t + 3c_0\eta\eta_\xi/2h + c_0\beta\eta_{\xi\xi\xi}/6)_\xi = \Omega^2\eta/2c_0 \ . \qquad (6)$$

Equation (6) was first derived in /6/ for internal oceanic waves ($\eta$ has the meaning of the perturbations of an isopycnic surface, i.e. a constant density surface). Later it was derived using different methods for surface and internal waves (see /5,7-9/). A similar equation neglecting HF dispersion ($\beta = 0$) was obtained for inertial gravity equatorial waves in the ocean /10/, for acoustic waves in a bent rod /26/ and for waves propagating in a randomly inhomogeneous medium /27/.

Let us underline straight away two simple integral consequencies of Eq.(6). Considering the localized perturbations, for which $\eta \to 0$ as $|\xi| \to \infty$ and intergrating (6) with respect to $\xi$, it is easy to show that

$$<\eta> \equiv \int_{-\infty}^{\infty} \eta(\xi,t)d\xi = 0 .$$

A similar conclusion holds also for space-periodic perturbations (the integral in this case is taken over the wave period). Another integral that follows form (6) has a sense of the wave energy: $<\eta^2> = \text{const}$.

Consider now the Cauchy problem for Eq.(6). Let the perturbation at the initial moment $t = 0$ be specified in the form $\eta(\xi) = \eta_0 f(\xi/\Lambda)$ where $\eta_0$ is the characteristic perturbation amplitude, $\Lambda$ is the characteristic size and f is the dimensionless function. Introducing new variables $x_n = \xi/\Lambda$, $t_n = 3tc_0\eta_0/2h\Lambda$ and $\tilde{\eta} = \eta/\eta_0$, Eq.(6) and the initial conditions can be represented in the dimensionless form

$$(\tilde{\eta}_t + \tilde{\eta}\tilde{\eta}_x + \tilde{\eta}_{xxx}/Ur)_x = \tilde{\eta}/S_0 , \quad \tilde{\eta}(x,0) = f(x) ,$$
$$Ur = 9\eta_0\Lambda^2/\beta h , \quad S_0 = 3\eta_0 c_0^2/h\Omega^2\Lambda^2 . \qquad (7)$$

The subscript n for the variables is omitted. It follows from (7) that the evolution of initial perturbations is determined, besides the form of the initial function f(x), by two dimensionless parameters: Ur (the Ursell parameter) and $S_0$ which relate the nonlinearity to HF and LF dispersion, respectively. The perturbations having identical forms and equal parameters Ur and $S_0$, evolve similarly.

Initial system (1) cannot be reduced to one equation for two-dimensional perturbations. However, we can obtain an approximate equation generalizing (2) for low-amplitude perturbations that are much longer along y and along x:

$$v_{tt} - c_0^2(v_{xx}+v_{yy}) + \Omega^2 v - \beta v_{ttxx}/3 = \left(\frac{v_t v_x}{\Omega+v_x}\right)_t + \frac{\Omega}{2}\left[\left(\frac{v_t}{\Omega+v_x}\right)^2\right]_x -$$
$$- \frac{c_0}{\Omega^2}\left[v_{xx}v_{ty} - v_{xy}v_{tx} + \frac{\Omega}{c_0^2}(2vv_{ty}+v_t v_y)\right] . \qquad (8)$$

This equation was derived in detail in /11/. Restricting the consideration to the quadratic nonlinearity for the waves travelling primarly along x and diffracting weakly in the transverse direction, we can obtain from (8) a two-dimensional variety of Eq.(6) for free surface perturbations:

$$(\eta_t + 3c_0\eta\eta_\xi/2h + \beta c_0\eta_{\xi\xi\xi}/6)_\xi = \Omega^2\eta/2c_0 - c_0\eta_{yy}/2 . \qquad (9)$$

Equations (6) and (9), in contrast to KdV and Kadomtsev-Petviashvily (KP) equations /12/, are, apparently, not completely integrable, to say nothing of more complex equations (2), (3) and (8) for which only simple approximate solutions can be constructed analytically. In such a situation, computer simulation is of great importance.

## 3. ONE-DIMENSIONAL WAVES

### 3.1. Steady-State Solutions

Consider first the steady-state periodic solutions of Eq.(2) that depend only on one "moving" variable x-Vt. We introduce new dimensionless variables where the wave period is $2\pi$ (the dimensional period is equal to $\Lambda$): $\zeta = 2\pi(x-Vt)/\Lambda$, $Z = (2\pi/\Lambda)\cdot$ $\cdot(v'/\Omega) \equiv \eta/h$. The Z-variable has a meaning of a dimensionless value of the free surface displacement. Equation (2) in these notations takes the form

$$\left[QZ'' - (1-1/U^2)Z + \frac{(3/2+z)z^2}{(1+z)^2}\right]'' = RZ\ , \qquad (10)$$

where $U = V/c_0$, $R = (\Lambda/2\pi)^2(\Omega/V)^2$ and $Q = \beta(2\pi/\Lambda)^2/3$; the prime denotes differentation with respect to $\zeta$. For linear perturbations, when the right-hand side can be neglected, Eq.(10) yields the dispersion relation (3) which can be written in the dimensionless variables $U^2 = (1-Q-R)^{-1}$. When describing the perturbation having low but finite amplitude, the right-hand side of (10) can be restricted to quadratic nonlinearity and Eq.(10) reduces to the form

$$\left[QZ'' - (1-U^2)Z + 3Z^2/2\right]'' = RZ\ . \qquad (11)$$

Equation (6) for steady-state periodic solutions reduces to the same form.

Integrating Eqs.(10) and (11) over the wave period, one can easily find that their solutions also have zero average values:

$$\langle z\rangle \equiv \left(\int_0^{2\pi} Zd\zeta\right)/2\pi = 0$$

although nonstationary perturbations within (2) may not possess this property. Note also that Eq.(10) and its approximate version (11) can be represented in the Hamiltonian form. For this purpose we introduce a canonical variable $p_1$ in the form $Z = p_1'$ and rewrite Eq.(10) in terms of this variable (which is proportional to the transverse velocity component v):

$$\left[Qp_1''' - (1-1/U^2)p_1' + \frac{(3/2+p_1')(p_1')^2}{(1+p_1')^2}\right]' = Rp_1\ . \qquad (12)$$

It can be readily verified that Eq.(12) has the first integral

$$H = \{Rp_1^2 + Q(p_1'')^2 - 2Qp_1'p_1''' + (p_1')^2\left[1/(1+p_1')^2 - 1/U^2\right]\}/2Q\ . \qquad (13)$$

With an appropriate choice of the canonical variables $q_1$, $q_2$ and $p_2$ together with the introduced variable $p_1$, this integral acquires the sense of a Hamiltonian:

$$\mathcal{H} = \left[Rp_1{}^2 + Qp_2{}^2 - 2Qq_1q_2 + q_2{}^2/U^2 - q_2 - 1/(1+q_2) + 1\right]/2Q \ . \tag{14}$$

After that, Eq.(12) can be represented in the form of a canonical Hamiltonian system with two degrees of freedom:

$$\begin{aligned} q_1{}' &= \partial\mathcal{H}/\partial p_1 = Rp_1/Q \ , \quad p_1{}' = -(\partial\mathcal{H}/\partial q_1) = q_2 \ , \\ q_2{}' &= \partial\mathcal{H}/\partial p_2 = p_2 \ , \quad p_2{}' = -(\partial\mathcal{H}/\partial q_2) = q_1 + \left[1-2q_2/U^2-1/(1+q_2)^2\right]/2Q \ . \end{aligned} \tag{15}$$

The first three equations in (15) are linear and only the fourth is a nonlinear equation. For low-amplitude perturbations it can be written in an approximate form:

$$p_2{}' \cong q_1 + q_2(1-1/U^2-3q_2/2)/Q \ . \tag{16}$$

The corresponding Hamiltonian has a form:

$$\mathcal{H} \cong \left[Rp_1{}^2 + Qp_2{}^2 - 2Qq_1q_2 + q_2{}^2(1/U^2-1+q_2)\right]/2Q \ . \tag{17}$$

The existence of the first integral (13) permits us to decrease the order of (12) but even after that (12) is a rather complicated equation that cannot be solved analytically.

The situation is simpler for sufficiently smooth waves when the HF dispersion is insignificant. In this case, the term with the fourth derivative in (10) can be omitted, i.e. we can take $Q = 0$. Then the second-order equation remains the solutions of which can be readily analyzed in the phase plane /4/. The first integral of this equation has the form

$$Z' = U^2(1+Z)^2(R\left[Z^2(1+Z)^2/U^2+2Z^2+6Z+3\right]+2C(1+Z)^2)^{1/2}(U^2-(1+Z)^3)^{-1} \ , \tag{18}$$

where C is the integration constant.

There is one equilibrium configuration $Z = 0$ in the phase plane (Fig.2) for which $C = -3R/2$. Elliptical trajectories near this equilibrium configuration correspond to linear harmonic oscillations. There is also a singular line $Z = U^{2/3}-1$ in the phase plane. The region of periodic motions is bounded by a singular trajectory in the plane (a solid curve in Fig.2) for which the derivative $Z'$ remains finite on the singular curve. From this it follows that $C = -1.5R\left[1-(U^{2/3}-1)^3/3U^2\right]$ for this trajectory. Waves of limiting configuration with a nonsmooth crest correspond to the motion along the singular curve (see Fig.3). The maximal and the minimal values of Z in this case are $Z_{max} = U^{2/3}-1$, $Z_{min} = (U^{2/3}-1)^2+U^{2/3}\ {}^{1/2}-U^{2/3}$ and the modulus of maximal derivation when $Z = Z_{max}$ is $|Z'|_{max} = U^{3/4}\ (U^{2/3}-1)R/3\ {}^{1/2}$. Note that $Z_{min}(U\to\infty) = -1/2$, i.e. the level of the free surface of limiting configuration waves does not fall lower than the half-depth of the fluid.

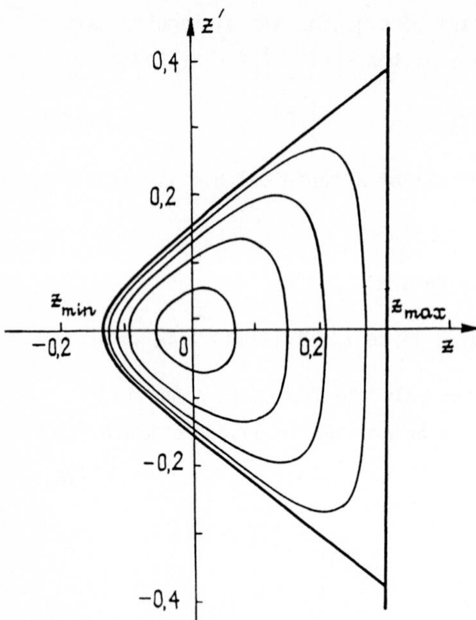


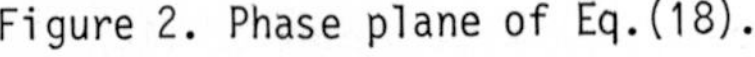
Figure 2. Phase plane of Eq.(18).

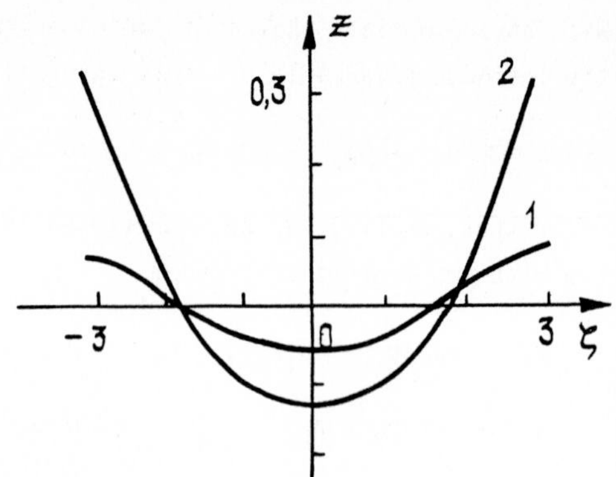


Figure 3. The shape of the solution of Eq.(18) in the form of a periodic nonlinear wave (1) corresponding to the motion along one of inner phase trajectories shown in Fig.2 and a periodic wave of limiting form (2) corresponding to the motion along the solid trajectory shown in Fig.2.

The solutions of Eq.(18) can be written in an implicit integral form, but the profile for limiting configuration waves can be expressed through elementary functions /4/. In the adopted normalization ($2\pi$ is the periodicity with respect to $\zeta$) parameter U - the dimensionless velocity of the wave - is not independent and is expressed through R. The R-U dependence for limiting-configuration waves is shown in Fig.4 by the solid curve. For comparison, Fig.4 shows also function $U = 1/\sqrt{1-R}$ (the broken curve) for small-amplitude sinusoidal linear perturbations.

For weakly nonlinear perturbations described by (11) with $Q = 0$ for which $U-1 \ll 1$, the limiting wave has a parabolic shape /6/: $Z = R(\zeta^2-\pi^2/3)/18$, $|\zeta| \leq \pi$. For such a wave we have $U \approx 1+\pi^2 R/18$, $Z_{max} = \pi^2 R/27$, $Z_{min} = -Z_{max}/2$, $|Z'|_{max} = \pi R/9$, $C = -3R/2 + \pi^6 R^4/39906$.

In the case $U < 1$ but $1-U \ll 1$ (for "slow" waves) analysis of the phase plane when $Q = 0$ easily shows /6/ that there are no periodic solutions. Limited solutions have the form either of finite-length pulses on a constant "pedestal" with derivative singularities at the ends /6/ or soliton-like pulses with asymptotic forms decreasing exponentially and, again, with derivative singularities at the front and back slopes of the pulse /26/. However, such solutions seem to be impossible in a complete equation with HF dispersion.

Let us now investigate the effect of HF dispersion on the structure of stationary waves. We shall consider the perturbations of a relatively low amplitude that are described by Eq.(11). This equation can be normalized so that it will contain only two independent parameters completely determining the structure of its solutions

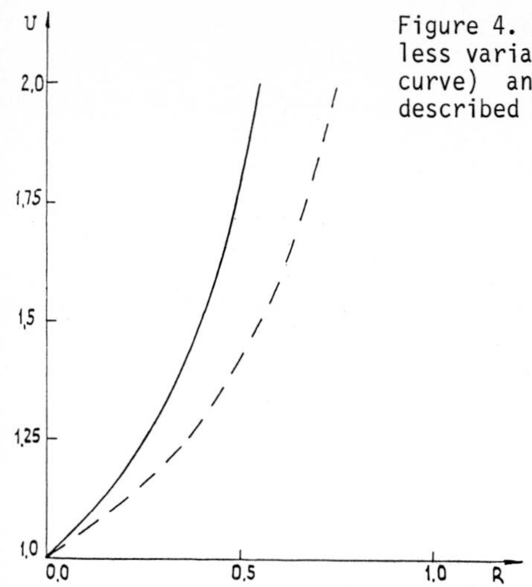


Figure 4. Dispersion dependences in the dimensionless variables for linear perturbations (dashed curve) and for limiting waves (solid curve) described by Eq. (18).

(while such a normalization is impossible in a more general equation (10)). Let us introduce the variable $W = 3ZU^2/(U^2-1)$ and write (11) in the form

$$(W''/Ur - W + W^2/2)'' = W/S_0 \,, \tag{19}$$

where $Ur = (U^2-1)/QU^2$ and $S_0 = (U^2-1)/RU^2$. We failed to find the solutions of this equation at finite Ur and $S_0$ analytically, therefore they were constructed numerically. We shall seek $W(\zeta)$ in the form of a Fourier series:

$$W(\zeta) = \sum_{k=1}^{N} a_k \cos k\zeta \,.$$

Substituting this series into (19), we obtain a system of nonlinear equations that was solved by the iterative Newton method; the number of harmonics was taken to be equal to 32, so that the relative contribution of the coefficient $a_{32}$ did not exceed 0.1% in the general solution (an alternative method of the numerical solution of (19) was proposed in /13/ but no computation results were presented).

Results of our calculations can be summed in the form of a "map of solutions" in the plane of the parameters Ur and $S_0$ (Fig.5). Note first of all that for sufficiently small perturbations, when the nonlinear term may be neglected, (19) yields a dispersion dependence that in our variables has the following form: $S_0 = Ur/(Ur+1)$. This relation is shown by a solid curve in Fig.5a. The steady-state solutions corresponding to the points on this curve are sinusoids (having a period $2\pi$ in our normalization). Using the perturbation theory, one can readily obtain corrections to the linear solution and, consequently, to the dispersion dependence:

$$S_0 = \frac{Ur}{Ur+1}\left[1 + \frac{\alpha^2 Ur^2}{6(Ur+1)(Ur+5)}\right] \,, \tag{20}$$

where $\alpha \ll 1$ is a small parameter proportional to the perturbation amplitude.

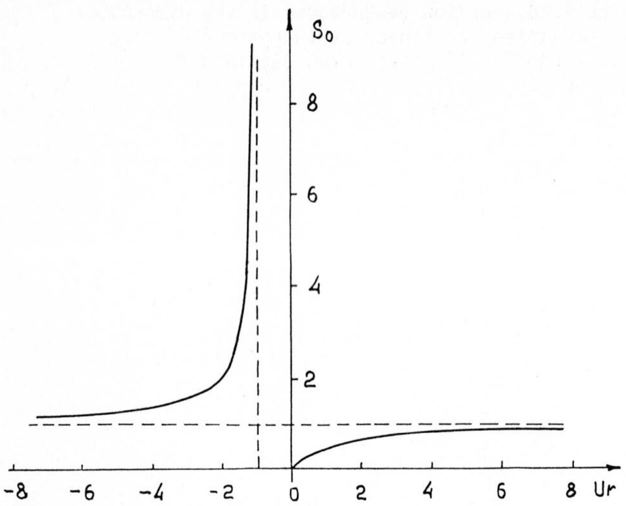


Figure 5a. The plane of the parameters Ur and $S_0$ for Eq.(19) and the structure of its steady-state solutions: $S_0$ versus Ur (solid curves) corresponding to linear solutions with a sinusoidal profile.

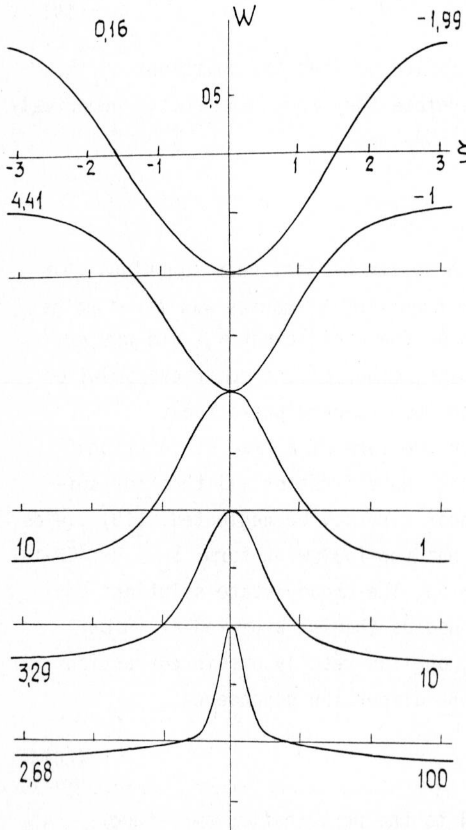


Figure 5b. The plane of the parameters Ur and $S_0$ for Eq.(19) and the structure of its steady-state solutions: the stationary wave profile for increasing Ur and fixed $S_0 = 2$; figures at the left indicate $|W|_{max}$ and those at the right indicate Ur.

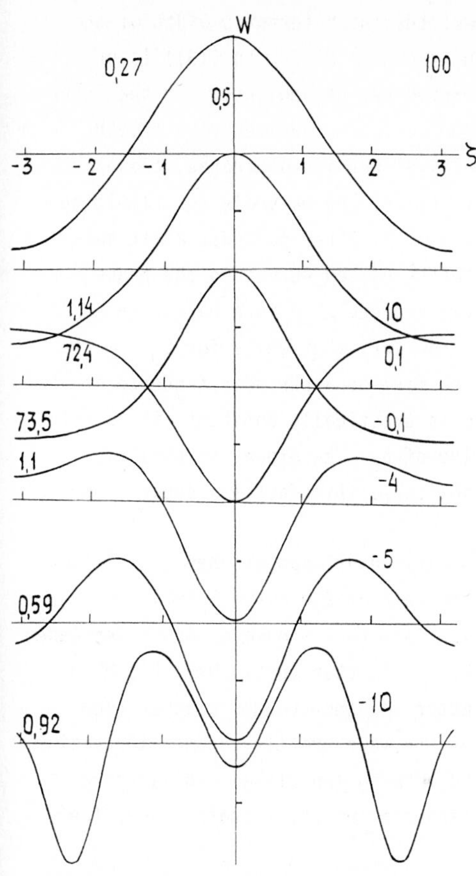


Figure 5c. The plane of the parameters Ur and $S_0$ for Eq.(19) and the structure of its steady-state solutions: the stationary wave profile for decreasing Ur and fixed $S_0 = 1$; see also the legend to Fig.5b.

Figure 5d. The plane of the parameters Ur and $S_0$ for Eq.(19) and the structure of its steady-state solutions: the stationary wave profile for decreasing Ur and fixed $S_0 = 0.001$. The last curve represents a stationary bisoliton that exists at the same values of the parameters Ur and $S_0$ as the single soliton (the previous curve); see also the legend to Fig.5b.

When $S_0 \to \infty$, the right-hand side of (19) can be neglected and the remaining equation is a steady-state form of a KdV equation the solution of which are well known. For Ur = -1 it is a sinusoid which transforms to a cnoidal wave with increasing Ur and then to a periodic sequence of solitons as $Ur \to \infty$. The solutions are physically meaningful only in the region $Ur > -1$, which is also seen from Eq.(20). For large and moderate values of $S_0 > 1$ the steady-state solutions, with the variation of Ur, behave in a similar fashion: they change from sinusoidal waves on the curve $S_0(Ur)$ to quasicnoidal waves as Ur increases. The stationary waves have flat crests and sharp troughs when $Ur < 0$ and, vice versa, sharp crests and flat troughs when $Ur > 0$ (Fig.5b). For sufficiently large Ur and fixed values of $S_0$ the wave crest has a shape similar to a KdV soliton, which is easily explained from qualitative considerations. Indeed, as Ur grows, the characteristic width of the crest decreases. This means that HF dispersion (the term $W^{IV}/Ur$ in (19)) is predominant in the wave crest region and LF dispersion can be neglected. In the intervals between the crests the wave profile, on the contrary, becomes very smooth, here the HF dispersion is negligible while the LF dispersion dominates, therefore, this section of the wave has a nearly parabolic shape. The parabola as a limiting solution of (19) when $Ur = \infty$ corresponds to $S_0 = \pi^2/9$. Other periodic limit solutions of (19) with $Ur = \infty$ (which correspond to the closed curves in the phase plane of Fig.2, that lie inside the solid curve) are enclosed in a narrow range of $S_0$ from 1 to $\pi^2/9$. It should be emphasized, however, that while for $S_0 = \infty$ the increase in parameter Ur gives a solution in the form of a periodic sequence of uncoupled KdV solitons between which the field is practically constant, it is not constant between the pulses for any finite value of $S_0$. Therefore, in this case there are no solutions in the form of single solitons. This fact was shown rigorously in /5/ for positive values of Ur.

The qualitative behaviour of solutions with varying Ur changes when $S_0 \leq 1$. Now physically reasonable solutions are found to the left of the curve $S_0(Ur)$. For $S_0 = 1$ and large values of Ur the wave shape is close to a sinusoid. As Ur decreases the wave crests sharpen and the wave troughs become flatter (until $Ur > 0$). When $Ur < 0$, then vice versa, the crests become flatter and the troughs sharper (see Fig.5c). As Ur decreases in the negative region the wave profile becomes increasingly more nonsinusoidal; however, in the wave period between two global minima there gradually appears another local minimum, then another and still another one. The wave acquires a complex shape (see Fig.5c).

A similar picture occurs also with smaller values of $S_0$ (see Fig.5d). Note that for $Ur < 0$ the "antisoliton" theorem /5/ is no linger valid, although the existence of solitons in this case was not proved either. Our computations show, however, that for small $S_0$ and sufficiently large negative Ur the wave profile is close to a periodic sequence of negative-polarity pulses ( a trough on the fluid surface) that are not coupled in fact. Each of these pulses can be considered as a soliton. Owing to nonmonotonic profiles of these solitons (the presence of local maxima) stationary bound states of two and more pulses can be formed in this case. An example is given in Fig.5d where the last curve represents a computer calculated stationary bisoliton. A still more complex and diverse picture of steady-state solutions occurs when $S_0 < 0$ (the "slow" waves mentioned above, for which $Ur < 1$, also belong here). We have not yet carried out a complete analysis of the solutions in this case, therefore we will not consider it here.

### 3.2. Dynamics of Nonstationary Perturbations

It is convenient to investigate nonstationary perturbations in the class of systems of interest on a model of a transmission line of the band filter type that

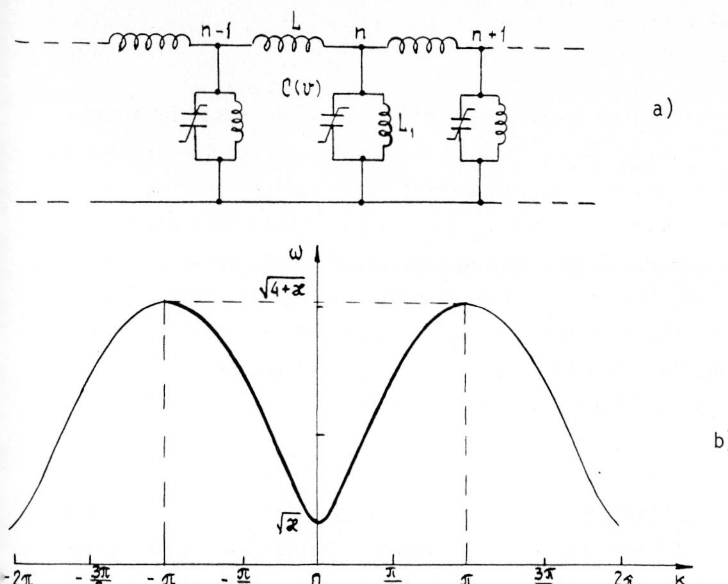


Figure 6. Scheme of a band-filter transmission line (a) and the corresponding dispersion curve (b).

contains nonlinear elements (Fig.6a). Assuming that the charge $Q_n$ is related to the voltage $v_n$ in a capacity of the n-th element of the line by a nonlinear function $Q_n = Q_0 \ln(1+v_n/V_0)$, one can easily obtain from the Kirchhoff equations the basic equation describing the propagation of perturbations in such a chain:

$$\frac{d^2}{dt^2} \ln(1+u_n) = u_{n-1} - 2u_n + u_{n+1} - \kappa u_n \,, \tag{21}$$

where $u_n = v_n/V_0$, $\kappa = L/L_1$ ($L$ and $L_1$ are the inductances shown in Fig.6a), the time is normalized to $(V_0/LQ_0)^{1/2}$. Note that for $\kappa = 0$ Eq.(21) reduces to a well-known, completely integrable Hamiltonian system: an "electric" variety of Toda chain /14/ which, in its return, reduces to a KdV equation for low-amplitude long perturbations.

Equation (21) yields a dispersion relation $\omega^2 = \kappa + 4\sin k/2$ which is plotted in Fig.6b. In view of spatial periodicity of the system, curve $\omega(k)$ is periodic with respect to k but only the sections corresponding to the first Brillouin zone (the more solid curve in Fig.6b) /15/ are physically meaningful. The figure demonstrates that within the first zone the dispersion curve for a long line is qualitatively similar to the dispersion curve for waves in a rotating fluid (cf. Fig.1) with $\beta > 0$. If $\kappa$ is sufficiently small, then in the long-wave range we have approximately $\omega \cong (\kappa+k^2-k^4/12)^{1/2} \cong k - k^3/24 + \kappa/2k$ which coincides with (5). In this case it is convenient to use the continuous approximation considering n as a continuous variable. Then for weakly nonlinear perturbations Eq.(21) easily yields an equation similar to Eq.(6):

$$(u_t+uu_x/2+u_{xxx}/24)_x = \kappa u/2 \ , \tag{6a}$$

where x = n-t. Equation (6a) can be transformed to (6) and vice versa, by substituting $u = 3c_0\eta/h(4c_0\beta)^{1/3}$, $x = \xi/(4c_0\beta)^{1/3}$ and $\kappa = \Omega^2(4c_0\beta)^{1/3}/c_0$. The differential-difference equation (21) can be used in numerical calculations for the investigation of the wave processes described by (6) and (6a). Note that PAPKO used such a transmission line for direct analog simulation of the nonlinear wave dynamics in a rotating fluid; the results were briefly reported in /16/.

Before starting the discussion of the results of numerical calculations of nonstationary waves, we would like to note that from (21) it follows that in the general case for linear spatially periodic perturbations the average value u

$$\langle u\rangle \equiv \sum_{n=1}^{\infty} u_n/N \ ,$$

where N is the number of the line elements in a period, oscillates in time with a frequency $\sqrt{\kappa}$, while within (6) and (6a) $\langle u\rangle = 0$. The oscillation amplitude is determined by the initial conditions. With the nonlinearity taken into account, the mean field generation occurs even when the initial value of this field is zero. The same holds for Eq.(2).

Let a periodic sequence of pulses the structure of which coincides with the KdV solitons (Fig.7a) be specified in the chain (21) when t = 0. When $\kappa = 0$ these pulses are stationary waves having zero mean value and travelling to the right along the chain. When $\kappa \neq 0$ (in our calculations $\kappa = 10^{-4}$), such an initial condition no longer corresponds to the wave travelling only to the right, there also emerges a small perturbation propagating to the left along the chain but in our case this this perturbation does not significantly affect the dynamics of the fundamental wave. The estimation of the parameters Ur and $S_0$ (see (7)) for the initial pulse with amplitude 0.02 and the characteristic half-width equal to 5 yields: Ur = 6 and $S_0 = 8$. It is apparent that LF and HF dispersion effects on the evolution of the pulse are quite comparable. Computations show that the trailing edge of the propagating pulse becomes steeper (Fig.7b) and a depression (gradually transforming into a long aperiodic tail) is formed behind it. Subsequently, the pulse amplitude decreases and the negative peak behind the pulse grows. Then the initial pulse is completely lost in nonstationary perturbations of different polarity (Fig.7b), however, in some time it is re-established again with almost initial amplitude (Fig.7d) on the background quasi-sinusoidal large- and small-scale perpurbations. The pulse has always a nearly KdV soliton shape with the appropriate amplitude. Then the process of the breaking and the formation of a new pulse is repeated. This phenomenon is similar to recurrence in a KdV equation /14/. Simultaneously, mean fields are generated in the chain due to nonlinearity. Value u averaged in space fluctuates in time with a frequency close to $\sqrt{\kappa}$ and the amplitude equals approximately to $4.7\cdot 10^{-4}$.

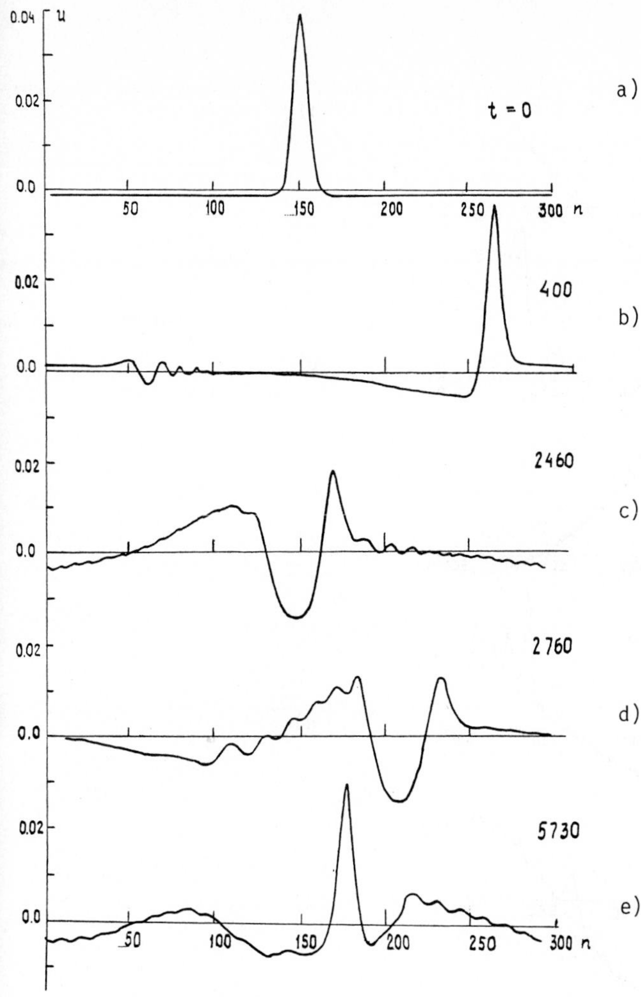


Figure 7. Evolution of a periodic sequence of pulses having a KdV-soliton shape at the initial moment.

Other computations used a parabolic profile as the initial condition corresponding to a stationary wave travelling to the right which is described by Eq.(11) with Q = 0 (Fig.8a). When $\kappa = 10^{-4}$, for such a perturbation $Ur = 2.7 \cdot 10^{-5}$ and $S_0 = 2.8 \cdot 10^{-2}$, i.e. the effect of the LF dispersion was predominant in this case all through the perturbation except the crest where the function is not analytical (we remind that the problem was solved with periodic boundary conditions with respect to n). The effect of the HF dispersion in this region caused intense generation of wave trains moving in opposite directions, with the shapes, again, similar to the KdV solutions (Fig.8b). It is clearly seen in Fig.8b that the perturbation can be divided into regions with predominant dispersions of their own: solitons are observed in the region with dominating HF dispersion and a parabolic wave profile is charac-

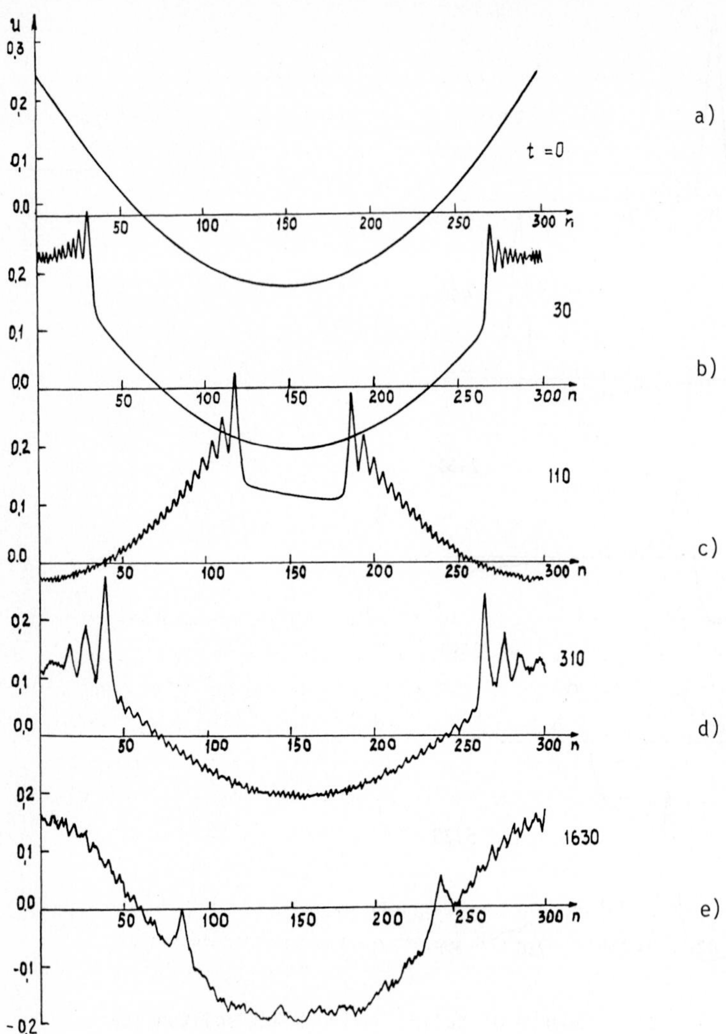


Figure 8. Evolution of a periodic wave with a parabolic initial profile described by Eq.(11) with Q = 0.

teristic of the regions with the dominating LF dispersion. Since the soliton velocity does not coincide with that of the parabolic wave, the solitons move along the parabola with an oscillating HF tail. Then they collide, the perturbations become extremely rough although the parabolic profile is well pronounced on the average (Fig.8c). Later the evolution pattern is the following: a LF wave with a nearly parabolic profile propagates along the chain, and solitons and nonstationary low-amplitude wave trains propagate on its background in opposite directions (Fig. 8d). This process is also accompanied with the generation of mean fields oscillating with the frequency $\sim\sqrt{\kappa}$ and the amplitude $\sim 8.7\cdot 10^{-2}$.

## 4. TWO-DIMENSIONAL WAVES IN A ROTATING FLUID

We shall consider weakly nonlinear perturbations which vary more along y (transverse to the primary motion) than along x. They can be described by the model Eq. (9) with appropriate boundary conditions which depend on the type of the problem of interest. Thus, when studying the waves in a rotating channel with the width L bounded by rigid walls along y, the boundary conditions have a form /8,9/:

$$\eta_y + \Omega\eta/c_0 = 0 , \quad \text{as } y = 0,L , \tag{22}$$

which corresponds to the normal velocity component v vanishing to zero on the walls. The average level of surface perturbations $\langle\eta\rangle$ may be not equal to zero in a channel. Indeed, integrating (9) with respect to the wave period $\Lambda$ along $\xi$ with (22) taken into account, we obtain the following expression /8,9/ (the integral for localized perturbations is taken in infinite limits):

$$\frac{1}{\Lambda}\int_0^\Lambda \eta(\xi,y,t)d\xi = M(t)\exp(-\Omega y/c_0) , \tag{23}$$

where M(t) can be interpreted as the wave "mass" taken at y = 0. For stationary waves M = const. This relation imposes stringent limitations on the transverse structure of the initial perturbations described by (9) and (22). For example, the initial condition in the experiments described in /17/ was a plane wave which is bounded in x (solitary wave) and has a uniform front along y. The evolution of such a perturbation cannot be considered, strictly speaking, within model Eqs.(9) and (22) (besides, the perturbations in /17/ were strong, which also contradicts the applicability conditions for (9)).

Results of numerical calculations made within the model (9), (22) were presented in /9/. The initial condition was specified as a KdV soliton with a plane front and an amplitude decreasing exponentially along y such that condition (23) was met. It was found that the wave front curvature grows with time in the x,y-plane, the sections with higher amplitude near one wall of the channel (y = 0) ran forward while the sections with low amplitude trailed behind (Fig.9). Weakly nonlinear waves were generated in this case behind the wave front. The wave profile near the boundary y = 0 was qualitatively similar to the KdV soliton profile but was much wider. The perturbation amplitude decreased exponentially along y in accord with the linear theory ($\eta \sim \exp(-\Omega y/c_0)$). An exponential decrease of amplitude, even if slower, was also observed along the curved front. In the course of propagation the wave as a whole was damped along x, which was evidently due to the emission of quasi-linear oscillatory waves that are seen in Fig.9. The formation of solitary waves is impossible in this case because the dispersion of the internal waves considered in /9/ is such that $\beta > 0$. The bending of the front can, apparently, be explained by the radiation wave damping, since dissipation in the medium results in the bending of the solitary wave front as was shown in /18/ for the Kelvin waves that are si-

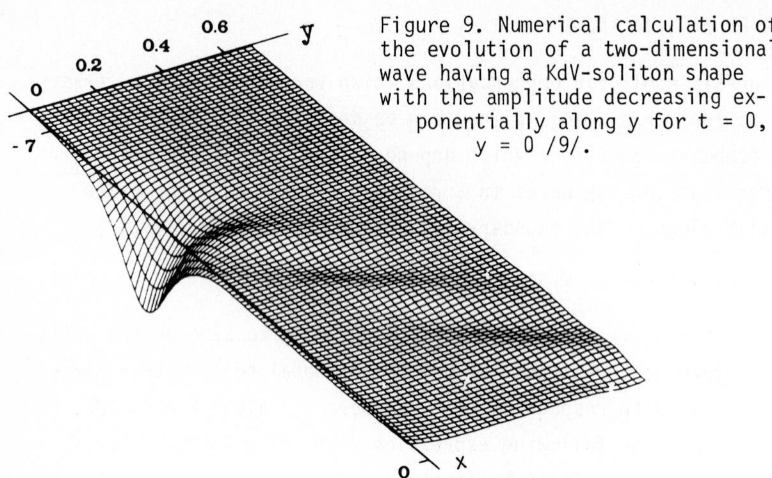


Figure 9. Numerical calculation of the evolution of a two-dimensional wave having a KdV-soliton shape with the amplitude decreasing exponentially along y for t = 0, y = 0 /9/.

milar to the waves studied here. The front curvature in this case depends on the nature of dissipation. These conclusions agree qualitatively with experimental results /17/ although the initial conditions in the latter case, as has already been mentioned, did not fit model (9), (22). The numerical computations /9/ also reproduced the conditions corresponding to the experiment /17/, i.e. a KdV soliton with a plane uniform front along y was taken at the initial moment. The computations have shown that the results were, in fact, the same as in the first case. This indicates that the violation of condition (23) is not very essential for the qualitative picture of the evolution of initial perturbations. This also explains the agreement between the results of experiments /17/ and numerical calculations.

### 4.1. Self-Modulation and Self-Focusing Instabilities of Harmonic Waves

The stability of plane stationary waves periodic along x relative to self-modulation and self-focusing effects was analyzed in /5/ within Eq.(11). Seeking the periodic solutions of this equation in the form of a series in the dimensionless wave amplitude $\alpha$, we shall write

$$\begin{aligned} Z &= \alpha Z_1(\zeta) + \alpha^2 Z_2(\zeta) + \alpha^3 Z_3(\zeta) + \dots , \\ 2(U-1) &= a_1 + \alpha^2 a_2 + \dots \end{aligned} \tag{24}$$

(note that the wave period within (11) is equal to $2\pi$). Then substituting (24) into (11) and equating the terms with equal powers of $\alpha$, we shall obtain the solution generalizing the Stokes solution /19/ to the case of a rotating medium:

$$\begin{aligned} Z &= \alpha \sin \zeta - \frac{\alpha^2 \cos 2\zeta}{R+4Q} - \frac{27\alpha^3 \sin 3\zeta}{16(R+4Q)(R+9Q)} + \dots , \\ 2(U-1) &= R - Q + \frac{\alpha^2 Q}{2(R+4Q)} + \dots . \end{aligned} \tag{25}$$

The second expression in (25) is the so-called nonlinear dispersion equation. It can be written in dimensional variables in the form (cf.(5)):

$$\omega = c_0 k + \frac{\Omega^2}{2c_0 k} - \frac{\beta c_0}{6}k^3 + \frac{9\eta^2\beta c_0{}^3 k^3}{4h^4(3\Omega^2+4\beta c_0{}^2 k^4)} \cdot \tag{26}$$

Equation (26) readily yields the stability condition for quasi-harmonic waves relative to longitudinal and transverse perturbations /2/. According to the Lighthill's criterion, the wave is stable with respect to self-modulation if

$$\left(\frac{\partial^2\omega}{\partial k^2}\right)_{\eta=0} \cdot \left(\frac{\partial^2\omega}{\partial \eta^2}\right)_{\eta=0} > 0 .$$

Using (26) we obtain

$$\left(\frac{\Omega^2}{c_0 k^3 - c_0 k}\right)\left(\frac{\beta c_0{}^3 k^3}{6\Omega^2+8\beta c_0{}^2 k^4}\right) > 0 . \tag{27}$$

If $\beta > 0$, the second multiplier is positive. Then the inequality (27) holds for $k < k_1 \equiv (\Omega^2/\beta c_0{}^2)^{1/4}$, i.e. in the region of sufficiently long waves. If $\beta < 0$, the first multiplier in (27) is positive. Then the inequality (27) is fulfilled when $k > k_2 \equiv (-3\Omega^2/4\beta c_0{}^2)^{1/4}$. From this it is clear that periodic waves are stable with respect to self-modulation in a narrow range $k_2 < k < k_1$ at any sign of $\beta$.

Let us now consider the transverse instability associated with the self-focusing. It is known /2/ that condition $(\partial\omega/\partial\alpha^2)_{\alpha=0} > 0$ is the criterion of such a stability. Equation (27) shows that this condition necessitates a positive second multiplier in (27). This multiplier is positive for all k when $\beta > 0$ and is positive only in the region of relatively short waves ($k > k_2$) when $\beta < 0$. Thus, large-scale waves with $k < k_1$ are unstable with respect to both self-modulation and self-focusing if there is a positive dispersion in a HF region ($\beta < 0$). This fact gives us grounds to velieve that solitons may exist both in one-dimensional (which was confirmed by numerical calculations in Sect.31) and two-dimensional cases in media with $\beta < 0$. Below we shall present results of calculations of the structure of two-dimensional solitons when $\beta < 0$.

## 4.2. Structure of Two-Dimensional Multisolitons

The investigations carried out in recent years showed that the presence of positive dispersion in the HF region of the spectrum in systems described by a generalized KP equation allows for the existence of localized solutions in the form of two-dimensional solitons and multisolitons, i.e. groups of steady-state coupled solitons /20/. The considerations presented above also favour the existence of such formations when $\beta < 0$. We shall seek them numerically using the stabilizing multiplier method that was first proposed in /21/.

Let us first rewrite (9) in a standard dimensionless form similar to (11) introducing the variables

$$Z = \eta/h\ , \quad \zeta = \xi 2\pi/\Lambda_x\ , \quad \theta = y2\pi/\Lambda_y\ ,$$

where $\Lambda_x$ and $\Lambda_y$ are the characteristic scales of perturbations along x and y. Then (9) will yield for the waves steadily propagating along $\xi$ (= $x-c_0t$) with a velocity $\delta V$, the following expression

$$(QZ_{\xi\xi} + 3Z^2/2 - 2\delta UZ)_{\xi\xi} = RZ - DZ_{\theta\theta}\ , \tag{28}$$

where $Q = \beta(2\pi/\Lambda_x)^2/3$, $U = \delta V/c_0$, $R = (\Lambda_x/2\pi)^2(\Omega/c_0)^2$ and $D = (\Lambda_x/\Lambda_y)^2$. Then, the steady-state solutions of (28) were sought in a spectral form by a known iteration scheme /20,21/. The use of a bell-shaped function as a starting one resulted, as we expected, in a two-dimensional soliton shown in Fig.10 for the following values of coefficients: Q = -0.25, D = 3, $\delta U$ = -1.5 and R = 30. A two-dimensional soliton is a depression on the surface of a fluid. As applied to water waves, such solitons can be realized only in a sufficiently flat rotating container with a depth smaller than 5 mm when capillary forces prevail over the gravity ones. Note, incidentally, that these conditions can presumably be implemented in the setups described in /22, 23/ where wave-vortex formations were investigated at much lower frequencies ($\omega \ll \Omega$) belonging to the oscillation spectrum of the Rossby waves. At first sight, these solitons are very much like two-dimensional KP solitons /21/ which are produced also in our case if there is no rotation (R = 0). However, there is a significant difference between them. An analytical expression for KP solitons is known /24/, namely:

$$Z(\zeta,\theta) = 8Q\,\frac{3Q/2\delta U - 2\delta U\theta^2/D - \zeta^2}{(3Q/2\delta U - 2\delta U\theta^2/D + \zeta^2)^2}\ , \tag{29}$$

which shows that this solution has power asymptotics $Z \sim \zeta^{-2},\theta^{-2}$ when $|\zeta|,|\theta| \to \infty$. In our case when $R \neq 0$, a two-dimensional soliton is much more compact and its field decreases in space exponentially. This fact follows directly from the linearized equation (28) that holds at large distances from the soliton centre where Z is sufficiently small. Indeed, when R = 0 and $Z^2$ is neglected, the asymptotic solution of this equation is determined by the eigenfunctions of the Laplace equation $\Delta Z = 0$ to which the linearized equation (28) reduces after the corresponding scale transformation. It is known that the power functions are the eigenfunctions of the Laplace operator. Therefore the dispersion term with the fourth derivative for such functions is negligibly small at $|\zeta|,|\theta| \to \infty$. When $R \neq 0$, then instead of the Laplace operator the linearized operator will contain in the main order the Helmholtz operator ($\Delta$ - R) the eigenfunctions of which are already exponents. At these functions, however, the dispersion term is not small in comparison to other terms, therefore it must also be included in the construction of the asymptotic form of the solution. The resulting scale of the decrease of a soliton field (the decrement) is anisotropic at large distances: the characteristic scale of the field variation along the soliton motion (along $\zeta$) does not coincide with the decrease in the transverse direction (along $\theta$).

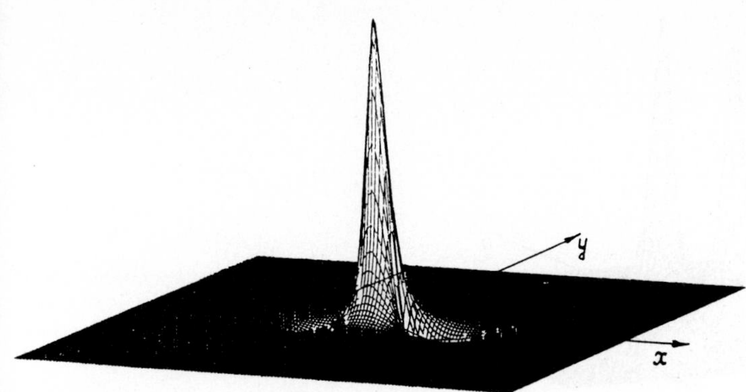


Figure 10. General form of a two-dimensional soliton in media with positive HF disperion.

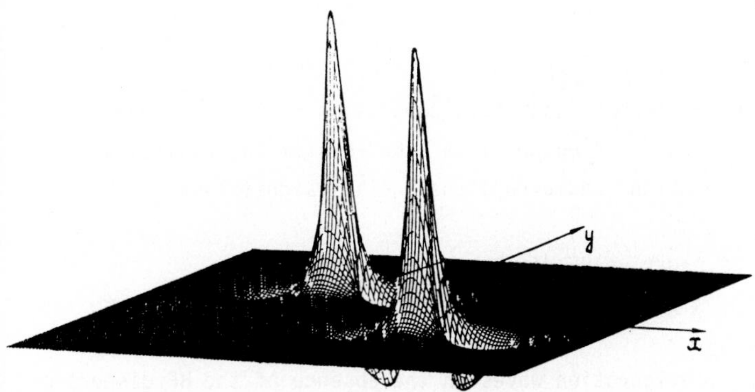


Figure 11. General form of a two-dimensional bisoliton moving at an angle to the line connecting its centres.

The analysis of the structure of single solitons (see Fig.10) shows that they have local field extrema also outside the centre. This gives us grounds to believe that there exist coupled states of two or more solitons /25/. We have performed the calculation with a two-humped starting function which confirmed the idea of the existence of such solutions in the form of stationary bisolitons. An example of such a bisoliton is presented in Fig.11. It is interesting that the solitons in the group do not move along $\zeta$ strictly one after another, as it was within the KP equation /20/, but their motion is such that the line connecting their centres makes an angle with the trajectory of the motion. We revealed that there exist several different solutions in the form of bisolitons with different angles of turn for the same parameters of Eq.(28). In particular, it is possible to construct a coupled pair of bisolitons moving side by side (Fig.12).

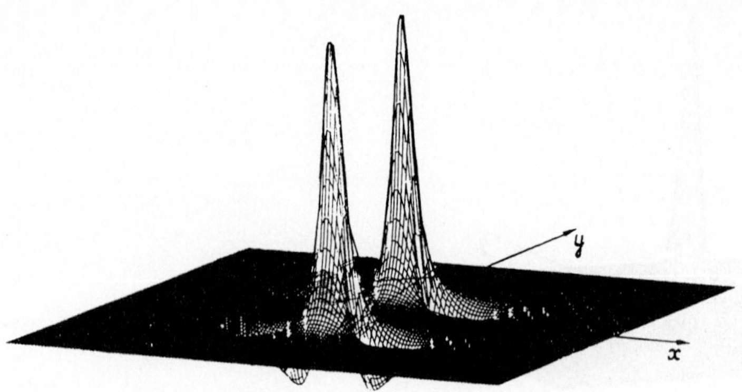


Figure 12. Two-dimensional bisoliton moving perpendicularly to the line connecting its centres.

## 5. CONCLUSION

The analysis presented in this paper leads to a conclusion that even a weak rotation (or other similar effects) introduced into a KdV-system may radically change the behaviour of the solutions and expand their class significantly. Then emerge the following new aspects:

1) the absence of solitons at negative HF dispersion;
2) the appearance of solitons with nonmonotonic asymptotic behaviour related to rotation at positive dispersion;
3) the existence of limiting-configuration waves in the absence of the HF dispersion even at weak nonlinearity;
4) the appearance of "two-scale" periodic waves with sharp soliton-like crests on the background low-frequency waves;
5) the mutual conversion of these two types of waves in the evolution process;
6) the formation of nonstationary soliton packets interacting with a LF component;
7) the formation of nonuniform waves with bended fronts in systems bounded or semi-bounded in the transverse direction (e.g., in channels);
8) the possibility of simultaneous meeting of the focusing and self-modulation conditions in a certain region of wavelength in systems with positive HF dispersion;
9) the existence of two-dimensional solitons with strong (exponential) localization in these systems;
10) possible formation of two-dimensional bisolitons (multisolitons, most probably, as well) in the form of steadily moving soliton pairs with different orientation relative to their trajectory.

All these various properties are, in fact, realized as a result of the change in the sign and magnitude of one parameter determined by rotation. We believe that this class of systems is worth further investigating using different "synergetic" methods: analytical, numerical and experimental.

*Acknowledgements*

The authors are grateful to L.A.Abramyan and I.P.Ryazantseva for help in numerical computations and also to N.B.Krivatkina and N.B.Mezentseva for their help in preparing the English translation of the text.

REFERENCES

1. O.M.Braun, I.I.Zelenskaya, Yu.S.Kivshar. Frenkel-Kontorova Model with Far Interaction. Preprint No.2, Inst. of Physics, Ukrainian Acad.Sci., Kiev, 1989, 36p. (in Russian).
2. V.I.Karpman. Nonlinear Waves in Dispersive Media. Pergamon Press, Oxford, 1975.
3. L.D.Landau, E.M.Lifshits. Hydrodynamics. Nauka, Moscow, 1988.
4. V.I.Shrira. On the long strongly nonlinear waves in rotating ocean. Izvestiya Acad.Nauk SSSR, FAO, 1986, 22, 4, 395-405 (in Russia ); Atmospheric and Oceanic Physics, 1986, 22, 4.
5. A.I.Leonov. The effect of Earth rotation on the propagation of weak nonlinear surface and internal long oceanic waves. Annals New York Acad.Sci., 1981, 373, 150-159.
6. L.A.Ostrovsky. Nonlinear internal waves in a rotating ocean. Okeanologiya, 1978, 18, 2, 181-191 (in Russian); Oceanology, 1978, 18, 2.
7. L.Y.Redekopp. Nonlinear waves in geophysics: Long internal waves. Lectures in Appl.Math., 1983, 20, 29-78.
8. R.G.Grimshaw. Evolution equations for weakly nonlinear long internal waves in a rotating fluid. Stud.Appl.Math., 1985, 73, 1, 1-33.
9. C.Katsis, T.R.Akylas. Solitary internal waves in a rotating channel: A numerical study. Phys.Fluids, 1987, 30, 2, 297-301.
10. S.V.Muzylev. Nonlinear equatorial waves in the ocean. Digest of Reports, 11 All-Union Congress of Oceanographers, Sevastopol, USSR, 1982, 2, 26-27 (in Russian).
11. V.I.Shrira. The propagation of long nonlinear waves in the layer of rotating fluid. Izvestiya Akad.Nauk SSSR, FAO, 1981, 17, 1, 76-81 (in Russian); Atmospheric and Oceanic Physics, 1981, 17, 1.
12. B.B.Kadomtsev, V.I.Petviashvili. On the stability of solitary waves in weakly dispersive media. Sov.Phys.-Dokl., 1970, 15, 539-541.
13. A.A.Zaitsev. Steady-state solution of the Ostrovsky equation. Digest of Reports, All-Union Meeting on Numerical Methods in the Problem of Tsunami, Krasnoyarsk, USSR, 1987, 58-60 (in Russian).
14. M.Toda. Theory of Nonlinear Lattices. Springer-Verlag, Berlin-Heidelberg-New York, 1981.
15. L.Brillouin, M.Parodi. Waves Propagation in Periodic Structures. Dover, 1953.
16. L.A.Ostrovsky. Nonlinear internal waves in the ocean. In: Nonlinear Waves, Nauka, Moscow, 1979, 292-323.
17. D.P.Renouard, Y.Ch.D'Hieres, X.Zhang. An experimental study of strongly nonlinear waves in a rotating system. J.Fluid.Mech., 1987, 177, 381-394.
18. Sh.M.Khasanov. The propagation of Kelvin's solitary wave in an absorbing medium. Izvestiya Akad. Nauk SSSR, FAO, 1989, 25, 307-311 (in Russian); Atmospheric and Oceanic Physics, 1989, 25, 3.
19. G.B.Whitham. Linear and Nonlinear Waves. John Wiley & Sons, New York et al, 1974.

20. L.A.Abramyan, Yu.A.Stepanyants. Structure of two-dimensional solitons in the context of a generalized Kadomtsev-Petviashvili equation. Izvestiya Akad.Nauk SSSR, Radiofizika, 1987, 30, 10, 1175-1180 (in Russian); Radiophysics and Quantum Electronics, 1987, 30, 10.

21. V.I.Petviashvili. Equation of an extraordinary soliton. Fizika Plasmy, 1976, 2, 3, 469-472 (in Russian); Plasma Physics, 1976, 2, 3, 257.

22. R.A.Antonova, B.P.Zhvanya, D.G.Lominadze, D.I.Nanobashvili, V.I.Petviashvili, Modeling of vortices in a homogeneous magnetized plasma in a shallow rapidly rotating liquid. Fizika Plasmy, 1987, 13, 11, 1327-1331 (in Russian); Plasma Physics, 1987, 13, 11.

23. S.V.Antipov, M.V.Nezlin, V.K.Radionov, A.Yu.Rylov, E.P.Snezhkin, A.S.Trubnikov, A.V.Khutoretsky. The properties of drift solitons in a plasma as inferred from model experiments on a rapidly rotating shallow water. Fizika Plasmy, 1988, 14, 9, 1104-1121 (in Russian); Plasma Physics, 1988, 14, 9.

24. S.V.Manakov, V.E.Zakharov, L.A.Bordag, A.R.Its, V.B.Matveev. Two-dimensional solitons of the Kadomtsev-Petviashvili equation and their interaction. Phys. Lett., 1977, 63A, 3, 205-206.

25. K.A.Gorshkov, L.A.Ostrovsky, V.V.Papko. Interactions and coupled states of solitons as classical particles. Sov.Phys.-JETP, 1976, 71, 2.

26. S.A.Rybak, Yu.I.Skrynnikov. Solitary waves in a thin constant-curvature rod. Akust.Zh., 1990, 36, 4, 548-553 (in Russian).

27. E.S.Benilov, E.N.Pelinovsky. On the theory of wave propagation in dispersionless nonlinear fluctuating media. Sov.Phys.-JETP, 1988, 94, 1, 175-185.

# A Note on the Mechanism of the Resonant Nonlinear Instability in Boundary Layers

*V.P. Reutov*

Institute of Applied Physics, USSR Academy of Sciences,
46 Ulyanov Str., 603600 Gorky, USSR

It has been shown that in the inviscid boundary-layer type flow a degenerate nonlinear instability arises, which can be interpreted as the nonlinear Landau growing of resonantly coupled waves.

The resonant nonlinear instability plays an important role in the breakdown of a laminar boundary-layer flow. Usually, its investigation is based on the formal theory of coupled waves of a viscous flow /1/. The occurrence of such an instability in an inviscid flow has not been discussed so far. The proof of the inviscid nature of the resonant instability is of interest from the viewpoint of the general theory of flow instability as well as in relation with the use of the ideal fluid approximation in the numerical simulation of perturbation development in the boundary layers /2/.

This paper investigates the possibility of nonlinear instability of a resonant wave triplet in an inviscid wall flow with the piece-wise linear velocity profile: $\bar{u} = y$ at $0 < y < 1$ and $\bar{u} = 1$ at $y > 1$ (here and further standard dimensionless variables ables are used). In such a flow three-dimensional perturbations of the form $\exp\left[i\alpha(x-ct) + i\beta z\right]$ have the real phase velocity $c = 1 - (1/2\bar{\alpha})\left[1-\exp(-2\bar{\alpha})\right]$, where $\bar{\alpha} = (\alpha^2+\beta^2)^{1/2}$. We shall consider the resonant interaction of a two-dimensional wave $(\alpha_1,0)$ with a pair of symmetric oblique waves $(1/2\alpha_1, \pm\beta_1)$. It can be shown that in this case the frequency resonance conditions are satisfied at $\beta_1/\frac{1}{2}\alpha_1 = \sqrt{3}$. Since the phase velocities of all waves are identical and equal to $c_1 = c|_{\alpha=\alpha_1}$, a single critical layer (CL), localized near the surface $y = y_c \equiv c_1$, arises in the flow. At small wave amplitudes the solution of the problem can be reduced to an analysis of the self-consistent evolution of a weakly nonlinear three-dimensional CL. Note that the dynamics of two-dimensional nonlinear CL in an inviscid fluid was investigated in /3-6/. The formation of a three-dimensional nonlinear CL at fixed wave amplitudes was considered in /7/.

The equations of the problem include the Euler equations, the boundary conditions at $y = 0$, $y \to \infty$ and the boundary conditions on the vorticity jump reduced to the level $y = 1$. The velocity components of the flow can be represented as $v_1 = u(y) + u$, $v_2 = v$, $v_3 = w$. To solve the problem, we shall make use of the method of multiple scales /8/. Let us introduce a small parameter $\varepsilon \ll 1$ and a coordinate $\xi = x - c_1 t$ instead of x. To obtain a two-term uniformly valid expansion it is sufficient to introduce one variable $\tau = \varepsilon^{1/4}t$ instead of t. The expansion for v is sought in the form:

$$v = \varepsilon \sum_{\alpha,\beta} af(y;\alpha,\beta)\exp(i\alpha\xi+i\beta z) + \text{c.c.} + \varepsilon^{5/4}v^{(5/4)} + \dots , \quad (1)$$

where $\varepsilon a(\tau\ \alpha,\beta)$ is the complex amplitude of the wave $(\alpha,\beta)$; f is the wave profile in the linear theory; the summation is performed over all the waves of the triplet. In our normalization $f = -i\bar{\alpha}\,\mathrm{sh}\bar{\alpha}y/\alpha\,\mathrm{sh}\bar{\alpha}y_c$ at $0 < y < 1$. Similar expansions for u, w and pressure p can conveniently be constructed by using the expressions for these variables through v and the nonlinear terms of the equations (see /7/). Denoting the complex amplitudes of the harmonics $\exp(i\alpha\xi+i\beta z)$ by ^ , we can represent the solution of the boundary problem for $\hat{v}^{(5/4)}$ in the region $0 < y < 1$ as

$$\hat{v}^{(5/4)} = A_{\pm}^{(5/4)}(\tau;\alpha,\beta)\frac{1}{\bar{\alpha}}\,\mathrm{sh}\bar{\alpha}\eta + B^{(5/4)}(\tau;\alpha,\beta)\mathrm{ch}\bar{\alpha}\eta , \quad (2)$$

where $\eta = y - y_c$, the coefficients $A_{\pm}^{(5/4)}$ refer to the regions $\eta > 0$ and $\eta < 0$, respectively. From the condition for uniform validity of the two-term expansion for v, the following equation for the wave amplitudes is derived:

$$\frac{da}{d\tau} = \alpha^2 G \cdot (A_{+}^{(5/4)} - A_{-}^{(5/4)}) , \quad (3)$$

where

$$G(\bar{\alpha}) = (1/4\bar{\alpha}^2)\left[1 - \exp(-2\bar{\alpha}c_1)\right]\{2\bar{\alpha}(1-c_1) - 1 + \exp\left[-2\bar{\alpha}(1-c_1)\right]\} .$$

To define the jump $A_{+}^{(5/4)} - A_{-}^{(5/4)}$, we shall make use of the method of matched asymptotic series /8/. After introducing the internal variable $Y = \eta/\varepsilon^{1/4}$ into the Euler equations we shall construct the internal series for the CL in the form

$v = V^{(1)}+\varepsilon^{5/4}V^{(5/4)}+\varepsilon^{3/2}V^{(3/2)}+\dots$ , $p = \varepsilon P^{(1)}+\varepsilon^{5/4}P^{(5/4)}+\varepsilon^{3/2}P^{(3/2)}+\dots$ ,

$u = \varepsilon^{3/4}U^{(3/4)}+\varepsilon U^{(1)}+\varepsilon^{5/4}U^{(5/4)}+\dots$ , $w = \varepsilon^{3/4}W^{(3/4)}+\varepsilon W^{(1)}+\varepsilon^{5/4}W^{(5/4)}+\dots$ .

Construction of the solution reduces to an alternate search of the terms of the internal series for p and v and an elimination of the arbitrariness by matching the external and internal series using van-Dyke criterion: $(\hat{p}_n{}^0)_m{}^i = (\hat{p}_m{}^i)_n{}^0$, $(\hat{v}_n{}^0)_m{}^i = (\hat{v}_m{}^i)_n{}^0$, where the indices 0 and i denote external and internal expansion and m and n are the numbers of terms in expansion. It should be sufficient to match the series with $n \leq 2$ and $m \leq 3$. The procedure of calculation of the internal series can be greatly simplified by taking into account the form of the desired dependence of a on $\tau$ beforehand, i.e. by introducing $a(\tau;\alpha,\beta) = \bar{a}(\alpha,\beta) \cdot \exp \gamma(\alpha,\beta)\tau$ , where $\gamma(1/2\alpha_1,\pm\beta_1) = \gamma > 0$ and $\gamma(\alpha_1,0) = 0$. In particular, the matching yields: $\hat{P}^{(1)} = a$, $\hat{V}^{(1)} = \bar{\alpha}^2 a/i\alpha$, $\hat{U}^{(3/4)} = i\beta^2 a/\alpha(\gamma+i\alpha Y)$, $\hat{W}^{(3/4)} = -i\beta a/(\gamma+i\alpha Y)$, $\hat{P}^{(3/4)} = -(\gamma\bar{\alpha}^2/i\alpha) \cdot$ $\cdot\, Ya - 1/2\bar{\alpha}^2 Y^2 a + D_1$, where in this order of matching function $D_1(\tau)$ is not determined explicitly. From the conservation laws for the x and z components of the momentum and the continuity equations we obtain the equation for $\hat{V}^{(3/2)}$ in the form

$$\left(\frac{\partial}{\partial\tau} + i\alpha Y\right)\frac{\partial\hat{V}^{(3/2)}}{\partial Y} - i\alpha\hat{V}^{(3/2)} = -\bar{\alpha}^2\hat{P}^{(3/2)} + \hat{F}^{(3/2)} , \quad (4)$$

where

$$F^{(3/2)} = \frac{\partial^2}{\partial\xi^2}(U^{(3/4)})^2 + \frac{\partial^2}{\partial\xi\partial Y}(U^{(3/4)}V^{(1)}) + 2\frac{\partial^2}{\partial\xi\partial z}(U^{(3/4)}W^{(3/4)}) +$$

$$+ \frac{\partial^2}{\partial\xi\partial Y}(W^{(3/4)}V^{(1)}) + \frac{\partial^2}{\partial z^2}(W^{(3/4)})^2 .$$

Equation (4) can be easily solved taking into account the above mentioned dependence of the wave amplitudes on $\tau$. From the condition $(v_2^{\,0})_3^{\,i} = (v_3^{\,i})_2^{\,0}$ we can find the sought jump $A_+^{(5/4)} - A_-^{(5/4)}$. For a two-dimensional wave, this jump is equal to zero, and, in accordance with (3), its amplitude is constant. If the amplitudes of oblique waves are assumed to be identical and the phase difference $\arg\bar{a}(1/2\alpha_1,\beta_1) + \arg\bar{a}(1/2\alpha_1,-\beta_1) - \arg\bar{a}(\alpha_1,0)$ is assumed to be equal to $\pi/2$, then the pair of equations (3) degenerates to one equation, from which the instability increment is found:

$$\bar{\gamma} = (\frac{\pi}{8}\,\alpha_1^{\,4}\beta_1^{\,2}G_1|a(\alpha_1,0)|)^{1/4} , \qquad (5)$$

where $G_1$ is the value of G at $\bar{\alpha} = (1/4\alpha_1^{\,2} + \beta_2^{\,2})^{1/2}$.

Thus, a degenerate nonlinear three-wave instability appears in the inviscid boundary-layer type flow. In this case, the amplitudes of oblique waves increase exponentially and the amplitude of a two-dimensional wave is constant. This instability is due to the energy withdrawal by the waves from the mean flow in the CL and can be interpreted as a nonlinear Landau growing of resonantly coupled waves. Unlike the known effect of linear Landau growing (damping) of two-dimensional waves, which arises in the presence of the vorticity gradient of the mean flow in the CL /4, 5/, the energy withdrawal from the flow at a resonant nonlinear instability is due to the combinative deformation of the vortex lines in a three-dimensional CL. The vorticity perturbations induced by such a deformation yield a jump of the Reynolds stresses across the CL. The nonlinear Landau growing takes place also in a viscous flow, since the elimination of the singularity by viscosity and nonstationarity leads to the formation of similar wave profiles in the CL.

It can be shown that a noticeable distortion of the wave amplitude variation in the course of time is started when the oblique wave amplitudes reach the magnitude $\sim\varepsilon^{3/4}$. An analysis also shows that with the reverse action of the flow in the CL on the wave amplitudes taken into account, it is impossible to construct the solution of the boundary-value problem containing a strongly nonlinear but thin CL. This indicates the high efficiency of the resonant nonlinear instability as a mechanism of the breakdown of the laminar flow in the boundary layers.

REFERENCES

1. A.D.D.Craik. Wave Interactions and Fluid Flows. Cambridge Univ. Press, Cambridge, 1985.
2. A.Leonard. In: Turbulent Shear Flows. V.2. Springer, Berlin, 1980.
3. J.L.Robinson. J.Fluid Mech., 1974, 63, 4, 723-752.

4. V.P.Reutov. Sov.Phys.-Atm. Oceanic Phys., 1980, 16, 12.
5. V.P.Reutov. J.Appl.Mech.Tech.Phys., 1982, No.4
6. V.P.Reutov. Sov.Phys.-Atm. Oceanic Phys., 1982, 18, 4.
7. V.P.Reutov. J.Appl.Mech.Tech.Phys., 1987, No.5
8. A.H.Nayfeh. Introduction to Perturbation Techniques. John Wiley & Sons, New York, 1981.

# Stability of Quasi-Two-Dimensional Shear Flows and Jets in Rotating Fluids

*F.V. Dolzhansky and D.Yu. Manin*

Institute of Atmospheric Physics, USSR Academy of Sciences, Moscow, USSR

The stability problems are briefly discussed based on the recent results obtained by the authors.

Two-dimensional flows as solutions of hydrodynamic equations consititute a special class of great variety and importance. However, real flows are never strictly two-dimensional. Either they are destroyed by three-dimensional instability, or some mechanism is present maintaining the flow two-dimensional. Such a mechanism is commonly provided by a hard flat boundary (top and/or bottom). If this is the case, the inevitable boundary ("external") friction essentially modifies two-dimensional flows in many aspects. It is shown /1/ that taking the external friction into account, we are able to explain the experimental results on the stability of the Kolmogorov flow. A general stability theory for the flows dominated by the external friction is constructed in /2/. Formally, such flows are governed by the equation

$$\begin{gathered}\partial_t \Delta\Psi + \left[\Delta\Psi, \Psi\right] = -\lambda\Delta\Psi + \nu\Delta^2\Psi + f\ , \\ u = \partial_y \Psi\ , \quad v = -\partial_x \Psi\ , \quad \Delta = \partial_x{}^2 + \partial_y{}^2\ , \\ \left[F, G\right] = \partial_x F \partial_y G + \partial_y F \partial_x G\ ,\end{gathered} \tag{1}$$

where $\Psi$ is the stream function, $\lambda$ the external friction decrement, $\nu$ the internal viscosity and f the external forcing.

Quasi-two-dimensional flows are characterized by two dimensionless similarity parameters which are the usual Reynolds number $Re_\nu = UD/\nu$ and the Reynolds number in terms of the external friction $Re_\lambda = U/\lambda D$, where U and D are the characteristic velocity and length scale of the flow in question, respectively. DOLZHANSKY /2/ has demonstrated that marginal stability curves preserve their shape with the vanishing viscosity ($\nu = 0$), but do not so when $\lambda = 0$. Moreover, nothing depends on $Re_\nu$ when $Re_\nu$ is reasonably large, so that the viscosity term in (1) can be neglected while the external friction $\lambda$ can not. This was verified in /3/ using both original experimental data and that available from literature. Here the stability problems are discussed in brief.

Research Reports in Physics Nonlinear Waves 3
Editors: A.V. Gaponov-Grekhov · M.I. Rabinovich · J. Engelbrecht

From the estimation of the right-hand-side terms in (1) it can be seen that the $\nu$ term is negligible compared with the $\lambda$ term for flows with $D \gg \sqrt{\nu/\lambda}$. On the other hand, even the $\delta$-function-type forcing f results in a flow with $D \sim \sqrt{\nu/\lambda}$. Thus, the viscosity again appears to be negligible.

There are two principal examples of quasi-two-dimensional systems: 1) a thin fluid layer on a hard flat bottom, and 2) rotating fluid at small Ekman numbers. In the first case $\lambda = 2\nu/H^2$, H being the fluid depth, and we have $\sqrt{\nu/\lambda} = H$. Hence the obvious inequality $D \gg H$ required for the flow to be quasi-two-dimensional implies that we can neglect the $\nu$ term in (1). The full three-dimensional numerical analysis of the thin layer flow was accomplished in /4/ using the Galerkin method. From the obtained results it follows that Eq.(1) does adequately describe the actual flow with D several times greater than H.

For the second case we have $\lambda = \sqrt{\Omega\nu}/H$ ($\Omega$ being the angular velocity of the system) which is the reverse Ekman time scale. The natural length scale $\sqrt{\nu/\lambda}$ equals to $H^{1/2}\nu^{1/4}\Omega^{-1/4} = \delta_s$ which is the Stewartson $E^{1/4}$-layer width.

A special example of quasi-two-dimensional flows is provided by large-scale zonal winds in atmospheres of the Earth and planets, the external friction being due to the Ekman boundary layer. It is essential that $Re_\nu$ is of the order $10^6$ for zonal winds in the Earth's atmosphere, while $Re_\lambda$ does not exceed the value of 40. The fact is a striking evidence of the quasi-two-dimensional nature of large-scale atmospheric flows and it also proves that their laboratory simulation is valid. The general equation governing these flows was obtained in /5/. A special case of this equation describing potential vorticity was derived in /6/, assuming quasi-geostrophicity and weak two-dimensional compressibility. The Ekman friction having been taken into account, the obtained equation appears in the form (1) but includes an additional term $\beta\partial_x\Psi$ due to the $\beta$-effect, i.e. the variance of the Coriolis parameter along the transverse coordinate. In /7/ the stability of zonal flows is investigated and the $\beta$-effect is shown to be as significant as the external friction for the Earth's atmosphere. Hence it is important to simulate the $\beta$-effect in a laboratory model of zonal flows, which is easily achieved with a rotating fluid of variable depth.

The stability of a jet in a laboratory model like that was investigated in /8/, the jet being produced using the sink/source method. However, in order to correctly interpret the experimental data, it is necessary to analyze the structure of the sink/source flow and demonstrate that it is in fact essentially two-dimensional. We shall outline the analysis following /9/.

A common approach to a problem of the kind (cf./10/) is to assume the Rossby number $\varepsilon = U/D\Omega$ small as well as the Ekman number $E = \nu/\Omega H^2$. The Navier-Stokes equations are linearized with respect to $\varepsilon$. The smallness of E makes it possible to separate the Ekman layer problem from the internal region problem, the standard solution of the former providing boundary conditions for the latter. The internal flow equation are of the form /10/

$$E\partial_y^2 u - 2\partial_z \chi = 0 , \quad E\partial_y^4 \chi + 2\partial_z u = 0 ,$$

$$2(\chi(y,0)-\chi_B(y)) = -E^{1/2}(u(y,0)-u_B(y)) , \tag{2}$$

$$2(\chi(y,1)-\chi_T(y)) = -E^{1/2}(u(y,1)-u_T(y)) .$$

Here u is the zonal velocity component (along the x-axis), $\chi$ the transverse stream function, $\chi_{B,T}$ and $u_{B,T}$ the boundary values at bottom and top boundaries, all lengths are non-dimensionalized by H. The posed problem is not formulated along the x-axis, but a rotationally symmetric one is also valid.

System (2) is now solvable with the aid of Fourier transform in y coordinate. Fourier integrals representing the solution can be used to obtain asymptotic (in E) series. These series, however, converge poorly in the region $y,z \to 0$ of sharp variance. Hence we prefer to analyze the asymptotic behaviour of solutions immediately from the Fourier integral representation. The result is shown in Fig.1 for the simplest case of the single linear sink (source). There are four domains in the (y,z)-plane with distinct asymptotic behaviour of the zonal velocity u (the domains are different for the stream function $\chi$). Asymptotics are as follows: in the domain I we have $u \sim$ const, in the II - $u \sim y/E^{1/4}$, in the III - $u \sim y/E^{1/4} \cdot z^{2/3}$, in the IV - $u \simeq E^{1/2}/y$. We do not present the results for the transverse flow since it can be seen from (2) that the sink/source problem is equivalent to the moving boundary problem usually considered (boundary conditions posed on $\chi$ or u, respectively). So the streamlines of the transverse flow can be readily outlined for the experimental set-up (Fig.2), and we do not need any precise fomula. The principal fact essential for the following is that $E^{1/3}$ Stewartson layer has no effect on the internal flow , except for the asymptotically small domain $z \leq \leq E^{1/8}$, $|y-l| \leq E^{3/8}$ (l is the coordinate of the sink/source). Beyond the domain the flow is two-dimensional, velocity being independent of z.

Let us now proceed to the experimental results /8/. The jet was produced in a cylindrical container mounted on a rotating table. Floating particles provided the flow visualization. The "topographic" β-effect was due to the sloping bottom and

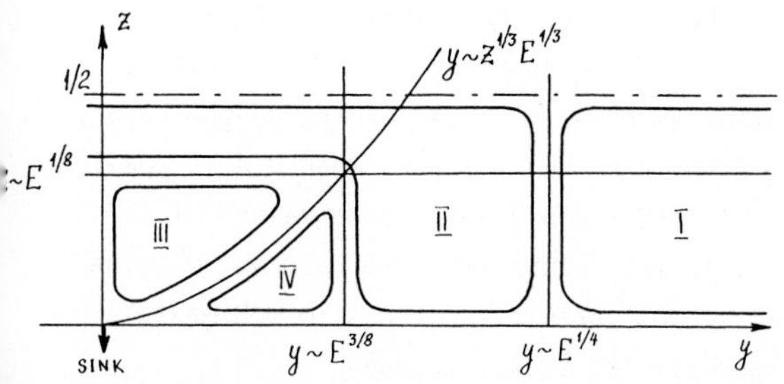


Figure 1. Domain of distinct asymptotic behaviour of the zonal velocity u.

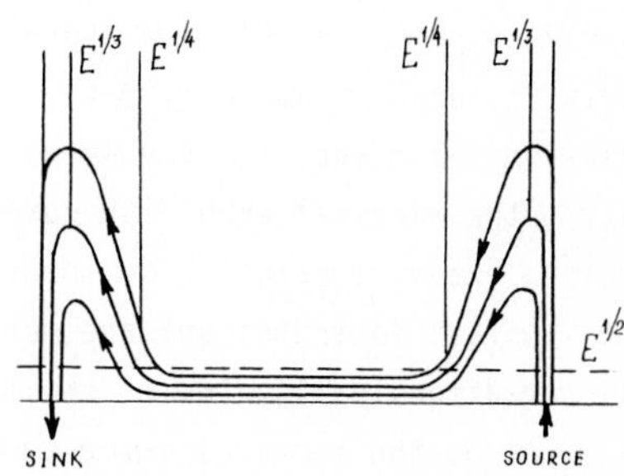


Figure 2. The structure of transverse flow for the experimental set-up.

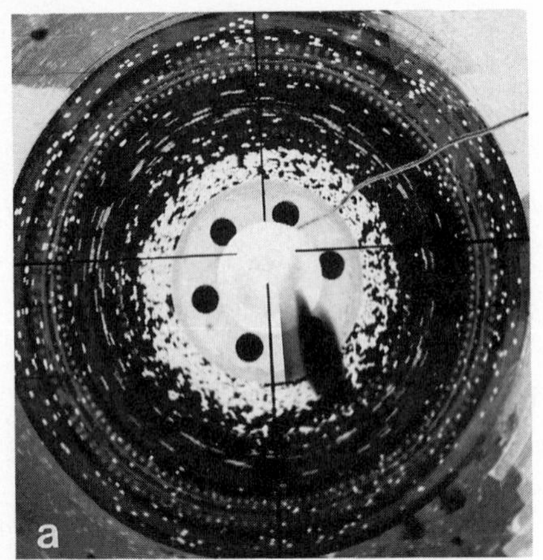


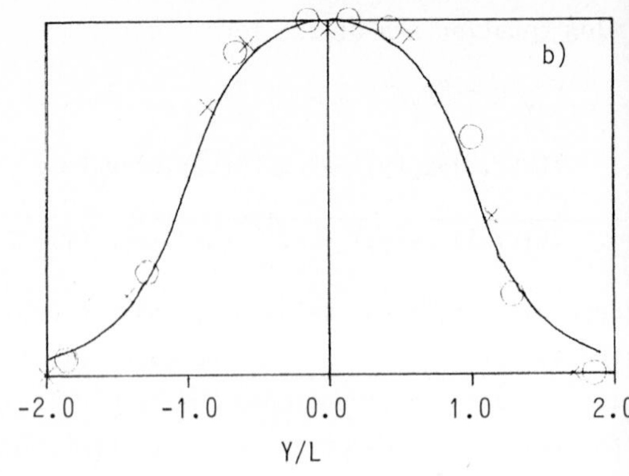


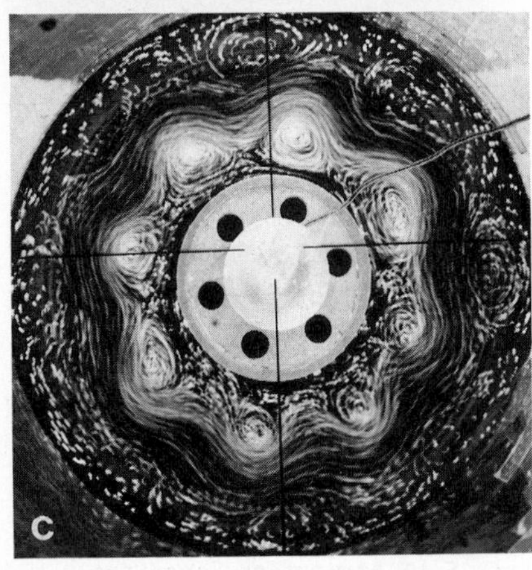


Figure 3. a) a stable jet flow; b) experimental (points) and theoretical (line) velocity profile; c) a supercritical flow.

parabolic free surface. A laminar jet with a Π-shaped velocity profile originated at small sink/source intensity (Fig.3a). The profile evaluated from a long-exposure photograph is shown in Fig.3b along with the theoretical one.

With the forcing increasing, the jet becomes unstable, and a regular eddy pattern occurs (Fig.3c). The experimental results can be represented by points in the $(D/d, Re_\lambda)$-plane. Here $D = \max(u(y))/\max|u'(y)|$ is the jet width, $d = \lambda/\beta$ a characteristic length scale /7/, $\beta$ the Coriolis parameter gradient, $\lambda$ the external friction decrement, $Re_\lambda$ the Reynolds number in terms of $\lambda$. According to the theory /2,7/, the marginal stability curve is of the form $Re_\lambda$ = const, when D/d is small and the external friction dominates. On the other hand, with large values of D/d, the $\beta$-effect dominates and the marginal stability occurs when $Re_\lambda = D/dm$, m being the non-dimensional value of the maximal second derivative of $u(y)$: $m = D^2\max(u'')/\max(u)$. So the marginal curve can be approximated by the formula

$$Re_\lambda^* \approx \sqrt{C^2+(D/dm)^2} \; . \qquad (3)$$

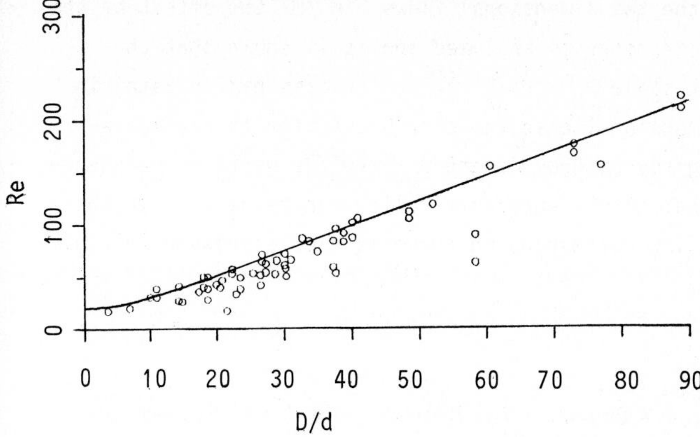


Figure 4. Observed stable flow regimes (points) with the stability curve (Eq.3).

Note that in order to plot the experimental data we only need from the theory the value of D.

The data processed in that way are shown in Fig.4. It can be seen that most of the data lies in the region where D/d is considerably large, which implies that the β-effect dominates. The points corresponding to stable but nearly marginal state make it possible to draw the curve (3) with C = 20, m = 0.42. The value of m differs considerably from the theoretical one $m_{th}$ = 2. The inconsistency is possible only due to the fact that u" is highly sensitive to small changes of u(y). Nevertheless, it follows from the results above that the theory correctly describes the velocity profile up to the first derivative.

The analysis of supercritical flow also confirms the conclusion. The flow is characterized by the number of vortices, n, from which nondimensional wavenumber is evaluated $\alpha = nD/R$, where R is the radius of the jet ($R \gg D$). The number of vortices, as well as $\alpha$, decreases as the forcing increases (this is very common for the flows of this kind). In fact, the value of $\alpha$ proves to be dependent almost uniquely on the supercriticality Re/Re* (Fig.5). The enlargement of vortices itself results from the well-known "inverse cascade" effect, i.e. the energy transfer to

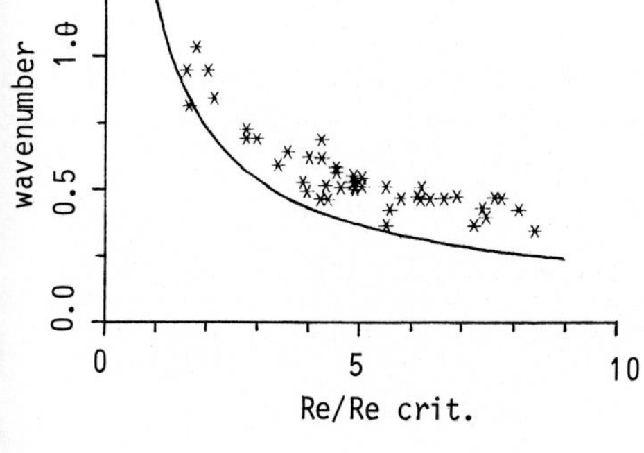


Figure 5. Dimensionless wavenumber vs. supercriticality: experiment (points) and theory (line).

small wavenumbers intrinsic to the two-dimensional flows. In /7/ the effect of the external friction on the energy transfer is analyzed and it is shown that the dissipation occurs at the length scale $D_\lambda \sim \sqrt{\varepsilon/\lambda^3}$ ($\varepsilon$ is the dissipation rate) in the same way as in the three-dimensional case the energy transfer to high wavenumbers is cut off at the Kolmogorov-Obukhov length $D_\nu \sim \sqrt[4]{\nu^3/\varepsilon}$. Next, a dependence of the minimum possible wavenumber on the supercriticality can be derived in the form /11/ $\alpha_{min} \sim (Re/Re^*)^{-3/4}$. The corresponding curve is also displayed in Fig.5.

REFERENCES


1. E.B.Gledzer, F.V.Dolzhansky, A.M.Obukhov. The Hydrodynamic-Type Systems and Their Applications. Moscow, Nauka, 1981, 366 p. (in Russian).
2. F.V.Dolzhansky. On the effect of external friction on the stability of parallel flows of the homogeneous incompressible fluid. Izvestiya AN SSSR, FAO, 1987, 23, 4, 348-355 (in Russian).
3. V.A.Krymov. Stability and supercritical regimes of quasi two-dimensional shear flows in the presence of the external friction (experiment). Izvestiya AN SSSR, MZhG, 1989, 2, 12-18 (in Russian).
4. V.A.Dovzhenko, A.M.Obukhov, V.M.Ponomarev. On the eddy generation in an azimuthially symmetric flow. Izvestiya AN SSSR, MZhG, 1981, 4, 27-36 (in Russian).
5. A.M.Obukhov. On the dynamics of stratified fluid. Sov.Phys.-Dokl., 1962, 145, 6, 1239-1242.
6. F.V.Dolzhansky. On the effect of Ekman layer on the stability of planetary waves Izvestiya AN SSSR, FAO, 1985, 21, 4, 383-390 (in Russian).
7. D.Yu.Manin. Stability of quasi two-dimensional shear flows dominated by $\beta$-effect and external friction. Izvestiya AN SSSR, FAO, 1989 (in press, in Russian).
8. D.Yu.Manin, Yu.L.Chernous'ko. An experimental investigation of the stability of a quasi two-dimensional jet generated in a rotating fluid by sinks and sources. Izvestiya AN SSSR, FAO, 1989 (in press, in Russian).
9. D.Yu.Manin. On the structure of the flow generated in a rotating fluid by point and linear sinks/sources. Izvestiya AN SSSR, FAO, 1989 (in press, in Russian).
10. H.Greenspan. Theory of Rotating Fluids. Cambridge University Press, 1969.
11. D.Yu.Manin. On the characteristic length scale of eddies in well-developed quasi two-dimensional fluid flows. Izvestiya AN SSSR, FAO, 1989 (in press, in Russian).

# On One Solution of the Nonlinear Problem of Electrodynamics

*A.G. Eremeev and V.E. Semenov*

Institute of Applied Physics, USSR Academy of Sciences,
46 Ulyanov Str., 603600 Gorky, USSR

An example of using the "rectangular" approximation of the nonlinearity characteristic in the nonlinear electrodynamic problem is analyzed. It is shown that this approximation in some cases permits one to define the general parameters of the produced structures, even if the equations cannot be solved analytically.

When powerful microwave radiation propagates in a nonlinear dissipative medium, structures characterized with essential reflection of incident electromagnetic waves can be formed. In this case, of practical importance are the general characteristics of the produced structures (reflection and absorption coefficients, velocity of propagation, etc.) and their dependece on incident radiation power. Within the framework of the theory, these characteristics can usually be determined only if the corresponding nonlinear problem of electrodynamics has been solved completely. There are no standard methods to solve exactly these problems, since the equations of nonlinear electrodynamics do not have the first integrals in the general case of absorbing medium. Therefore, it is important to develop various approximate methods of investigating such problems. For example, in a low-absorption medium it is possible to use averaged equations for the parameters of quasiperiodic solutions. In /1,2/ it is shown that in the last case there is the first integral like an adiabatic invariant permitting one to reduce the solution of the problem with arbitrary nonlinearity to quadratures. At strong nonlinearity, when the dielectric permittivity perturbation $\delta\varepsilon$ of the medium is a fast-growing function of the high-frequency electric field amplitude A, the approximation using the "rectangular" characteristic of nonlinearity seems to be very attractive: $\delta\varepsilon = 0$ if $A < A_c$, $\delta\varepsilon \to \infty$ if $A > A_c$. Such an approximation makes it possible to reduce the initial direct problem of electrodynamics to the inverse problem that is much more simple, i.e. to find the spatial distribution of $\delta\varepsilon(\vec{r})$, which ensures the equality $A = A_c$ in the region where $\delta\varepsilon \neq 0$. Specifically, as is shown in /3/, this problem can be reduced to a linear partial differential equation with a constant coefficient and, therefore, the non-one-dimensional solution can be constructed. Particularly, this was a case of plasma (the dielectric permittivity $\varepsilon = 1 - n - i\nu n$), where the electron density n depends locally on electric field amplitude. In /4/ this method is used to calculate the stationary structure of the microwave discharge arising in the crossed beams of electromagnetic waves. The results of this method were verified numerically and analytically by investigating the problem in the case of "nonrectangular" characteristivs of nonlinearity /2/, as well as experimentally /5/. The method of "rectangular" characteristic of nonlinearity and its applications are reviewed in ample detail in /6/.

Below it will be shown that using the approximation of "rectangular" characteristic of nonlinearity it is possible, in some cases, to define the general parameters of the produced structures, even if the equations of the inverse problem cannot be solved analytically. We give an example of a solution of such a problem, which arises in calculating the dynamics of a nonequilibrium high-pressure microwave discharge in the pre-breakdown electromagnetic wave field.

The plasma produced in this case is characterized with strong absorption of electromagnetic waves. Therefore, in the equation for the complex amplitude of the HF electric field $E \cdot \exp(i\omega t)$ we can neglect the perturbation of the real part of dielectric permittivity:

$$\frac{\partial^2 E}{\partial x^2} + (1 - i\sigma)E = 0 . \tag{1}$$

The nonlinearity is nonlocal in this problem, since the dimensionless conductivity $\sigma$ of the plasma depends only on the field amplitude $A = |E|$ but also on the gas temperature T: $\sigma = \sigma(A,T)$; the isobaric heating of the gas is due to the ohmic absorption of the field energy in the plasma. With thermal conductivity of the gas and gas motion due to the thermal spread taken into account, the equations in dimensionless variables are the following:

$$\rho(\frac{\partial T}{\partial t} + v\frac{\partial T}{\partial x}) = \sigma A^2 + \frac{\partial^2 T}{\partial x^2} . \tag{2}$$

The gas density $\rho$ and the gas velocity v satisfy the continuity equation:

$$\frac{\partial \rho}{\partial t} + \frac{\partial}{\partial x}(\rho v) = 0 . \tag{3}$$

Thus, using the isobaricity conditions ($\rho T = 1$), the equation for gas velocity can be written as:

$$\frac{\partial v}{\partial x} = \sigma A^2 + \frac{\partial^2 T}{\partial x^2} . \tag{4}$$

The thermal conductivity of gas leads to the heating of the gas layers neighbouring the discharge thereby promoting the ionization of these layers and moving the discharge front opposite to the incident electromagnetic wave, since the electric field amplitude behind the plasma layer is weakened. Thus, there occurs the well-known phenomenon - the stationary propagation of the ionization wave opposite to (towards) the incident electromagnetic radiation.

Assuming that the gas heating in the discharge is strong enough ($T \gg T_0$), we shall neglect the unperturbed gas temperature $T_0$ putting $T \to 0$ at $x \to -\infty$. In a coordinate frame bound to the front of the stationary ionization wave we have: $\rho v = u = \text{const}$ , because of (3), and Eq. (2) can be integrated once:

$$S = S_0 + \frac{dT}{dx} - uT \; . \tag{5}$$

Here $S = (E\partial E^*/\partial x - E^*\partial E/\partial x)/(2i)$ is the electromagnetic energy flux density, $S_0 = A_0^2(1-R^2)$ is its value at $x \to -\infty$, from where a wave of amplitude $A_0$ is incident, R is the unknown reflection coefficient of the incident electromagnetic wave from the discharge plasma.

To investigate the basic regularities of stationary wave ionization, it is convenient to pass over from Eq. (1) to two real equations with respect to the field amplitude A and the energy flux density S:

$$\frac{dS}{dx} = - A^2 \; , \tag{6}$$

$$\frac{d^2A}{dx^2} - \frac{S^2}{A^3} + A = 0 \; . \tag{7}$$

The solution of Eqs. (5)-(7), which corresponds to the ionization wave, should satisfy the following boundary conditions: at $x \to -\infty$ the temperature turns to zero, $S \to S_0$ and the amplitude has periodic oscillations from the minimum $A_0(1-R)$ to the maximum value $A_0(1+R)$; at $x \to +\infty$ the gas temperature reaches the finite value $T_m$, $A \to A_m$, $S \to S_m = A_m^2$. In general, the problem appears to be re-determined, thus enabling one to find the unknown values of u, R, $A_m$ and $T_m$ at a given value of $A_0$.

Using the rectangular approximation of the nonlinear characteristic, we should put $\sigma = 0$, $AT < A_c$ at $x < 0$, $\sigma > 0$, $AT = A_c$ at $x > 0$. On the leading edge of the ionization wave (at point $x = 0$) we should demand the continuity of the values of A, T, $\partial A/\partial x$ and $\partial T/\partial x$ (or S). In the region $x > 0$ this approximation reduces system (5)-(7) to one equation of the second order:

$$\frac{d^2T}{dx^2} - \frac{2}{T}\left(\frac{dT}{dx}\right)^2 - T + \frac{T^5}{A_c^4}\left(S_0 - uT + \frac{dT}{dx}\right)^2 = 0 \; . \tag{8}$$

It is hardly possible to solve this equation. Nevertheless, we can obtain relations between parameters u, R, $A_m$ and $T_m$. Indeed, according to the condition $AT = A_c$ at $x > 0$ we have: $A_mT_m = A_c$. Therefore, from (5) it follows that

$$A_0^2(1-R^2) = uT_m + A_c^2/T_m^2 \; . \tag{9}$$

On the other hand, analyzing the spatial distribution of function F:

$$F \equiv \frac{d^2}{dx^2}(AT) = \frac{S^2}{A^3}T - AT + uA\frac{dT}{dx} - \sigma A^3 + 2\frac{dA}{dx}\frac{dT}{dx} \; , \tag{10}$$

we can draw the following conclusions. Since $F = F_+ \equiv 0$ at $x > 0$, $F = F_- \leqq 0$ at $x \to 0$ and $x < 0$, and $F_- - F_+ = \sigma(0)A^3(0) \geqq 0$, we have: $\sigma(0) = 0$ and $F_- = 0$. Using the solutions of Eqs. (5)-(7) in the region $x < 0$, where $\sigma = 0$:

$$T = T(0)\exp(ux) \; ,$$

$$A^2 = A_0^2(1+R^2) + 2RA_0^2 \sin(2x+\phi) \; ,$$

where $\phi$ is the integration constant, it is easy to show that the conditions $F_- = 0$ and $\frac{d}{dx}(AT)\big|_{x=0} = 0$ will be satisfied simultaneously only if

$$R = u/(1 - \sqrt{1+u^2}) \ . \tag{11}$$

A qualitative analysis of the solution of Eq. (8) in the phase plane of the variables T, $T_x = dT/dx$ (see Fig.1) indicates that at small values of u ($u < u_c \sim 1$) the ionization wave is characterized by two free parameters: amplitude $A_0$ and velocity u (the range of permissible values is shown in Fig.2). Therefore, expressions (9) and (11) completely determine the sought general characteristics of a stationary ionization wave at given $A_0$ and u. The corresponding phase trajectory must come from the point $T = T(0) = A_c(1+u^2)^{1/4}/\sqrt{S_0}$, $T_x = uT(0)$ to an equilibrium state of a "stable node" type, the position of which is determined by the lesser root $T_m$ of Eq. (9). The arbitrariness in choosing the values of u and $S_0$ is due to the fact that a continuous set of trajectories enters the "node".

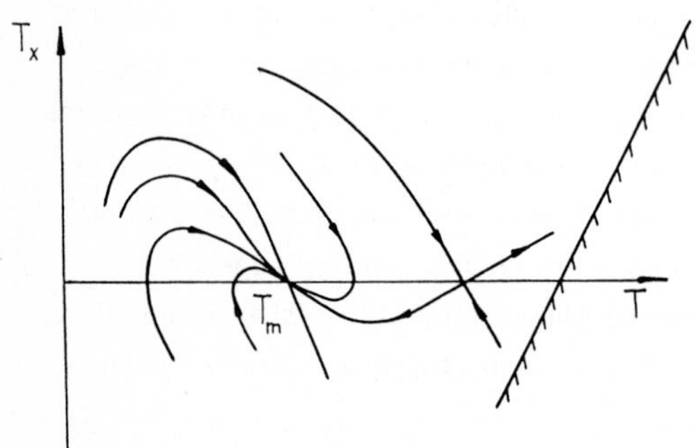


Figure 1. Phase portrait of Eq. (8) in the range of parameter values $3(u/2)^{2/3} < S_0A_c^{-2/3} < (u+w)w^{-1/3}$ where $w = (u+\sqrt{u^2+4})/4$. The hatched region corresponds to the negative values of the energy flux density of the electromagnetic field S.

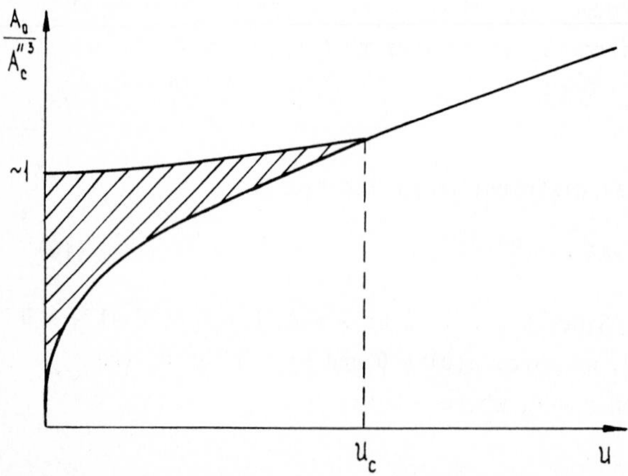


Figure 2. The region of the existence of solutions in the plane $u, A_0$.

In the two-parametric totality of ionization waves, a particular role is played by those which have the largest velocity for a given amplitude $A_0$ of the incident electromagnetic wave. Such waves are most likely to occur in general. Let us consider the dynamics of smooth monotonic perturbations of gas temperature behind the front of the ionization wave. Neglecting the thermal conductivity influence and the electromagnetic wave reflection (i.e. putting $S = A^2$) and assuming, as before, $AT = A_c$, we obtain, in view of (2)-(4) and (6), the following heat transfer equation for a simple wave:

$$\frac{\partial T}{\partial t} + (uT_m + S_m - 3A_c^2/T_m^2)\frac{\partial T}{\partial x} = 0 , \qquad (12)$$

where u, $T_m$ and $S_m$ are the parameters of the stationary ionization wave. It is easy to see that the thermal perturbations behind the front of the ionization wave, moving at a velocity different from the maximum one, overtake the ionization wave (since $uT_m < 2A_c^2/T_m^2$) and "push" it. Thus, we can construct a qualitative picture of the initial state evolution to a stationary wave with the maximum velocity $u_m$, the amplitude dependency of which is given by

$$A_0^2 = \frac{3}{2} A_c^{2/3}(\frac{u_m}{2})^{2/3}(\sqrt{1+u^2} + 1) . \qquad (13)$$

The reflection coefficient for this wave is, as before, determined by formula (11), and expression (9) can be reduced essentially:

$$T_m = (2A_c^2/u_m)^{1/3} . \qquad (14)$$

This means that for this ionization wave the absorption coefficient of the incident electromagnetic radiation is exactly twice the transmission coefficient, i.e. $S_m = S_0/3$.

Unlike the "slow" waves, the ionization wave is completely determined by amplitude $A_0$ at $u > u_c$. The phase trajectory represents a separatrix of a "saddle", the position of which depends on the larger root of Eq. (9); this predetermines the unambiguous relationship between u and $A_0$ (see Fig.2). To calculate the dependence $u(A_0)$ and to define exactly the value of $u_c$, Eq. (8) should be solved numerically. Note, however, that for a "fast" ionization wave with $u > u_c$ we cannot construct a qualitative picture of evolution. Therefore, the realization of such a wave is doubtful.

## REFERENCES

1. E.I.Yakubovich. On general properties of averaged polarizabilities in the nonlinear quasioptics. Sov.Phys.-JETP, 1969, 56, 2, 676-682.

2. V.B.Gil'denburg, V.E.Semenov. The stationary structure of a nonequilibrium high-frequency discharge in electromagnetic wave fields. Nonlinear Waves. Structures and Bifurcations. Nauka, Moscow, 1987, p.376-383.

3. V.B.Gil'denburg. Nonequilibrium high-frequency discharge in electromagnetic wave fields. Nonlinear Waves. Propagation and Interaction. Nauka, Moscow, 1981, p.87-96.

4. V.E.Semenov. The multilayer structure of discharge in the self-consistent field of two quasioptical beams of electromagnetic waves. Fizika Plazmy, 1984, 10, 3, 562-567.

5. A.L.Vikharev, V.B.Gil'denburg, A.O.Ivanov, A.N.Stepanov. Microwave discharge in crossed beams of electromagnetic waves. Fizika Plazmy, 1984, 10, 1, 165-172.

6. A.L.Vikharev, V.B.Gil'denburg, A.V.Kim et al. Electrodynamics of a nonequilibrium high-frequency discharge in wave fields. High-frequency discharge in wave fields. Inst.Appl.Phys., USSR Acad.Sci., Gorky, 1988, p.41-135.

# Part II

# Quantum Physics, Physics of the Solid State

# Phenomenological Quantum Electrodynamics of Active Media and Superradiance

*V.V. Kocharovsky and Vl.V. Kocharovsky*

Institute of Applied Physics, USSR Academy of Sciences,
46 Ulyanov Str., 603600 Gorky, USSR

In this paper phenomenological quantum electrodynamics (PQED) is developed for collective coherent processes in macroscopic samples of active media. Superradiance (SR) in the inverted two-level medium is considered as the example. The remarkable phenomenon of macroscopic manifestation of quantum fluctuations observed in superradiance experiments is analyzed.

## 1. INTRODUCTION: MACROSCOPIC QUANTUM FLUCTUATIONS

The present paper is devoted to the theory of macroscopic quantum fluctuations. The question is about a macroscopic system which has been driven to an unstable equilibrium state. It is the instability process initiated by internal microscopic fluctuations that leads to a subsequent decay of the unstable state and an amplification of quantum fluctuations up to the macroscopic level. The ensemble of such experiments starting from physically identical initial conditions displays large fluctuations. The variances of observables are of the same orders of magnitude as their mean values. The principal feature is the impossibility to predict the result of an individual experiment.

There are a lot of such processes. First of all, nonequilibrium phase transitions in lasers, ferromagnets, ferroelectrics, etc. In this paper we appeal to the analogous example from electrodynamics, namely the optical SR in a two-level medium /1/.

Macroscopic quantum effects result from the coherent addition of microscopic interactions in a multi-body system of electrons, photons, ions, etc. Such is the origin of a macroscopic current of Cooper electron pairs in superconductors, for example. The effects of quantum behaviour of macroscopic degrees of freedom from the particular class. It involves the zero-point motion of the crystal as a whole and the zero-point current of the circuit, for example. There is no doubt theoretically that the laws of Quantum Mechanics are the same for micro- and macro-degrees of freedom /2/. The direct experiments on macroscopic quantum tunnelling involving the Josephson junction confirm this assertion /3/.

The choice of unstable macro-systems for investigating macroscopic quantum phenomena is not accidental. Quantum fluctuations in stable systems exist but are tiny and nonobservable. Moreover, the evolution of stable systems is usually close to the evolution of linear harmonic oscillators, for which quantum dynamics of mean variables as well as of its higher moments coincides with the classical one /4/.



Research Reports in Physics Nonlinear Waves 3
Editors: A.V. Gaponov-Grekhov · M.I. Rabinovich · J. Engelbrecht

It is natural to describe the quantum fluctuations of unstably systems by direct quantization of macroscopic collective variables proceeding from its classical equations. Deriving quantum properties of every macroscopic many-body system in a framework of macroscopic quantum electrodynamics (QED) of interacting vacuum modes and particles is a complicated problem. As a rule, the solutions require simplifying assumptions which may be obstacles afterwards for applications. In this paper we illustrate the approach of PQED to the theory of processes pointed out on the SR example. It is essential that well known conceptions of classical electrodynamics (CED) of continuous media are used in PQED naturally /5,6/.

## 2. CLASSICAL ELECTRODYNAMICS OF SUPERRADIANCE AND DISSIPATIVE INSTABILITY

Let us consider an active volume V filled with $\overline{N}V$ two-level molecules, $\overline{N}V=\int_V N(\vec{r})d^2\vec{r}$, inverted by a short pump pulse at the initial moment t = 0. The stored energy $\hbar\omega_0\overline{N}V$ will be emitted spontaneously on the molecular transition with optical frequency $\omega_0$, dipole moment d and homogeneous broadening $T_2^{-1}$. For molecular concentration, N being large enough, the powerful coherent pulse is radiated within short duration $\tau$ and delay $t_d$: $\tau \ll t_d \ll T_2 \lesssim T_1$ (Fig.1). Such a SR differs substantially from a noncoherent spontaneous emission of isolated molecules as well as from a quasistationary superluminescence, i.e. an amplified spontaneous emission. In a point sample, thw length of which is $L \ll \lambda = 2\pi c/\omega_0$, the SR increment ($\tau^{-1} \sim \overline{N}V/T_1$) and power ($Q_{SR} \sim Q_{spont}\cdot\overline{N}V$) are $\overline{N}V \gg 1$ times greater than that of the noncoherent spontaneous emission: $T_1^{-1} = 4d^2\omega_0^3/3\hbar c^3$ and $Q_{spont} = \overline{N}V\hbar\omega_0/T_1$. SR was proposed by DICKE /7/ in 1954 and has been observed in optics in 1973 /8/.

One can say that SR arises due to the build-up correlation between moelcular dipoles which tend to radiate coherently as united macroscopic dipole. However, to comprehend the nature of collective molecular behaviour it is necessary to analyze mechanisms

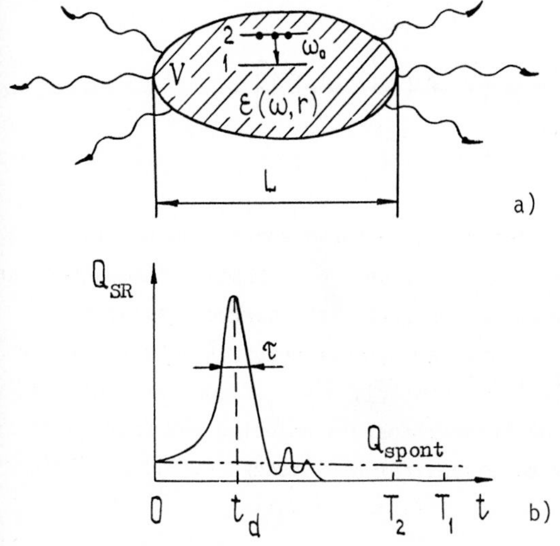


Figure 1. Superradiance of the active sample: (a) geometry, (b) power oscillogram $Q_{SR}(t)$.

of instability and quantum fluctuations of collective degrees of field and polarization freedoms in the active medium. It is the way of macroscopic CED and QED /1/.

CED of SR starts with the Maxwell equations for electric $\vec{E}$ and magnetic $\vec{B}$ fields in the medium

$$\mathrm{rot}\,\vec{E} = -c^{-1}\partial\vec{B}/\partial t\ , \quad \mathrm{rot}\,\vec{B} = c^{-1}\partial(\vec{E}+4\pi\vec{P})/\partial t + 4\pi\sigma c^{-1}\vec{E}\ , \tag{1}$$

and material equations for average polarization $\vec{P}$ per unit volume. The ohmic conductivity $\sigma$ of the "background" medium is taken into account in (1). The polarization of two-level medium is described by the oscillator equation /9/:

$$\partial^2\vec{P}/\partial t^2 + 2T_2^{-1}\partial\vec{P}/\partial t + (\omega_0^2+T_2^{-2})\vec{P} = \omega_c^2\vec{E}/4\pi\ ; \quad \omega_c^2 = -8\pi d^2\Delta N\omega_0/\hbar\ . \tag{2}$$

"The cooperative frequency" $\omega_c$ depends on population difference $N = N_2 - N_1$ (per unit volume) governed by the action of field $\vec{E}$ on current $\partial\vec{P}/\partial t$:

$$\partial\Delta N/\partial t = 2\hbar^{-1}\omega_0^{-1}\vec{E}\partial\vec{P}/\partial t\ . \tag{3}$$

The properties of plane waves $\vec{E} = (1/2)\vec{E}\exp(-i\omega t+i\vec{k}\vec{r}) + \mathrm{c.c.}$ are determinated by the permittivity $\varepsilon(\omega,\vec{r}) \approx 1 + i4\pi\sigma/\omega - \omega_c^2(\vec{r})/2\omega_0(\omega-\omega_0+i/T_2)$, according to (1) and (2) in the resonance approximation $\omega \approx \omega_0$. Such an approach fixes the dispersion $\omega(\vec{k}) = \omega' + i\omega''$ of transverse *normal waves*, i.e. photons in the medium: $\omega^2\varepsilon = c^2k^2$. They are known as polaritons in solid state physics and are the real collective excitations which determinate the local evolution of field and polarization in the medium. In the two-level model the polariton spectrum consists of two branches /10/

$$\omega_{e,p} = \omega_0 - iT_2^{-1} + \frac{1}{2}\left[ck-\omega_0+i(T_2^{-1}-2\pi\sigma)\right]\{1\pm(1+\omega_c^2/\left[ck-\omega_0+i(T_2^{-1}-2\pi\sigma)\right]^2)^{1/2}\}, \tag{4}$$

known as electromagnetic and polarization waves. Their energies in the inverted medium ($\omega_c^2 < 0$, $\Delta N > 0$) have opposite signs: $w_p \leq 0 \leq w_e$. The energy density and the power loss of linear waves are deduced from the energy balance law of the system (1) and (2) in which $N = \mathrm{const}$ /10/:

$$\frac{\partial w}{\partial t} = -Q - \frac{c}{4\pi}\,\mathrm{div}\,\vec{E},\vec{B}\ ; \quad w = \frac{\vec{E}^2+\vec{B}^2}{8\pi} + \frac{2\pi}{\omega_c^2}\left[\left(\frac{\partial\vec{P}}{\partial t}\right)^2 + \left(\omega_0^2+\frac{1}{T_2^2}\right)\vec{P}^2\right],$$

$$Q = \frac{8\pi}{\omega_c^2T_2}\left(\frac{\partial\vec{P}}{\partial t}\right)^2 + \sigma\vec{E}^2\ . \tag{5}$$

The perpetual problem, as usual, is to distinguish between energy and the loss in CED of dispersive media /6/. Using Eqs.(1) and (2) and $\Delta N = \mathrm{const}$ in the Lagrangian form with the Rayleigh dissipative function, we can verify the correctness of the division (5) in the two-level model. Let us consider a homogeneous ($\vec{k} = \mathrm{Re}\,\vec{k}$) linear polarized wave $\vec{P} = \vec{P}_\perp(t)\cdot\sin(\vec{k}\vec{r})||\vec{E} = c^{-1}k^{-2}\,\vec{k},\vec{B}\,\sin(\vec{k}\vec{r})$, $\vec{B} = \vec{B}_\perp(t)\cos(\vec{k}\vec{r})$. That is enough if a space flow of the energy is not interesting. Introducing the Lagrangian (L) and Rayleigh ($\Phi$) functions, we obtain equations for amplitudes $P_\perp$, $B_\perp$:

$$\frac{d}{dt}\frac{\partial L}{\partial \dot{P}_\perp} - \frac{L}{\partial P_\perp} = -\frac{\partial \Phi}{\partial \dot{P}_\perp}, \quad \frac{d}{dt}\frac{\partial L}{\partial \dot{B}_\perp} - \frac{\partial L}{\partial B_\perp} = -\frac{\partial \Phi}{\partial \dot{B}_\perp}; \tag{6}$$

$$L = \frac{\dot{B}_\perp^2/c^2k^2 - B_\perp^2}{8\pi} + \frac{2\pi}{\omega_c^2}\left[\dot{P}_\perp^2 - (\omega_0^2 + \frac{1}{T_2^2})P_\perp^2\right] + \frac{P_\perp \dot{B}_\perp}{ck}, \quad \Phi = \frac{\sigma}{2c^2k^2}\dot{B}_\perp^2 + \frac{4\pi}{\omega_c^2 T_2}\dot{P}_\perp^2 \tag{7}$$

(the dot denotes $\partial/\partial t$). They lead to the well known relations $H = \dot{P}_\perp \partial L/\partial \dot{P}_\perp + \dot{B}_\perp \partial L/\partial \dot{B}_\perp - L$ and $dH/dt = -2\Phi$ which prove the "truthfulness" of energy and loss formulae (5): $w = H$, $Q = 2\Phi$ .

Only one of the two normal waves is unstable. For strong dissipation, when $2\pi\sigma > T_2^{-1}$, it is the polarization wave the negative energy of which becomes more negative due to the field dissipation ($Q_p > 0$). This means that the wave amplitude increases with the rate /10/ $\omega_p'' = -Q_p/2w_p > 0$, and the *dissipative instability* occurs. When $2\pi\sigma < T_2^{-1}$, the maser instability of the electromagnetic wave is realized: $\omega_e'' = -Q_e/2w_e > 0$, because of its positive energy and negative loss ($Q_e < 0$). In the limit of zero relaxation and dissipation ($Q = 0$), the normal waves, possessed of growth rates $\omega_{e,p}'' \neq 0$, are waves of zero energy: $w = -Q/2\omega'' = 0$. They grow due to energy exchange between partial oscillations of polarization and field. Such a dynamical "swinging" of two subsystems which have opposite signs of energies is a limiting case of a dissipative instability. In both cases an instability arises due to an interaction between dynamical and/or dissipative systems which have opposite signs of energies.

CED treatment of SR problem intends to solve the Maxwell-Bloch equations (1)-(3) subject to the appropriate boundary and fluctuating initial conditions in the active sample (Fig.1) /1,11/. At the linear stage, $\Delta N = N$, the SR problem can be reduced to $(\varepsilon\omega^2/c^2 - \vec{\nabla}\times\vec{\nabla}\times)\vec{E}(\omega,\vec{r}) = \vec{C}(\vec{r})$ for the temporal Laplace-transformated complex amplitude of the field. This equation without RHS ($\vec{C} = 0$) has a set of intrinsic solutions $\vec{E}_m(\vec{r})$ which satisfy the outgoing emission condition. These are the *natural modes* with discrete spectrum $\omega_m = \omega_m' + i\omega_m''$. An initial condition $\vec{C} \neq 0$ fixes the solution inside a sample in the form of the sum of natural modes $\sum_m a_m(\omega)\vec{E}_m(\vec{r})$ with definite amplitudes $a_m(\omega)$. The field outside the sample is determined by their emission from the sample surface $S_0$. In the presence of incident external radiation, the solution includes waves with continuous spectrum transmitted through a sample.

There are two types of natural modes: positive-energy (m,e)-modes and negative-energy (m,p)-modes /1,10,11/. Their properties are analogous to those of electromagnetic and polarization waves in an unbounded medium. The dissipation due to emission through the sample boundaries is added to $\sigma$ only. For a quasi-one-dimensional cylindrical sample of finite length L and of small cross area $S \lesssim L$ this dissipation includes diffraction through sides $\sigma_{dif} \sim c\lambda/6\pi S$ and radiation through ends of sample $\sigma_{rad} = (c/4\pi L)\ln R^{-1}$, determined by the reflection coefficient R.

The instability dynamics of natural modes and normal waves as well as resulting SR have been analyzed for some models: unidirectional SR in the active ring (Fig.2), SR of counterpropagating waves in the one-dimensional layer, SR in the sphere, etc.

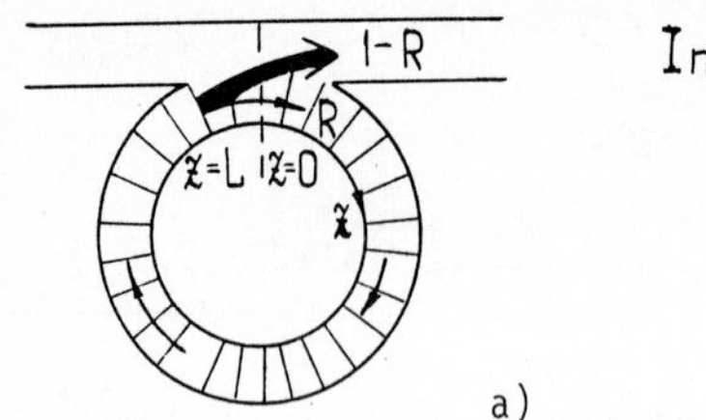


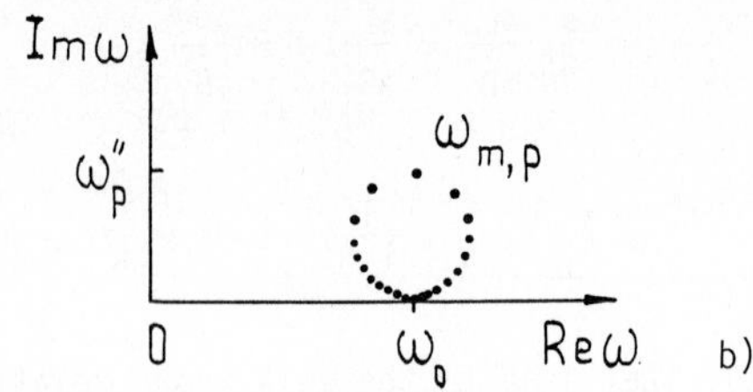


Figure 2. Model of a unidirectional superradiance in an active ring with feedback: (a) geometry of the light propagation, (b) discrete spectrum of the natural modes.

/1,11/. The former is the simplest and is described by the following reduced Maxwell-Bloch equations (1)-(3) for slowly varying inversion $\Delta N$ and complex field E and polarization P amplitudes of linear polarized waves $\sim \exp(-i\omega_0 t + i\omega_0 z/c)$:

$$\left[\frac{\partial}{\partial t} + c\frac{\partial}{\partial z} + 2\pi(\sigma+\sigma_{dif})\right]E = 2\pi i\omega_0 P\ , \quad (\frac{\partial}{\partial t} + \frac{1}{T_2})P = \frac{i\omega_c^2}{8\pi\omega_0}E\ ,$$

$$\frac{\partial \Delta N}{\partial t} = \mathrm{Im}(E^*P)/\hbar\ . \tag{8}$$

Setting $P = -idN\sin\psi$, $\Delta N = N\cos\psi$, $E = (\hbar/d)\partial\psi/\partial t$, in the limit of vanishing relaxation, $T_2^{-1} = 0$, one obtains the sine-Gordon equation for the Bloch angle $\psi = \mathrm{Re}\,\psi$:

$$\left[\partial/\partial t + c\partial/\partial z + 2\pi(\sigma+\sigma_{dif})\right]\partial\psi/\partial t = \frac{1}{4}\Omega_c^2 \sin\psi;\quad \Omega_c = |\omega_c|_{t=0}\ . \tag{9}$$

In the absence of dissipation, it possesses Hamiltonian $H = \frac{1}{2}(\partial\psi/\partial t)^2 + \frac{1}{4}\Omega_c^2(\cos\psi - 1$ with a canonical coordinate $\psi$ and momentum $\pi = \partial\psi/\partial t + (c/2)\partial\psi/\partial z$. If there is no feedback, this model describes an instability existing within the whole spectral inetrval up to the resonance: $|ck-\omega_0| < \Omega_c$. Its analysis confirms the conclusion above about the instability origin due to interaction between partial waves of opposite signs of energies. Here we have $H = \left[(c\partial\psi/\partial z)^2 - (\Omega_c\psi)^2\right]/8 < 0$ for $(\psi \neq 0, \pi = 0)$-wave and $H = \frac{1}{2}\pi^2 > 0$ for $(\psi = 0, \pi \neq 0)$-wave. According to Eq.(9), unidirectional SR of waves with a contiunous spectrum is realized. Its linear asymptotics gives $\psi \sim$ $\sim \exp(\Omega_c\sqrt{tL/c})$. A feedback $R \neq 0$ selects natural modes with a discrete spectrum and suppresses other components of radiation. Neglecting a retardation in the short ring, $L\omega_c/c \ll \ln R^{-1}$, $\ln \psi_{t=0}^{-1}$, i.e. ignoring the decay electromagnetic modes, we have the spectrum $\omega_{m,p} = \omega_0 + i\omega_p''/(1+4\pi im/\ln R^{-1})$, where $\omega_p'' = \Omega_c^2/8\pi\sigma_{rad}$, $m = 0,\pm1,\ldots$ (Fig.2 cf. (4)). In this case SR is the dissipative instability of negative-energy (m,p)-modes caused by the radiation dissipation $\sigma_{rad}$.

A link between SR and dissipative instability takes place for a three-dimensional sample as well. In the case of vanishing local loss Q, for example, it is the positive energy flow $\Sigma_{rad} = (c/4\pi)\int_{S_0}\left[\vec{E},\vec{B}\right]\cdot d\vec{s} > 0$ through the boundary $S_0$ which causes the decrease of negative energy $W = \int_V w d^3\vec{r} < 0$ and gives rise to a dissipative instability with the growth rate $\omega'' = -(\Sigma_{rad} + \int_V Q d^3\vec{r})/2W > 0$, calculated according to Eq.(5) as per $\omega'' = (dW/dt)/2W$.

The energy stored in the active sample is emitted as *SR of discrete spectrum modes* /1/ if the reflection coefficient, or a feedback, is greater than a critical one: $R > R_{cr} \sim (M/\bar{N}V)^{1/4} \ll 1$. Here M represents the number of natural modes of instability the growth rates of which are close to the maximum one. Otherwise, discrete modes have no time for standing out and SR occurs in the regime analogous to *SR of continuous-spectrum waves* whenever $R \to 0$.

Dicke's original concept of SR as an aperiodic collective spontaneous relaxation of excited molecules with emission of the whole energy stored in the sample, corresponds most closely to a single-pulse SR regime produced by the dissipative instability of polarization waves or modes. With decreasing dissipation, $2\pi(\sigma+\sigma_{dif}+\sigma_{rad}) < T_2^{-1}$, this instability is replaced by its limiting variant, i.e. by an instability due to an interaction between two dynamic subsystems, for example, partial oscillators of polarization and field, which have opposite signs of energies. More detailed CED treatment of SR is outlined in /1,11/.

## 3. QUANTUM ELECTRODYNAMICS OF TRANSPARENT DISPERSIVE MEDIA

Normal waves and natural modes are the basis of CED as well as its quantum generalization, i.e. PQED. The last one intends to describe quantum-statistical properties of macro-fields in continuous media. PQED, unlike microscopic QED, starts with classical (nonoperator) equations for local values of macroscopic fields and polarization in a medium /5,6,9/, not with equations for quantum interactions between individual particles and photons in vacuum. After the former equations are reformulated to Hamiltonian form and subsequently canonically quantized, it is immediately possible to study the quantum statistics of collective matter-field excitations, which turn into quantum oscillators of natural modes and photons in the medium.

PQED originated from GINZBURG /6/ in 1940 in connection with the theory of the Čerenkov radiation and was elaborated by many authors later on. To treat PQED in transparent linear media, we use , as a rule, the Hamiltonian method of the field expansion $E = i \sum_{\vec{k},j} (a_{\vec{k}j}\vec{g}_{\vec{k}j} - a_{\vec{k}j}^{+}\vec{g}_{\vec{k}j}^{*})\omega_{\vec{k}j}/c$, where $\vec{g}_{\vec{k}j} = (2\pi\hbar c^2/\omega_{\vec{k}j}V)^{1/2}\vec{e}_{\vec{k}j}\exp(i\vec{k}\vec{r})$ are modes (normal waves) of a quantization volume $V \to \infty$, filled with medium. Mode frequencies are real, $\omega_{\vec{k}j} > 0$, and the permittivity is an even function of frequency, $\varepsilon_{\gamma\sigma}(\omega,\vec{k}) = \varepsilon_{\gamma\sigma}(-\omega,\vec{k})$, because of the absence of relaxation and reversibility of medium polarization equations. The creation $a_{\vec{k}j}^{+}$ and annihilation $a_{\vec{k}j}$ operators of *photons in the medium* obey canonical commutation relations

$$a_{\vec{k}j}, a_{\vec{k}'j'}^{+} = \sigma_{\vec{k}\vec{k}'}\sigma_{jj'} , \qquad a_{\vec{k}j}^{+}, a_{\vec{k}'j'}^{+} = a_{\vec{k}j}, a_{\vec{k}'j'} = 0 . \tag{10}$$

The free-field Hamiltonian $H_0 = \sum_{\vec{k},j} \hbar\omega_{\vec{k}j}(a_{\vec{k}j}^{+}a_{\vec{k}j} + 1/2) \equiv \sum_{\vec{k},j} (p_{\vec{k}j}^2 + \omega_{\vec{k}j}^2 q_{\vec{k}j}^2)/2$ is the sum of energies of normal field-oscillators which are characterized by coordinates $q_{\vec{k}j} = (\hbar/2\omega_{\vec{k}j})^{1/2}(a_{\vec{k}j}^{+}+a_{\vec{k}j})$ and momenta $p_{\vec{k}j} = i(\hbar\omega_{\vec{k}j}/2)^{1/2}(a_{\vec{k}j}^{+}-a_{\vec{k}j})$. Eigenfunctions $\vec{g}_{\vec{k}j}$ are normalized with respect to one-quantum energy:

$$\frac{1}{4\pi}\int_V \left\{\frac{\omega^2}{c^2}\sum_{\gamma,\sigma=1} \frac{d(\omega\varepsilon_{\gamma\sigma})}{d\omega}(\vec{g}_{\vec{k}j}{}^*)_\gamma(\vec{g}_{\vec{k}j})_\sigma + |\vec{k},\vec{g}_{\vec{k}j}{}^*|^2\right\}_{\omega=\omega_{\vec{k}j}} d^3\vec{r} = \hbar\omega_{\vec{k}j} . \qquad (11)$$

The Bose-Einstein quantization (10), instead of the Fermi-Dirac one, is accounted by the initial fact of a coherent macro-field existence. It means that a lot of elementary excitations are condensed in the same quantum state, so that Fermi statistics is ruled out according to the Pauli exclusion principle /14/.

The Maxwell equations in nonlinear media with external currents can be reformulated to the Hamiltonian equations and quantized too. For a systematic treatment of PQED of the transparent media, see /6,15-18/ and references therein.

## 4. QUANTUM THEORY OF DISSIPATIVE INSTABILITY

According to CED (see Sect.2), SR and analogous processes are associated with a dissipative instability of interacting modes which have opposite signs of energies. Therefore, before discussing PQED of active media, one has to deal with the quantum theory of a dissipative instability /1/. At first, let us treat its limiting variant, i.e. the *dynamical dissipative instability* of two coupled oscillators characterized by opposite-sign quanta $-\hbar\omega_1^{(0)} < 0$ and $\hbar\omega_2^{(0)} > 0$. It is described by the sign-indefinite Hamiltonian quadratic in the creation-annihilation operators $a_j^+$ and $a_j$ of partial oscillators ($j = 1,2$) obeying canonical commutation relations like (10):

$$H = -\hbar\omega_1^{(0)}a_1^+a_1 + \hbar\omega_2^{(0)}a_2^+a_2 + \frac{\hbar}{2}(\eta a_1a_2+\eta^*a_2^+a_1^+) , \quad da_j/dt = a_j,H /i\hbar . \qquad (12)$$

If higher order terms are taken into account, the Hamiltonian would be nonnegative-definite and the analysis of the latter, the nonlinear stage of instability would be possible. Nevertheless, for many problems, for example SR, macro-fluctuations are formed in a linear stage already, and then a quadratic Hamiltonian is sufficient. Variant (12) describes, for example, the instability in the inverted two-level medium (see (4)) due to interaction ($\eta = \Omega_c \gg T_2^{-1}, 2\pi\sigma$) between partial polarization ($\omega_1^{(0)} = \omega_0$) and field ($\omega_2^{(0)} = ck$) oscillators in the "single-mode" SR model. In a resonance approximation, the oscillators with opposite-sign energies are coupled by operators $a_1a_2$, $a_2^+a_1^+$ different from the traditional ones $a_1^+a_2^+$, $a_2^+a_1^+$ used in quantum optics and laser theory for coupling oscillators with identical-sign energies /19-21/. The omitted antiresonant terms are vanished by averaging and are responsible for weak effects like the Bloch-Siegert frequency shift only.

Complex transformation of creation-annihilation operators

$$\tilde{a}_1^+ = a_1^+-a_2\eta/2(\omega_1-\omega_1^{(0)}), \quad \tilde{a}_2 = -i\eta\left[a_1^+\eta^*/2(\omega_1-\omega_2^{(0)}-a_2\right]:$$

$$: 2(|\eta^2|-(\omega_1^{(0)}-\omega_2^{(0)}))^{1/2} \tag{13}$$

leads to noncommuted normal oscillators with Hamiltonian

$$H = \hbar\omega_1\tilde{a}_2^{+}\tilde{a}_1^{+} + \hbar\omega_2\tilde{a}_1\tilde{a}_2 + \hbar\omega_1^{(0)} ;$$

$$\omega_{1,2} = \frac{1}{2}\left[\omega_1^{(0)} + \omega_2^{(0)} \pm i(|\eta^2| - (\omega_1^{(0)}-\omega_2^{(0)}))^{1/2}\right]. \tag{14}$$

They obey *cross commutation relations* /16,21/, $\left[\tilde{a}_1^{+},\tilde{a}_2^{+}\right] = 1$, $\left[\tilde{a}_j^{+},\tilde{a}_{j'}\right] = 0$ $(j = 1,2;$ $j' = 1,2)$, different from the canonical ones (10). The Heisenberg equations of motion for new, increasing and decaying, oscillators are decoupled: $\dot{\tilde{a}}_1^{+} = -i\omega_1\tilde{a}_1^{+}$, $\dot{\tilde{a}}_2 = -i\omega_2\tilde{a}_2$. Hereafter we suppose the growth rate to be $\omega_1'' \equiv \mathrm{Im}\,\omega_1 > 0$.

The Hermitian Hamiltonian (12), as well as the equivalent one (14), is boundless from below. Its eigenfunctions have infinite norms, i.e. do not belong to ordinary Hilbert space. The last fact invalidates the standard proof that eigenvalues are complex /22/ ($\omega_{1,2}'' \neq 0$). It is exactly what is necessary for the description of instability.

The simple solutions of decoupled equations allow us to study the process statistics completely. According to (12), the difference between partial oscillator numbers of quanty $n_{1,2} = a_{1,2}^{+}a_{1,2}$ is conserved: $d(n_1-n_2)/dt = 0$. Therefore, the dynamical dissipative instability is the process of mutual exchange of quanta between oscillators. To find a dynamical probability distribution $\rho(n,t)$ of a photon number n of either oscillator is enough. This problem has to be solved by the method of the characteristic function

$$\theta(u,t) = \mathrm{Sp}\{\rho_0 e^{iun(t)}\} = \int_{-\infty}^{\infty} \rho(n,t)e^{iun}dn, \quad \rho(n,t) = \frac{1}{2\pi}\int_{-\infty}^{\infty}\theta(u,t)e^{-iun}du . \tag{15}$$

It is connected with $\rho(n,t)$ by the Fourier transformation, which can be calculated by means of a quantum averaging of $\exp(iun) = \sum_{q=0}^{q}(iu)^q n^q/q!$ with respect to inital density operator $\rho_0$. The normal ordering of operators, $n^q \equiv (a^+a)^q = \sum_{j=0}^{q} D_q^{\,j}(a^+)^j a^j$, is suitable, cf. the statistical Bloch-de-Dominicis theorem and the Wick theorem /22/. Here $D_{q+1}^{j} = D_q^{\,j-1}\theta(j) + jD_q^{\,j}$; $D_q^{\,q} = 1$, $D_q^{\,0} = \sigma_{0,q}$, and $D_q^{\,1} = \theta(q)$ is unit step function.

It is the spontaneous creation of pairs of quanta from the vacuum state (Fig.3a), i.e. the growth of spontaneous fluctuations of initially deexcited coupled oscillators ($\rho_0 = |0,0\rangle\langle 0,0|$, $\rho(n,t=0) = \sigma(n)$), that represents just a quantum result. In this case one finds

$$\mathrm{Sp}\{\rho_0 n^q\} = \sum_{j=0}^{q}(j!)D_q^{\,j}\bar{n}^j = P_0(t)\sigma_{0,q} + \sum_{n=1}^{\infty}P_n(t)n^q$$

$$\theta(u,t) = \sum_{q=0}^{\infty}\frac{(iu)^q}{q!}\sum_{j=0}^{q}(j!)D_q^{\,j}\bar{n}^j , \tag{16}$$

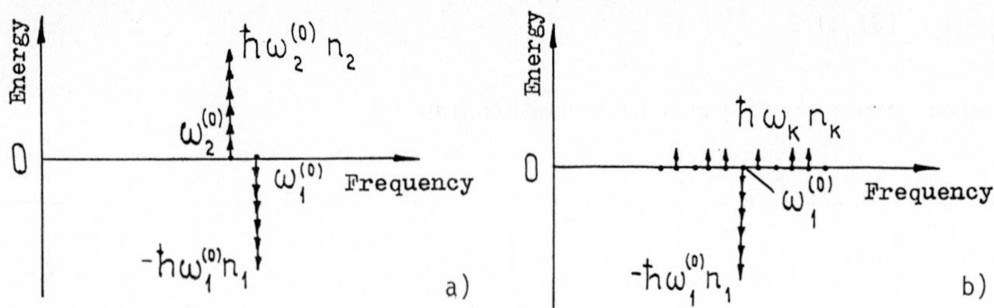


Figure 3. Diagram of spontaneous creation of photons from the vacuum initial state in a process of dissipative instability due to interaction between dynamical negative-energy oscillator ($\hbar\omega_1^{(0)}$) and (a) dynamical positive energy oscillator ($\hbar\omega_2^{(0)}$) or (b) the reservoir of positive-energy oscillators ($\hbar\omega_k$).

where the mean photon number increases starting from zero: $\bar{n}(t) = |\eta/2\omega_1''|^2 \mathrm{sh}^2(\omega_1''t)$. The probability of n-quantum excitation at the moment t is $P_n(t)$, i.e. $\rho(n,t) = \sum_{m=0}^{\infty} P_m(t)\sigma(n-m)$. Retaining in (16) terms up to the second order ($D_q^2 = \theta(q-1)(2^{q-1}-1)$) and calculating (15), it is easy to find the probabilities of one- and two quantum excitations $P_1(t) = \bar{n}(1-2\bar{n})$, $P_2(t) = \bar{n}^2$ at the beginning of the instability, when $\omega_1''t \ll 1$ and $\bar{n} \ll 1$. Later, when $t \gg 1/\omega_1''$, $\bar{n} \gg 1$ and the discreteness of n is irrelevant, only the highest order terms $j = q$ and $j = q-1$ in (16) ($D_q^{q-1} = (q-1)q/2$) are significant. They give the smooth distribution, $\rho(n,t) = \exp(-n/\bar{n})(1-1/\bar{n}+n/2\bar{n}^2)/\bar{n}$, correct up to $n \ll \bar{n}^2$. Its asymptotics for $t \to \infty$,

$$\rho(n,t) = \frac{1}{\bar{n}}\exp(-n/\bar{n}), \quad n \geq 0;$$

$$\int_0^\infty \rho(n,t)dn = 1 , \quad \bar{n}(t) = n_{eff}\exp(2\omega_1''t) , \tag{17}$$

corresponds to the Gaussian distribution of oscillator amplitudes.

Thermal fluctuations, at a temperature T, results in the same asymptotics (17), but $\bar{n}$ is greater:

$$n_{eff} = |\eta/4\omega_1''|^2\left[\mathrm{cth}(\hbar\omega_1^{(0)}/2\kappa_B T) + \mathrm{cth}(\hbar\omega_2^{(0)}/2\kappa_B T)\right]/2 . \tag{18}$$

Let us consider another variant of a dissipative instability produced by an interaction between the negative-energy dynamical open subsystem, namely the $a_1$-oscillator, and the positive-energy dissipative subsystem, namely the (thermal) *reservoir* consisting of an infinite number of $b_k$-oscillators /1/. The appropriate Hamiltonian,

$$H = -\hbar\omega_1^{(0)}a_1^+a_1 + \sum_k \hbar\omega_k b_k^+ b_k + \frac{\hbar}{2}\sum_k(\beta_k a_1 b_k + \beta_k^* b_k^+ a_1^+) , \tag{19}$$

produces coupled equations for the creation and annihilation operators of the partial oscillators with canonical commutators like (10):

$$\dot{a}_1^+ + i\omega_1^{(0)}a_1^+ = \frac{i}{2}\sum_k \beta_k b_k , \quad \dot{b}_k + i\omega_k b_k = -\frac{i}{2}\beta_k^* a_1^+ . \tag{20}$$

A weak coupling limit ($1 \gg |\beta_k|/\omega_1'' \to 0$) combined with a large reservoir limit, implying continuous $\omega_k$-frequency spectrum ($\sum_k \ldots \approx \int \ldots g(\omega)d\omega$), leads to a temporal irreversibility /20/. The correctness of a reservoir model and, generally speaking, of a phenomenological approach applied to quantum dynamics of a primary classical system is grounded on the fact that the macroscopic results are essentially independent of a choice of microscopic parameters /20/ $g(\omega)$ and $\beta_k = \beta(\omega)$.

The formulated problem has the analytical solution in the Weisskopf-Wigner approximation. Indeed, the Laplace transform of (20), solving the obtained algebraic system and substituting $\omega - \omega_1$ in return for its determinant $\det = \omega - \omega_1^{(0)} + \sum_k |\beta_k^2|/4(\omega-\omega_k)$, results in

$$a_1^+(t) = \{a_1^+(0) - \frac{1}{2}\sum_k \frac{\beta_k b_k(0)}{\omega_1-\omega_k}\left[1 - \exp(i(\omega_1-\omega_k)t)\right]\}\exp(-i\omega_1 t) \ . \tag{21}$$

Here the observed, renormalized by a reservoir, complex frequency of the normal dynamical oscillator is introduced:

$$\omega_1 = \omega_1^{(0)} + \Delta\omega_1' + i\omega_1'' \ ; \qquad \Delta\omega_1' = \mathrm{V.p.}\int_{-\infty}^{\infty} \frac{|\beta(\omega)|^2 g(\omega)}{4(\omega-\omega_1^{(0)})}\, d\omega \ ,$$

$$\omega_1'' = \frac{\pi}{4}\, g(\omega_1^{(0)})|\beta(\omega_1^{(0)})|^2 \ll \omega_1^{(0)} \ . \tag{22}$$

Again, from solution (21) it follows that a dissipative instability develops spontaneously even from a deexcited vacuum state: $\bar{n}_1(t) = \exp(2\omega_1''t) - 1$, $t \geq 0$ (Fig.3b). The method of the characteristic function (15) gives the evolution of a photon number distribution again. The asymptotics (17) remains correct for spontaneous as well as thermal fluctuations, if one takes $n_{eff} = \mathrm{cth}(\hbar\omega_1^{(0)}/2\kappa_B T) \geq 1$. Note that the analysis stated above is consistent with a widely used in laser theory /19/ method based on introducing the operator Langevin noise source. In our case, the RHS of the first equation (20) plays its part.

Quantum theory, developed for model examples (12) and (19), can be generalized for any type of a dissipative instability, including both an interaction between dynamical oscillators of various signs of energies and an irreversible removing of their energies by reservoirs. The coupling of dynamical oscillator with a reservoir, which has the same or opposite sign of energy, renormalizes its frequency analogous to (22) and describes a relaxation or an incoherent amplification accordingly. For the polarization oscillator in the two-level medium, this is the relaxation $T_2^{-1}$. For the field mode, this is the ohmic positive dissipation $\sigma$ or the negative dissipation $\sigma_a$, realized in lasers.

## 5. PHENOMENOLOGICAL QUANTUM ELECTRODYNAMICS OF ACTIVE MEDIA

Let us extend PQED ideas to the collective processes in active media /1/. It is not a trivial problem because PQED is commonly accepted for transparent media only, where Im $\omega_{\vec{k}j} = 0$ (Sec.3). As for absorbing media, where Im $\omega_{\vec{k}j} < 0$, one has to quantize a damped oscillator which has a positive-definite energy /20,21/. The situation in active media, where Im $\omega_{\vec{k}j} > 0$, differs qualitatively from it. As it was pointed out in /10,23/, the quantum fluctuation growth in active media, during a SR process, for example, must be considered as the instability arising due to an interaction between quantum oscillators (modes or waves) possessed of opposite-sign energies. Such an instability can, broadly speaking, be called by the term "dissipative". Indeed, with regard to a distinguishable dynamical subsystem of unstable oscillators, the other oscillators, in any event, play a part of a dissipative subsystem taking away the energy from the first one (Sec.2). It involves in the quantum theory the Hermitian Hamiltonian operator which is not sign-definite in the linear approximation (Fig.3). This approach gives a general scheme of the quantization and the description of the fluctuation evolution from micro- up to macro-level. A frequency-spatial dispersion, a nonlinearity, an inhomogeneity, an inhomogeneity, an anisotropy, and external currents in the medium can be taken into account too.

In a principal outline, the quantization procedure of PQED in active linear media is as follows. First of all, properties and frequencies $\omega_j(\vec{k})$ of normal waves must be elucidated. Then all relaxation and dissipation constants will be taken zeroes, and Hamiltonian equations for dynamical field oscillators in the medium will be obtained. By this step, there will be pairs of oscillators with complex-conjugated frequencies $\Omega_{\alpha 1,2}^{(0)} = \Omega_{\alpha}^{(0)'} \pm i\Omega_{\alpha}^{(0)''}$ and individual stationary oscillators with real frequencies $\Omega_{\beta}^{(0)}$. Every pair will be converted into a coupled system of partial oscillators with opposite-sign energies $-\hbar\omega_{\alpha 1}^{(0)}$ and $+\hbar\omega_{\alpha 2}^{(0)}$, analogous to (12). The amplitudes of obtained oscillators will be normalized with respect to one-quantum energy; cf. (11). Further, every dynamical partial oscillator ($\Omega_{\beta}^{(0)}$, $-\Omega_{\alpha 1}^{(0)}$, $+\omega_{\alpha 2}^{(0)}$) will be coupled with the pair of partial oscillator reservoirs, possessing opposite-sign energies, so that frequencies $\Omega_{\alpha}^{(0)}$ and $\Omega_{\beta}^{(0)}$ renormalized by reservoirs will coincide with the primary ones $\omega_j$; cf. (22).

At last, the theory is quantized by replacing all canonical coordinates and momenta of all partial oscillators by the Bose operators; cf. Sec.3. Dynamical oscillators with $\Omega_{\alpha}^{(0)''} \neq 0$ will then satisfy cross commutation relations (Sec.4). The last means that the sign-indefinite Hamiltonian in active media, unlike the Hamiltonian of positive-energy oscillators in transparent media, is not diagonalized like an analog of the Bogolyubov transformation /22,24/ with conserved commutation relations. This fact is significant in order to pass the limits of the transparent medium PQED and to quantize field in active and absorbing media. After all, PQED analysis of unstable macro-field oscillators reduced to the quantum theory of a dissipative instability.

In inhomogeneous and bounded media one has to take into account a change in the spatial structure of eigenmodes $\vec{g}_j(\vec{r})$ as well as a coexistence of discrete-spectrum natural modes and continuous-spectrum waves (Sec.2). Further generalization of PQED for describing nonlinear and nonstationary media is associated with momentary linearizing of field equations in the medium and with introducing the current normal and partial oscillators (cf. momentary Bogolyubov transformation in quantum field theory /24/).

It is important to convert classical field equations in the medium into the form of the Heisenberg equations for field and polarization operators with local commutation relations. The last ones are determined, in any case, by the local properties of normal waves, by the analogy with a quantization procedure outlined above. It is clear in geometrical optics /9/ . Unlike the momentum $\vec{k}$-representation, the spatial $\vec{r}$-representation gives the Heisenberg equations the convenient form of partial differential equations /18/, according to the quantum-classical correspondence principle. One can use well-developed methods of classical wave theory to solve the Heisenberg equations for slowly varying macro-field amplitudes in the quantum theory. This attractive opportunity will promote the application of PQED to the analysis of quantum-statistical phenomena in amplifiers, lasers, etc. Close approaches are employed in the problems of the coupled excitón-photon propagation in the kinetic equation approximation /25/, of the stimulated parametric scattering on the basis of the operator paraxial equation /15/, and of SR /1,12,13,23,26-28/.

## 6. SUPERRADIANCE MACROSCOPIC QUANTUM FLUCTUATIONS

Let us consider SR of discrete modes (Sec.2) to illustrate PQED of active media /1,23/. Quantum fluctuations of SR pulse parameters are macroscopic and not predictable from shot to shot. Statistics is concerned with an ensemble of shots, or pulses, all originating from identically prepared systems. Note that in every individual shot the system evolves from a quantum noise to a deterministic classical trajectory during the linear stage already /12,13/. This is because of the existence of the macroscopic system. For the sake of simplicity, we shall take into account only those modes which have growth rates $\omega_m''$ ($m = 1,\dots,M$), close to their maximum $1/2\tau$. According to Sec.4, the numbers of quanta $n_m$ of different modes are independent random quantities with the same asymptotic distribution (17). Therefore, the distribution of the total number of quanta

$$q = \sum_{m=1}^{M} n_m \text{ is } \rho(q,t/\tau) = (q/\bar{n}_1)^{M-1} \, (M-1)!\bar{n}_1{}^{-1}\exp(-q/\bar{n}_1) \, .$$

The probability of the SR pulse to be emitted during the interval $(0,t_d)$ is equal to the probability of the total number of quanta q to exceed the half number of inverted molecules $\bar{N}V/2$ at the moment $t_d$. To determine the delay time $t_d$ like that, the linear approximation of SR is used, in a manner similar to /12,13,26/. The condition stated reads

$$\int_0^{t_d/\tau} f(t_d'/\tau)d(t_d'/\tau) = \int_{\bar{N}V/2}^{\infty} \rho(q,t_d/\tau)dq.$$

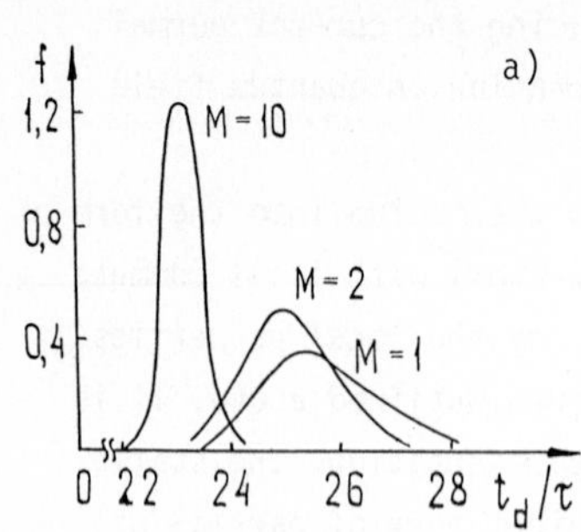


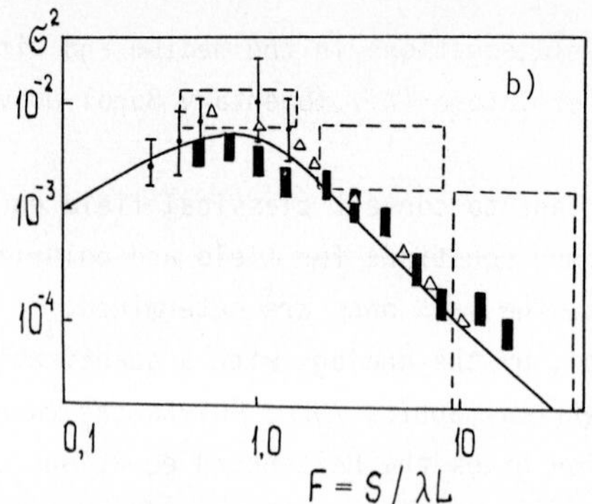


Figure 4. Dependence of the delay time statistics from the sample form. (a) Probability distribution (23) for $u=10^{11}$. (b) Relative dispersion squared $\sigma^2$ of delay time $t_d$ versus the Fresnel number $F=S/\lambda L$ for cylindrical samples ($\overline{N}\lambda L^2=10^8$, L=const): solid line - envelope of (24), dashed rectangles - experiment /29/, triangles, hatched rectangles and vertical segments - calculations /26,31,32/, respectively.

Differentiating it with respect to $t_d$ gives the searching distribution of delay time /23/

$$f\left(\frac{t_d}{\tau}\right) = \frac{u^M}{(M-1)!}\exp\left(-M\frac{t_d}{\tau} - ue^{-t_d/\tau}\right) ; \quad u \equiv \frac{\overline{N}V}{2n_{eff}} >> 1 . \tag{23}$$

According to (23), the dependence of SR statistics on sample form exists and is essentially the dependence of number M of unstable modes on sample form (Fig.4a). Number M is determined by solving a corresponding CED problem. For example, for a sphere of radius $a >> \lambda$ one has /10/ $M \sim (\omega_0 a/c)^2 >> 1$; for a cylinder of Fresnel number $F = S/\lambda L$ estimates /12/ give $M \sim (F^2+1+1/F)/3$. Increasing M leads to reducing the mean delay time as $\overline{t}_d = \tau\ln(u/M)$ and, in agreement with experiments /29,30/, to decrease fluctuations as (Fig.4b)

$$\sigma^2(M) \equiv \overline{t_d^2} - (\overline{t}_d)^2 / (\overline{t}_d)^2 = \left(\frac{\pi^2}{6} - \sum_{m=1}^{M-1}\frac{1}{m^2}\right) / \ln(u/M)^2 \simeq 1/M\ln(u/M)^2 . \tag{24}$$

The observed statistics of SR polarization ellipse fluctuations /12,27/ is also explained by PQED of active anisotropic media. Experimentally one is dealing with a distribution of a random angle $\beta \in (0,\pi/2)$ determined according to equation $tg^2\beta = I_y/I_x \quad n_y/n_x$. Intensities $I_{x,y}$, or numbers of quanta $n_{x,y}$, are measured by two detectors with fixed orthogonal linear polarization directions x and y (Fig.5).Let

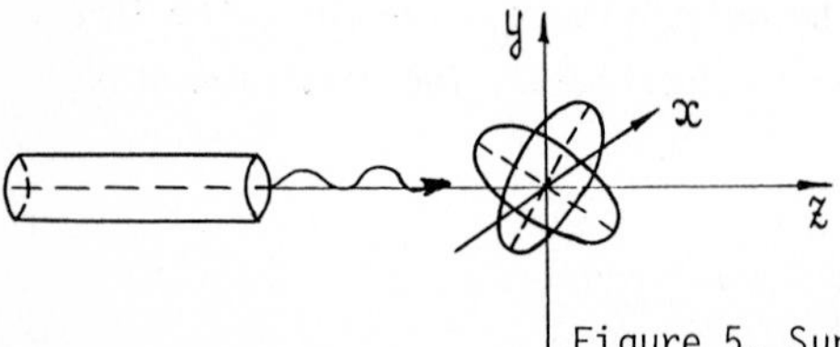


Figure 5. Superradiance of two elliptically polarized modes.

us consider the SR of two unstable modes with close growth rates $\omega_1'' \approx \omega_2''$ and arbitrary polarization ellipses, characterized by the polarization coefficients $K_1$ and $K_2$. A polarization coefficient of a mode field is determined by the ratio $K = E_y/E_x = -i\,ctg(\chi+i\theta)$ of complex field amplitude components $E_x$, $E_y$ /33/. It gives both the ellipticity th$\theta$ and the polar angle $\chi$ between the long axis of the ellipse and the y axis. Numbers of quanta of both modes are distributed independently, according to (17). Therefore, it is easy to refind the delay time statistics (23), (24) with M = 2 and to calculate distributions of the ratio $n_2/n_1$ and the angle $\beta$:

$$f(\frac{n_2}{n_1}) = T/(T+n_2/n_1)^2 ,$$

$$f_\beta = \frac{tg\beta}{\cos^2\beta}\frac{Ts}{\pi|K_2-K_1|^2}\left[\int_0^{2\pi}\left|\frac{K-K_2}{K-K_1}\right|^3 (T+s^2\left|\frac{K-K_2}{K-K_1}\right|^2)^{-2}d\psi\right]_{|K|=ctg\beta} . \qquad (25)$$

where $T = \bar{n}_2/\bar{n}_1 \sim \exp\left[2(\omega_2''-\omega_1'')t\right]$, $s = \left[(1+|K_1^2|)/(1+|K_2^2|)\right]^{1/2}$ and $K = |K|\exp(i\psi)$. If modes are linearly and orthogonally polarized, i.e. $K_1 = K_2^{-1} = 0$, the result (25) reduces to $f_\beta = T\sin 2\beta/(T\cos^2\beta + \sin^2\beta)^2$ (Fig.6). The obtained distributions $f_\beta$ are consistent with observations /27/, including appreciable and not fully reproducible changes of histograms due to weak variations of experimental conditions. The last conclusion is connected with the exponentially strong amplification of weak anisotropy and gyrotropy manifestations in SR (Fig.6b).

Papers /26,27,31,32/ refer to experiments on SR of continuous-spectrum waves in rarefield gases, where $L \lesssim c/\Omega_c$ and the reflection at boundaries is very weak. The last is not the case for SR of discrete-spectrum modes in resonators and crystals /34,35/. Nevertheless, in both SR regimes the behaviour of the SR statistics, as the anisotropy (Fig.6) and the scale of a cylindrical sample (Fig.4b) change, is similar. Microscopic QED was developed for the case of unidirectional SR of linear polarized continuous-spectrum waves in the absence of dissipation and relaxation /13/. It can be shown that active medium PQED treatment outlined in the present paper immediately leads to the same results by means of direct quantization of the Maxwell-Bloch equations (8). If two differently polarized waves ($K_1 \neq K_2$, Fig.5) are taken into account, then macro-fluctuations of the total SR polarization ellipse submit to the statistics (25), where $T = \bar{I}_2/\bar{I}_1$ (Fig.6). The reason of the last result is that the unidirectional SR intensity of the continuous-spectrum waves /28,36/ and the SR number of quanta of the discrete-spectrum modes obey the same asymptotical statistics (17).

The SR of continuous-spectrum waves in a three-dimensional sample with a large Fresnel number $F = S/\lambda L >> 1$ are formed by statistically uncorrelated filaments (cf. van Cittert-Zernike theorem) /26/. Every filament emission is confined within a solid angle $\sim\lambda^2/S$, so that the number of such "diffraction modes" is $M \sim F^2$. Obviously, one can repeat the treatment of the SR delay time statistics for another SR regime by means of the substitution discrete mode dynamics and statistics of SR for unidirectional SR waves. It was carried out in /36/ and the results are analogous to Fig.4.

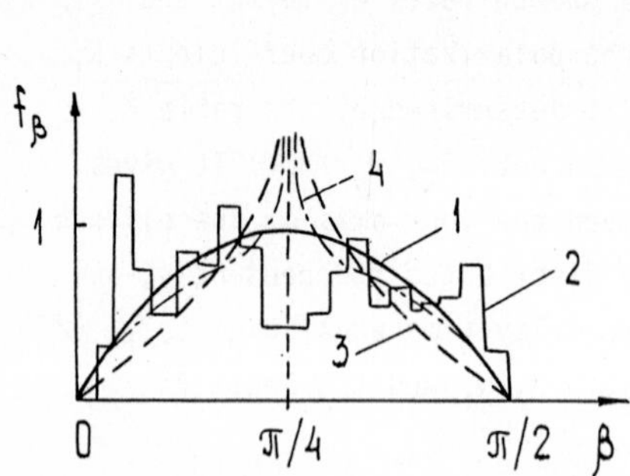


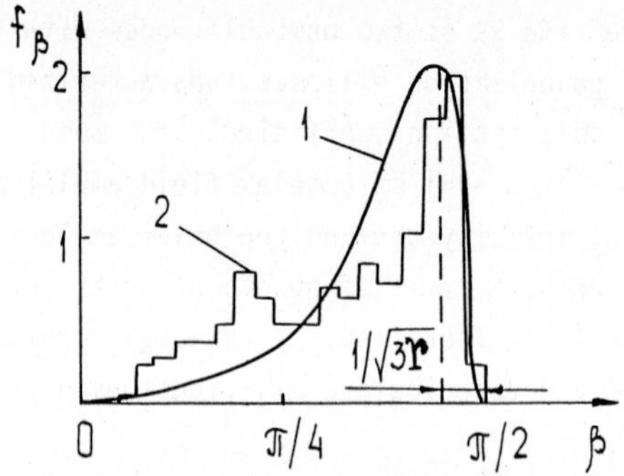


Figure 6. Probability distribution of the angle β, characterizing the orientation and eccentricity of the superradiance polarization ellipse of two modes with linear orthogonal polarizations, for (a) polarization degeneration ($\omega_1'' = \omega_2''$, T=1) and (b) weak anisotropy determined by the linear $\beta=\pi/2$-polarized pump radiation ($\omega_1'' \neq \omega_2''$, T=10). 1 - linear PQED theory (25); 2 - example of experimental histogram /27/; 3 and 4 - linear and nonlinear theory following /27/.

## CONCLUSION

Normal waves and natural modes, i.e. the collective excitations of an active medium sample, are the carriers of quantum properties. Their quantum fluctuations are amplified up to the macroscopic level during a process of a dissipative instability of interacting waves and/or modes which have opposite signs of energies. This is the origin of macroscopic quantum fluctuations in collective transient processes, like SR, which are not predictable from shot to shot. In order to treat quantum properties of macro-systems naturally and simply, one can use the direct phenomenological quantization of normal waves and natural modes, i.e. quantize the classical equations of electrodynamics by PQED method. In such a way, the quantum problem is studied just at these approximations and conditions being adequate to the classical ones. For instance, one can analyze the quantum statistics of unidirectional SR with the help of the same, but quantized sine-Gordon equation (9). On the whole, PQED is suitable for the correct treatment of various SR aspects, for example, geometry, inhomogeneity, and anisotropy of an active medium, reflections, generation of discrete-spectrum modes and continuous-spectrum waves, their nonlinear interaction, etc.

Macroscopic PQED conceptions are useful in the comparison of diverse collective coherent processes and for linking them into general science of wave processes in active media /1,37,38/.

## REFERENCES


1. V.V.Zheleznyakov, V.V.Kocharovsky, Vl.V.Kocharovsky. Usp.Fiz.Nauk (Sov.Phys.-Uspekhi), 1989, 159; Preprints No.227 and 228, Inst. of Appl.Phys., USSR Acad.Sci., Gorky 1989, 116p. (in Russian).
2. A.J.Leggett. Suppl. Progr.Theor.Phys., 1980, 69, 80.
3. M.H.Devoret et al. Helvetica Physica Acta, 1988, 61, 622.

4. V.V.Dodonov, V.I.Man'ko, V.N.Rudenko. Kvantovaya Elektronika, 1980, 7, 2124 (in Russian).

5. L.D.Landau, E.M.Lifshitz. Electrodynamics of Continuous Media. Pergamon, Oxford, 1960.

6. V.L.Ginzburg. Theoretical Physics and Astrophysics. Pergamon, Oxford, 1979.

7. R.H.Dicke. Phys.Rev., 1954, 93, 99.

8. N.Skribanowitz et al. Phys.Rev.Lett., 1973, 30, 309.

9. V.M.Fain. Photons and Nonlinear Media. Sovetskoye Radio, Moscow, 1972 (in Russian).

10. V.V.Zheleznyakov, V.V.Kocharovsky, Vl.V.Kocharovsky. Sov.Phys.-JETP, 1984, 60, 897.

11. V.V.Zheleznyakov, V.V.Kocharovsky, Vl.V.Kocharovsky. In: Nonlinear Waves 2, ed. by A.V.Gaponov-Grekhov, M.I.Rabinovitch, J.Engelbrecht. Springer, Berlin, Heidelberg, 1989, p.136.

12. M.Gross, S.Haroche. Phys.Rep., 1982, 93, 301.

13. F.Haake et al. Phys.Rev., 1979, A20, 2047; 1981, A23, 1322.

14. M.I.Kaganov, I.M.Lifshitz. Quaziparticles. Znanie, Moscow, 1976 (in Russian).

15. D.N.Klyshko. Photons and Nonlinear Optics. Nauka, Moscow, 1980 (in Russian).

16. V.P.Oleynik, I.V.Belousov. Problems of Quantum Electrodynamics of Vacuum, Dispersive media and Strong Fields. Shtiintza, Kishinev, 1983 (in Russian).

17. M.Hillery, L.D.Mlodinow. Phys.Rev. 1984, A30, 1860.

18. I.Abram. Phys.Rev., 1987, A35, 4661.

19. H.Haken. Laser Light Dynamics. North-Holland, Amsterdam, 1985.

20. R.Glauber, V.I.Man'ko. Zh.Eksp.Teor.Fiz. (Sov.Phys.-JETP), 1984, 87, 790 (in Russian).

21. H.Dekker. Phys.Rep., 1981, 80, 1.

22. N.N.Bogolyubov, N.N.Bogolyubov (Jr.). Introduction in Quantum Statistical Mechanics. Nauka, Moscow, 1984 (in Russian).

23. V.V.Kocharovsky, Vl.V.Kocharovsky. Opt.Comm., 1985, 53, 245.

24. A.A.Grieb, S.G.Mamaev, V.M.Mostepanenko. Vacuum Quantum Effects in Strong Fields. Energoatomizdat, Moscow, 1988 (in Russian).

25. A.S.Davydov, A.A.Serikov. Phys.st.sol.(b), 1973, 56, 351.

26. J.Mostowski, B.Sobolewska. Phys.Rev., 1984, A30, 1392.

27. A.Grubellier et al. J.Phys., 1986, B19, 2959; 1981, B14, L177.

28. F.Haake, R.Reibold. Phys.Rev., 1984, A29, 3208.

29. Q.H.P.Vrehen, J.J.der Weduve. Phys.Rev., 1981, A24, 2857.

30. K.Nattermann et al. Opt.Comm., 1986, 57, 212.

31. P.D.Drummond, J.H.Eberly. Phys.Rev., 1982, A25, 3446.

32. E.H.Watson et al. Phys.Rev., 1983, A27, 1427.

33. V.V.Zheleznyakov, V.V.Kocharovsky, Vl.Kocharovsky. Sov.Phys.-Uspekhi, 1983, 26, 877.

34. L.Moi et al. Phys.Rev. 1983, A27, 2043.

35. A.Schiller, L.O.Schwan, H.D.Schmid. J.Luminescence, 1987, 38, 243; 1988, 40&41, 541.

36. R.Reibold. Phys.Lett., 1986, A115, 325.

37. V.V.Zheleznyakov, V.V.Kocharovsky, Vl.V.Kocharovsky. Radiophys.Quant.Electron., 1986, 29, 830.

38. Ya.B.Zeldovich et al.Radiophys.Quant.Electron., 1986, 29, 761.

# Light Amplification by a Three-Level Atomic System Without Population Inversion

*Ya.I. Khanin and O.A. Kocharovskaya*

Institute of Applied Physics, USSR Academy of Sciences,
46 Ulyanov Str., 603600 Gorky, USSR

The possibility of light amplification in a three-level medium in the absence of population inversion is predicted. This requires the excitation of a coherent superposition of the lower level and partial occupation of the upper levels.

## 1. INTRODUCTION

It is well known that an atomic system can produce light amplification when transitions from the upper to the lower energy level prevail over the opposite transitions. When one deals with a two-level system, the only way is to create a state with population inversion (i.e. the upper level is more occupied than the lower one). In a three-level medium an alternative variant is possible. This variant is associated with the possibility for destructive interference of optical transitions. Specifically, in the $\Lambda$-scheme such a coherent superposition state of the two lower levels can be realized so that the transition probability from these to the upper level vanishes /1-2/. Meanwhile, the upward transition probability (to a state different from this coherent superposition) is not excluded. The rate of upward transitions is proportional to the upper level population. Therefore, even a small population is sufficient for light amplification.

The coherent bleaching phenomenon was investigated earlier /2/. It was shown that a finite layer of the absorbing three-level medium becomes transparent to powerful light because of the induced transfer of atoms into the coherent superposition state mentioned above /1-2/. It is obvious that the light cannot bring the atomic system into a state ensuring self-consistent amplification. Meanwhile, the inversionless amplification is possible when an external source is used for the coherent superposition formation.

This effect can be realized in various physical situations including both narrow and broad optical lines, both pulse and continuous radiation. Here we consider the propagation of ultrashort pulse in a three-level medium of $\Lambda$-configuration, where the homogeneous broadening of optical lines exceeds the pulse spectrum and the latter overlaps the splitting frequency: $T_2^{-1} \gg \tau_p^{-1} \gg \omega_{21}$. Under these conditions the pulse interacts resonantly with both optical transitions at once, so that the usual (for the Raman problems) distinction between two carrier frequencies spaced by the splitting frequency is meaningless here. Due to the last circumstance this situation is



Research Reports in Physics Nonlinear Waves 3
Editors: A.V. Gaponov-Grekhov · M.I. Rabinovich · J. Engelbrecht

radically different from the well-known case of simulation propagation. The latter is realized under the opposite conditions /3/ $T_2^{-1} << \tau_p^{-1} << \omega_{21}$.

## 2. EQUATIONS OF INTERACTION OF ULTRASHORT PULSE WITH THREE-LEVEL MEDIUM

If the field interacts resonantly with both optical transitions at once, the density matrix equations and the wave equation for the field $E = (1/2)\mathcal{E}\exp(-i\omega_p t + ikz) + $ + c.c. in the slowly varying amplitude approximation take the form:

$$\partial\sigma_{31}/\partial t + i(\omega_{31}-\omega_p)\sigma_{31} = i\mathcal{E}\left[\mu_{32}\rho_{21} + \mu_{31}(\rho_{11}-\rho_{33})\right]/2\hbar - \sigma_{31}/T_2 \ ,$$

$$\partial\sigma_{32}/\partial t + i(\omega_{32}-\omega_p)\sigma_{32} = i\mathcal{E}\left[\mu_{31}\rho_{21}^* + \mu_{32}(\rho_{22}-\rho_{33})\right]/2\hbar - \rho_{32}/T_2 \ ,$$

$$\partial\rho_{21}/\partial t + i\omega_{21}\rho_{21} = i(\mu_{32}^*\sigma_{31}\mathcal{E}^* - \mu_{31}\sigma_{32}^*\mathcal{E})/2\hbar - \rho_{21}/\tau_2 \ ,$$

$$\partial\rho_{11}/\partial t = i(\mathcal{E}^*\mu_{31}^*\sigma_{31} - c.c.)/2\hbar + R_1 \ ,$$

$$\partial\rho_{22}/\partial t = i(\mathcal{E}^*\mu_{32}^*\sigma_{32} - c.c.)/2\hbar + R_2 \ ,$$

$$\partial\rho_{33}/\partial t = i\left[\mathcal{E}(\mu_{31}\sigma_{31}^* + \mu_{32}\sigma_{32}^*) - c.c.\right]/2\hbar + R_3 \ ,$$

$$\partial\mathcal{E}/\partial z + c^{-1}\partial\mathcal{E}/\partial t = 4\pi i\omega_p N c^{-1}(\mu_{31}^*\sigma_{31} + \mu_{32}^*\sigma_{32}) \ . \qquad (1)$$

Here $\sigma_{31}$ and $\sigma_{32}$ are the complex amplitudes of the off-diagonal density matrix elements $\rho_{31} = \sigma_{31}\exp(-i\omega_p t)$, $\rho_{32} = \sigma_{32}\exp(-i\omega_p t)$, $\mu_{31}$, $\mu_{32}$ are the dipole matrix elements, N is the atom density, c is the velocity of the light; $R_1$, $R_2$, $R_3$ are some functions of level populations, describing pumping and relaxation processes; $\tau_2$ is the relaxation time of low frequency (LF) coherence (off-diagonal element of the density matrix $\rho_{21}$). In the approximation of the broadband optical transitions as compared with the radiation spectrum width, the $\omega_{31} - \omega_p$ and $\omega_{32} - \omega_p$ detunings, and the cooperative and Rabi frequencies, the optical polarization follows the field adiabatically. Then the first two equations of system (1) can be reduced to algebraic relations. Expressing the amplitudes $\sigma_{31}$ and $\sigma_{32}$ we substitute them to the remaining equations. At the same time one should take into account that pumping and relaxation processes are too slow to noticeably change the state of the medium during the ultrashort pulse $\tau_p << \omega_{21}^{-1}$, $\tau_2$, $T_1^{\ell}$ ($T_1^{\ell}$ are the relaxation times of populations $\ell = 1,2,3$). As the result we have a self-consistent set equations

$$\partial u/\partial t = -\left[2\eta n + (1+\eta^2)u\right]\sigma I/2\hbar\omega_p \ , \qquad (2a)$$

$$\partial n/\partial t = -3\ (1+\eta^2)n - (1-\eta^2)n + 2\eta u\ \sigma I/2\hbar\omega_p \ , \qquad (2b)$$

$$\partial\tilde{n}/\partial t = (1-\eta^2)n - (1+\eta^2)\tilde{n}\ \sigma I/2\hbar\omega_p \ ,,, \qquad (2c)$$

$$\partial I/\partial z + c^{-1}\partial I/\partial t = -\ (1+\eta^2)n - (1-\eta^2)\tilde{n} + 2\eta u\ \sigma I N \ . \qquad (2d)$$

This set describes the intensity transfer through the three-level medium taking into account the LF coherent effects. Here $\sigma = 4\pi\mu_{31}^2\omega_p T_2/c\hbar$ is the cross-section of the 3-1 transition, $u = \mathrm{Re}\,\sigma_{21}$, $\eta = \mu_{31}/\mu_{32}$. According to the normalization $\rho_{11} + \rho_{22} + \rho_{33} = 1$, the populations of all three levels are expressed through the half-sum $n = (n_{23}+n_{13})/2$ and the half-difference $\tilde{n} = (n_{23}-n_{13})/2$ of population differences for optical transitions in the following way:

$$\rho_{33} = (1-2n)/3, \quad \rho_{22} = \tilde{n} + (1+n)/3, \quad \rho_{11} = -\tilde{n} + (1+n)/3 . \tag{3}$$

## 3. BEHAVIOUR OF THE ATOMS UNDER THE ULTRASHORT PULSE ACTION

The medium evolution under the action of an ultrashort pulse is described by the solution of the first three equations (2):

$$\begin{pmatrix} u \\ n/\sqrt{3} \\ \tilde{n} \end{pmatrix} \equiv \vec{x} = \sum_{\alpha=1}^{3} c_\alpha \vec{x}^{(\alpha)} \exp(-\lambda_\alpha \zeta) \; ; \quad c_\alpha = \vec{x}_0 \vec{x}^{(\alpha)} ; \quad \vec{x}_0 \equiv \begin{pmatrix} \dot{u}_0 \\ n_0/\sqrt{3} \\ n_0 \end{pmatrix} . \tag{4}$$

Here $\vec{x}_0$ is the vector of the medium state before the pulse arrival,

$$\zeta = \sigma \int_{-\infty}^{t} I(t')dt'/2, \quad \lambda_1 = 0, \quad \lambda_2 = 4(1+\eta^2), \quad \lambda_3 = 1 + \eta^2,$$

$$\vec{x}^{(1)} = \frac{\sqrt{3}}{2} \begin{pmatrix} -2\eta/(1+\eta^2) \\ 1/\sqrt{3} \\ (1-\eta^2)/(1+\eta^2) \end{pmatrix}, \quad \vec{x}^{(2)} = \begin{pmatrix} -\eta/(1+\eta^2) \\ -\sqrt{3}/2 \\ (1-\eta^2)/2(1+\eta^2) \end{pmatrix}, \quad \vec{x}^{(3)} = \begin{pmatrix} (1-\eta^2)/(1+\eta^2) \\ 0 \\ 2\,/(1+\eta^2) \end{pmatrix},$$

$$\vec{x}^{(\alpha)}\,\vec{x}^{(\beta)} = \delta_{\alpha,\beta} . \tag{5}$$

The existence of a zero eigenvalue means that there is a set of states $c_1\vec{x}^{(1)}$ in which the medium does not interact with the field. The optical polarization (the r.h.s. of the last equation in (4)) in these states turns to zero. According to (4), it is the state $c_1\vec{x}^{(1)}$ into which the atomic system passes under the action of a rather strong field when $\lambda_2\zeta(\tau_p) >> 1$.

Equations (4), (5) and (3) give in particular the change in the upper level population under the field action described by

$$\rho_{33}^{(0)} - \rho_{33} = C_2\, 1-\exp(-\lambda_2\zeta) \;, \quad C_2 = -\eta u_0/(1+\eta^2) - n_0/2 + (1-\eta^2)\tilde{n}_0/2(1+\eta^2). \tag{6}$$

The pulse will decrease the population of the upper level for a variety of initial states defined by the condition $C_2 > 0$.

## 4. INVERSIONLESS AMPLIFICATION

Using Eqs. (2b), (2d) and (6) we find the equation for radiation intensity transfer /4/:

$$\partial I/\partial z + c^{-1}\partial I/\partial t = 2I\sigma N(1+\eta^2)C_2\exp(-\lambda_2\zeta) . \tag{7}$$

Integration with respect to pulse duration gives the law of pulse energy transfer:

$$W_p(z) = W_p(0) + \hbar\omega_p N \int_0^z (\rho_{33}^{(0)} - \rho_{33}(\tau_p))dz; \quad W_p \equiv \int_0^{\tau_p} I(t')dt' . \tag{8}$$

The pulse can be amplified at the expense of the medium energy stored by atoms at the upper level. The last one according to (6) takes place under the condition $C_2 > 0$.

In a particular case where the 3-2 transition is forbidden, i.e. $\eta = 0$, we have: $2C_2 = n_{13}$ and Eq. (7) coincides with the intensity transfer in an inertialess ($T_1 \ll \tau_p$) two-level medium in the rate equation approximation /5/. Its solution describes the well-known processes of pulse amplification in an inverted medium and of pulse absorption in a noninverted medium.

In general, the solution of Eq. (7) has a similar form:

$$I(t,z) = \frac{I_0(t-z/c)}{1-\{1-\exp(-\sigma N\lambda_2\int_0^z C_2 dz)\}\exp -\lambda_2\zeta_0(t-z/c)} \; ; \; \zeta_0(t) = \sigma\int_{-\infty}^{t} I_0(t')dt'/2\hbar\omega_p, \tag{9}$$

where $I_0(t)$ is the pulse intensity at the input to a three-level medium (z=0). On principle, the coefficient $C_2$ in a three-level medium depends not only on the population difference in optical transitions but also on the LF coherence. Owing to this fact, the pulse can grow even in the inversionless medium (at $u_0 < 0$) and, vice versa, decrease in spite of inversion population (at $u_0 > 0$).

Inversionless amplification occurs in a wide range of initial conditions of the medium, which is determined, besides $C_2 > 0$, by the following inequalities:

$$0 \leqq \rho_{33}^{(0)} \leqq \rho_{11}^{(0)}, \rho_{22}^{(0)}; \quad |\rho_{21}^{(0)}| \leqq \sqrt{\rho_{11}^{(0)}\rho_{22}^{(0)}};$$

$$Sp(\rho^{(0)})^2 \equiv \sum_{i=1}^{3} \rho_{ii}^{(0)2} + 2|\rho_{21}^{(0)}|^2 \leqq 1 . \tag{10}$$

The first one means that the inversion is absent. The last two are needed for positive definiteness of the density matrix $\hat{\rho}$.

It should be emphasized that even a small population of the upper level ($\rho_{33}^{(0)} \ll 1/3$) is sufficient for inversionless amplification if the maximum value of the LF coherence is excited: $-u_0 \lesssim (\rho_{11}^{(0)}\rho_{22}^{(0)})^{1/2}$. According to (7), the gain will be the same as in a two-level system with the inversion equal to $\rho_{33}^{(0)}$.

The coherence $\rho_{21}$ can be excited, for example, using a resonant microwave field in the form of a $\pi/2$ pulse. The latter transform the atoms into a state with the maximum value of the LF coherence $\mathrm{Im}\rho_{21}$, and equal populations of sublevels /6/. After the field action the Bloch vector corresponding to a LF transition will be rotating in the plane ($u$, $\mathrm{Im}\rho_{21}$) with a frequency $\omega_{21}$ (see (2a)). The result of optical pulse transformation in a three-level system will depend on the phase of the LF oscillations just before the ultrashort pulse. To achieve the maximum attainable (at a given $\rho_{33}^{(0)}$) gain, the optical pulse has to reach the medium exactly when $u < 0$ is negative and maximum in module. For this purpose the optical pulse delay with respect to the microwave pulse has to be a quarter or three quarters of the period $2\pi/\omega_{21}$ depending on the sign of $\tilde{n}^*$, where $\tilde{n}^*$ is $\tilde{n}$ before the microwave pulse arrival. According to the conservation of the Bloch vector, the increase in this initial population difference $\tilde{n}^*$ leads to a greater coherence /6/ $|u_0| = |\tilde{n}^*|$. For this increase one of the sublevel must be depleted, for example, by cooling the medium or using a resonant field, selective in frequency or polarization. Introducing $|u_0| = |\tilde{n}^*|$ into the inequality $C_2 > 0$ and taking into account that the $\pi/2$ pulse does not change the total population of sublevels, we find the necessary condition for inversionless amplification: $\rho_{33} > \min(\rho_{11}^*, \rho_{22}^*)$. Thus, the upper level must be more populated than one of the sublevels before the $\pi/2$ pulse. Note that the latter does not mean that the amplification is possible before the LF coherence excitation. Indeed, because of the large width of the homogeneous lines ($T_2^{-1} >> \omega_{21}$), the field interacts with both optical transitions at once and at $\rho_{33}^{(0)} < 1/3$ the absorption in the noninverted transition prevails. The excitation of low-frequency coherence by a $\pi/2$ pulse in the case $\eta = 1$ excludes this absorption and leads to the same gain as for the inverted transition.

Another way of atom operation before the pulse in the presence of an adjacent level 3' (Fig. 1) is to use two monochromatic fields with frequencies $\omega_{31}$ and $\omega_{32}$ (at $(T_2')^{(-1)} << \omega_{21}$) or a pulse train (at $T_2^{-1} >> \omega_{21}$).

The maximum value of the LF coherence excited by these methods is determined by the equilibrium population of the upper level: $u_0 = (3\rho_{3'3'}^{(0)}-1)/2$. Let us substitute it into the inequality $C_2 > 0$. Then the necessary condition of amplification takes the form $\rho_{33}^{(0)} > \rho_{3'3'}^{(0)}$. This means that the operating level must be more occupied than the adjacent one.

Partial population of the upper level needed for the amplification can be provided by noncoherent optical pumping or other traditional methods.

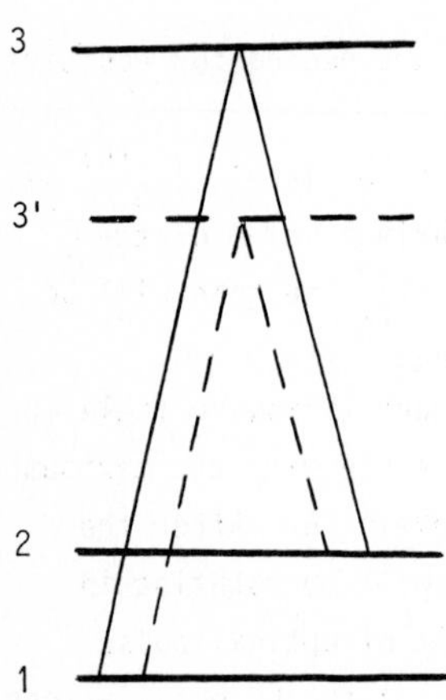


Figure 1. Λ-scheme (levels 1 and 2 are optically coupled with the upper level 3). The dashed line means adjacent level 3' by means of which the resonant optical pumping can be provided in the transitions 1-3', 2-3'.

## 5. CONCLUDING REMARKS

We have discussed the possibility of the inversionless amplification in the process of light propagation through the three-level medium. The main attention is paid on the case of ultrashort light pulse satisfying the conditions $T_2^{-1} \gg \tau_p^{-1} \gg \omega_{21}$ when the pulse interacts resonantly with both optical transitions at once. It is pointed out that the inversionless amplification can take place likewise in the case of narrow homogeneous optical lines when atoms interact with two monochromatic fields the frequencies of which $\omega_a \approx \omega_{31}$ and $\omega_b \approx \omega_{32}$ satisfy the condition $\omega_a - \omega_b = \omega_{21}$.

The amplification due to the parametric instability arises here if the rather great LF coherence ($|\sigma_{21}| > \sqrt{n_{13}n_{23}}$) is excited by the action of some external source. If one uses for this purpose two resonant fields interacting with the adjacent transitions (Fig. 1), the maximum value of the LF coherence will be achieved when the pump power satisfies the condition $I \gg I_0 = c\hbar^2/8\pi\mu'^2 T_2'\tau_2$. Here $\mu'$ is the dipole momentum, $T_2'$ is the transverse relaxation time for the adjacent transitions 1-3' and 2-3'. The condition of amplification takes the form: $\rho_{33}^{(0)} > \rho_{3'3'}^{(0)}$, which is the same as for the ultrashort pulse. The gain is determined by the population difference $\rho_{33}^{(0)} - \rho_{3'3'}^{(0)}$ and in the case $\rho_{33}^{(0)} - \rho_{3'3'}^{(0)} \ll \rho_{11}^{(0)} - \rho_{33}^{(0)}$ it is equal to $k \approx g(\rho_{33}^{(0)} - \rho_{3'3'}^{(0)})$, where $g = 2\pi\mu^2 T_2(\omega_a+\omega_b)/c\hbar$.

The analysis of nonlinear regimes of amplification taking into account pumping exhaustion and the influence of the amplificating fields on the LF coherence is of interest.

At last, both in the case of broad and narrow optical lines the investigation of a nonlinear interaction field with the three-level (Fig. 2) system taking into account the feedback (when atoms are placed in a resonator) is an open problem. Specifically the determination of a threshold and the parameters of inversionless generation is of interest.

The predicted effect of the inversionless amplification can be used to obtain the generation in those quantum transitions where the population inversion is difficult to reach.

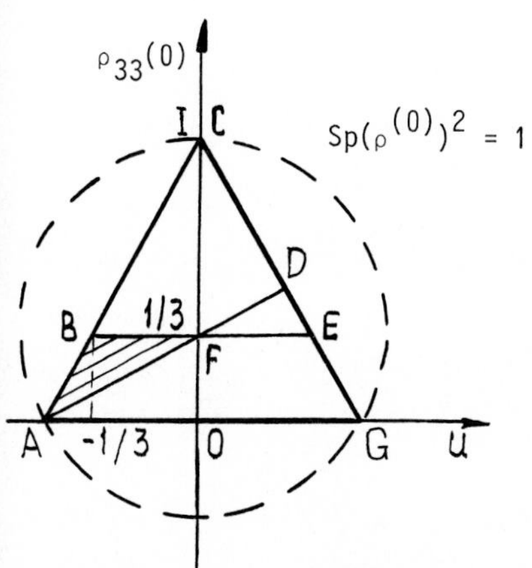


Figure 2. The region of physically allowed initial states of a three-level system (ACG) in the case $\mathrm{Im}\rho_{21}^{(0)}=0$, $\tilde{n}_0=0$. The letters mean the regions of inversionless amplification (ABF), amplification in inverted medium (BCDF) and absorption in inverted medium (DEF).

This effect can be realized probably in the sodium vapor. The pumping may be achieved by the dye-laser radiation and $D_1$ and $D_2$ lines of sodium can be used as operating and adjacent transitions. The excitement of the LF coherence at the superfine splitting of the ground state (1.77 GHz) is possible.

REFERENCES


1. Orriols G. Nonabsorption resonances by nonlinear coherent effects in a three-level system. Nuovo Cim., 1979, B53, No.1, 1-24.
2. Kocharovskaya O.A., Khanin Ya.I. Population trapping and coherent bleaching of a three-level medium by a periodic train of ultrashort pulses. Sov.Phys.-JETP, 1986, 63, 945-949.
3. Konopnicki M.J., Eberly J.H. Simultaneous propagation of short different wavelength optical pulses. Phys. Rev., 1981, A24, No.5, 2567-2583.
4. Kocharovskaya O.A., Khanin Ya.I. Coherent amplification of ultrashort pulse in a three-level medium without population inversion. Sov.Phys.-JETP Lett., 1988, 48, No.11, 581-584 (in Russian).
5. Frantz L.M., Nodvik J.S. Theory of light pulse propagation through light amplifier. J.Appl.Phys., 1963, 34, 2346-2356.
6. Allen L., Eberly J.H. Optical Resonance and Two-Level Atoms. Wiley, New York, 1975.

# On the Nonlinear Critical State Dynamics in Type II Superconductors

*I.L. Maksimov*

Physics Department, Gorky University, 23 Gagarin Ave., 603600 Gorky, USSR

The nonlinear dynamics of the critical state in type II superconductors is investigated taking into account the effects of dispersion and dissipation. Conditions under which a nonlinear thermomagnetic wave propagates inside the superconductor are found: the structure and velocity of the wave front are determined. It is shown that the wave profile is stable with respect to small thermal and electromagnetic disturbances. However, the mutual interaction of two "positive-energy" nonlinear waves may cause an explosive critical state dynamics in superconductors.

## 1. INTRODUCTION

The dynamics of the critical state in type II superconductors is not yet completely described. Meanwhile, the existence of the inherent critical state instability - the magnetic flux jump /1/ - makes this problem considerably interesting. Both the emergence condition and the instability dynamics have been investigated earlier in the linear approximation with respect to small thermal or electromagnetic perturbations /2,3/. The obtained instability criteria are in sufficiently good agreement with the corresponding experimental data (see e.g. /4/ and references therein). The further stage of an instability development has been discussed mainly in connection with so-called phenomena of limited flux jumps /2,4-6/. The earlier attempts to investigate the critical state dynamics in quasi-nonlinear approximation /2,5/ were not sufficiently well-grounded and self-consistent.

In the present paper the nonlinear critical state dynamics is investigated taking into account the effects of dispersion and dissipation. It is found that the thermomagnetic shock wave exists in a superconductor; the estimation is obtained for the velocity and the width of shock wave front. The stability problem of the found nonlinear wave is discussed; it is shown that solitary shock wave is stable with respect to small variations of the wave profile. The explosive behaviour of the thermomagnetic disturbances of large magnitude has been discovered to occur on the final stage of the magnetic flux penetration inside the sample. It is shown that the nonlinear interaction between two thermomagnetic shock waves could be responsible for the explosive critical state dynamics in superconductors.

Research Reports in Physics Nonlinear Waves 3
Editors: A.V. Gaponov-Grekhov · M.I. Rabinovich · J. Engelbrecht

## 2. CRITICAL STATE MODEL

The evolution in time and the space distribution of temperature T, electrical $\vec{E}$ and magnetic $\vec{H}$ fields in superconductor are described by the heat diffusion equation and the Maxwell equations:

$$\nu\dot{T} = \nabla(\kappa\nabla T) + \vec{j}\cdot\vec{E} \tag{1}$$

$$\nabla\times\vec{H} = \frac{4\pi}{c}\vec{j} \tag{2}$$

$$\nabla\times\vec{E} = -\frac{1}{c}\frac{\partial\vec{H}}{\partial t} \tag{3}$$

where $\nu = \nu(T)$ and $\kappa = \kappa(T)$ are the heat capacity and the heat conductivity coefficients, respectively. The current-voltage characteristics of the superconductor in flux flow (and/or flux creep) regime should be taken in the following form:

$$\vec{j} = \vec{j}_0(T,\vec{E}) = \vec{j}_c(T) + \vec{j}_1(\vec{E}) \ . \tag{4}$$

Here $j_c$ is the critical current density, $j_1(\vec{E})$ is the current density connected with the magnetic flux motion. In a wide temperature interval $j_c$ can be described as a linear function on T:

$$j_c = j_0 - a(T-T_0) \tag{5}$$

where $a = dj_c/dT \cong j_0/(T_c-T_0)$ is the thermal heat softening coefficient of the magnetic flux pinning force, $T_c$ and $T_0$ are the critical and the equilibrium temperatures of the superconductor, respectively. To simplify the further mathematical analysis of the problems, we have chosen the so-called Bean critical state model /7/, i.e. we neglect thereafter the critical current density dependence upon the magnetic field intensity H. It is important that function $j = j_0(E)$ being practically linear at $E \gg E_0$, is characterized by large differential conductivity $\sigma(E) = dj_1/dE$ in the flux creep region $E \leq E_0$ /8/. As it was shown experimentally in /9,10/ dependence $j_1(E)$ can be approximated by function $j_1(E) = j_1\, \ell n(E/E_0)$. For more detailed discussions of the applicability area of the used approach see /4/.

## 3. NONLINEAR THERMOMAGNETIC WAVE

Let us first consider the flat semiinfinite sample ($x \geq 0$) under adiabatical thermal conditions placed in an external electrical field $E = (0,E_e,0)$. We shall seek the solution of the system (1)-(5) in the "travelling wave" form:

$$E_y = E(\xi), \quad H_z = H(\xi), \quad T = T(\xi) \quad \xi = x - vt \ .$$

Using such a substitution, one can reduce the order of the system of differential equations (1)-(5) (naturally, the boundary conditions inside the sample volume $E(\infty) = 0$, $E'(\infty) = 0$, $T(\infty) = T_0$ are to be taken into account). The relationship $j = -(c^2/$

$/4\pi v)E_\xi'$ immediately follows from equations (2)-(3) and the electrodynamic boundary condition at $\xi \to +\infty$. Relationship $E = Hv/c$, reflecting the transformation of coordinates chosen, also naturally follows from equation (3). Finally, one finds the nonlinear equation to describe the electrical field distribution inside the superconductor:

$$E_{\xi^2}'' + \frac{v}{D_m(E)}\left[E_\xi' + \frac{va}{\sigma(E)\kappa(T)}\int_{T_0}^{T}\nu(T)dT\right] - \frac{a}{2\kappa(T)}\quad E^2 = 0\ . \tag{6}$$

Dependence $T = T(E,E_\xi')$ is defined by expressions (2)-(5):

$$T(E,E_\xi') = T_0 + \frac{1}{a}\ j_0(T_0,E) + \frac{c^2}{4\pi v}\ E_\xi'\quad . \tag{7}$$

The boundary conditions on the sample surface ($\xi = -\infty$) for the wave with the fixed amplitude of electrical field have the form: $E(-\infty) = E_e$, $E'(-\infty) = 0$, $T'(-\infty) = 0$.

The analysis of the phase plane $(E,E_\xi')$ of equation (6) shows that there are two equilibrium points: 1) $E = 0$, $E_\xi' = 0$; $T = T_0$ and 2) $E = E_e$, $E_\xi' = 0$; $T = T^* = T(E_e,0)$. The separatrix joining those two equilibrium points represents the solution of the shock wave type with amplitude $E_e$. Note that within the linear voltage-current characteristics region ($E \geq E_0$) in the case of a weak heating ($T^* - T_0 << T_c - T_0$) equation (6) coincides with the first integral of the stationary Korteweg-de Vries-Burgers equation, which has a solution of the shock wave type /11/. The "eigenvalue" $v_E$ in the problem has to be found from the requirement that the general solution $E = E_v(\xi)$ should coincide with separatrix $E = \overline{E}_{vE}(\xi)$. Finally, the velocity of an "E-wave" is determined by the expression:

$$v_E = cE_e\left(8\pi\int_{T_0}^{T^*}\nu(T)dT\right)^{-1/2} \tag{8}$$

where $T^* = T(E_e,0)$ and $H_e = E_e \cdot c/v_e$ is the magnetic field on the sample surface induced by the external curl electrical field.

The width of the shock wave front $\Delta\xi$ can be found by investigating the asymptotical behaviour of the solution $\overline{E}(\xi)$ in a close vicinity of the equilibrium points /11/. Simple calculations give the estimation for $\Delta\xi$:

$$\Delta\xi \cong \frac{2}{v}\cdot\left(\frac{\kappa}{v} + D_m\right)\Big|_{T=T^*,E=E_e}\quad . \tag{9}$$

It is seen that the width of the wave front is determined mainly by the largest one among the thermal $D_{th} = \kappa/\nu$ or magnetic $D_m = c^2/4\pi\sigma(E)$ diffusion coefficients.

## 4. STABILITY OF THE NONLINEAR WAVE

To consider the stability of the nonlinear stationary wave found with respect to small perturbations of the wave profile, one should solve a rather complicated system of linear partial differential equations with the variable coefficients. The complete investigation of the wave stability problem in the general case is not available because of the absence of the analytical solution of equation (6). Nevertheless, this problem permits an analytical investigation for a certain specified dependence $j_0(E)$ in the weak heating limit $((T^*-T_0)/(T_c-T_0) << 1)$. Namely, we shall consider the situation when $\partial j_0/\partial E \equiv \sigma_0$ = const at $E \to 0$ /12/. On the basis of the linear approch it has been ascertained that the dynamics of thermomagnetic disturbances in type II superconductors depends essentially on the ratio $\mu = D_{th}/D_m$. In the case $\mu >> 1$, magnetic diffusion is slow compared with the thermal diffusion process: therefore the effect of dispersion on the wave dynamics is neglible. Correspondingly, equation (6) coincides with the first integral of the Burgers equation /13/. The solution of (6) is:

$$\overline{E}(\xi) = 0.5E_e\left[1 - \tanh \frac{\nu V}{2\kappa} \xi\right],$$

here $\nu \equiv \nu(T_0)$. For the materials with the high Ginsburg-Landau parameter ("hard" superconductors) $\mu$ is usually small: $\mu >> 1$; therefore one can neglect the heat redistribution inside the sample. Surprisingly, the "travelling wave" solution of equations (1)-(5) in this case has practically the same appearance:

$$\overline{E}(\xi) = 0.5E_e\left[1 - \tanh \frac{\nu V\mu}{2\kappa} \xi\right].$$

In order to describe these two situations simultaneously, we shall use below the following interpolation for $E = \overline{E}(\xi)$:

$$\overline{E}(\xi) = 0.5E_e\left[1 - \tanh(\frac{\nu V}{2\kappa} \frac{\mu}{1+\mu} \xi)\right] \tag{10}$$

which is asymptotically valid for both cases considered. In order to find the wave stability condition, we must linearize the system of equations (1)-(5) with respect to small thermal $\delta T = T - \overline{T}(\xi) = \delta T(\xi)\cdot\exp(\lambda t)$ and electromagnetic $\delta E = E - \overline{E}(\xi) = \delta\overline{E}(\xi) \cdot \exp(\lambda t)$; $\delta H = H - \overline{H}(\xi) = \delta\overline{H}(\xi) \cdot \exp(\lambda t)$ perturbations of the initial wave profile $\overline{T}(\xi)$, $\overline{E}(\xi)$, $\overline{H}(\xi)$ defined by relations (7) and (10). By performing simple calculations and introducing the dimensionless variable z and increment

$$z = \frac{\nu V}{2\kappa} \frac{\mu}{1+\mu} \quad , \quad \Lambda = \lambda \cdot \frac{4\kappa}{\nu V^2} \frac{\mu}{1+\mu} \quad ,$$

we get the equation for function $\Psi(z) = \cosh z \cdot \delta\overline{E}$ :

$$\Psi'' + \left[-\Lambda(2-\tanh z) - 1+ \frac{2}{\cosh^2 z}\right]\Psi = 0 . \tag{11}$$

Here we have used the fact that relatively slow perturbations with $\Lambda \leqq 1$ are dominating near instability threshold (/3/; note also that for hard superconductors $\Lambda \sim \mu$) and consequently, the terms of the order $\Lambda^2$ and higher in equation (11) were neglected. Formally, equation (11) can be interpreted as the Schrödinger equation for the quantum particle in the nonsymmetrical potential well (note that the profile of the well depends on particle "energy"). Physically, the odd term in equation (11) describes the drift effect of the travelling wave on the dynamics of small disturbances of the wave profile. The solution of (11) limited at $\xi \to +\infty$ has the form:

$$\Psi(z) = (1-\tanh z)^{p-0.5} \cdot (1+\tanh z)^{q-0.5} \cdot F(\alpha,\beta,\gamma; \frac{1-\tanh z}{2})$$

where

$$p = \frac{1}{2}\Big[ (1+3\Lambda)^{1/2} + 1\Big]; \quad q = \frac{1}{2}\Big[ (1+\Lambda)^{1/2} + 1\Big];$$

$$\alpha = 1 + p + q, \quad \beta = \alpha - 3, \quad \gamma = 2q .$$

Using well-known transformation formula for the hypergeometric function $F(\alpha,\beta,\gamma\ x)$ (see, e.g. /14/), it is easy to show that the asymptotical behaviour of the electrical field perturbation $\delta E$:

$$\delta\bar{E}_\Lambda = \delta\bar{E}_\Lambda(0)(1-\tanh z)^p(1+\tanh z)^q F(\alpha,\beta,\gamma; \frac{1-\tanh z}{2})$$

at $z \to -\infty$ is the following:

$$\delta\bar{E}_\Lambda\Big|_{z\to -\infty} \sim \exp\,(1+q-2p)z \quad .$$

Since $1+q-2p = 0.5\ 1+\sqrt{1+\Lambda} - 2\sqrt{1+3\Lambda} \leq 0$ for all $\Lambda \geq 0$, then $\delta\bar{E}_{\Lambda\geq 0} \to \infty$ at $z \to -\infty$. Using analogous speculations it is easy to show that another linearly independent solution of equation (11) has the same singularity at $z \to +\infty$ provided $\Lambda \geq 0$. Hence, the limited solutions of equation (11) are possible only for $\Lambda \leq 0$. It means that finite thermomagnetic perturbations of the initial wave profile should eventually decrease in the course of the wave propagation. Thus, the found shock wave is stable against small profile disturbances. Another source of disturbances can arise from the shock waves interaction, for example, during magnetic flux jump. The solution of this problem requires more detailed nonlinear analysis of critical state dynamics. One possible approach based on the found asymptotically exact solution will be discussed below.

## 5. EXPLOSIVE DYNAMICS OF THE CRITICAL STATE IN SUPERCONDUCTORS

Let us now consider the plane superconducting sample of thickness 2b ($0 \leq |x| \leq b$) with the symmetrical current and electrical field $\vec{E} = (0,E,0)$ distribution relative to the plane $x = 0$. We will not limitate ourselves by searching for a certain self-similar solution of the problem, as we did in Sections 3 and 4, but we shall try to find an exact one. In general, the complete analytical investigation of such an essentially nonlinear system of equations (1)-(5) is not available. Therefore, we shall consider the latest stage of critical state dynamics, when the thermomagnetic disturbances become well developed, i.e. reach a considerable magnitude. For such disturbances with large amplitude $T \leq T_c$, $E \gg E_0$ the current-voltage characteristics of superconductor can be chosen in the linear form /3/:

$$j = j_0 - a(T-T_0) + \sigma E$$

and the temperature dependence of the heat capacity coefficient can be neglected:

$\nu(T) \cong \nu(T_c) \equiv \nu$. In order to present more clear physical picture of thermomagnetic instability and, consequently, thermomagnetic shock waves evolution, we shall consider the most simple case of hard superconductors ($\mu << 1$). For these materials, as it is well known, the flux jump is of adiabatical character, i.e. instability grows with heat flux being practically fixed /4/. Since the heat redistribution inside the sample is neglible, then one can omit the first term in the r.h.s. of equation (1). By excluding the temperature variable T from system (1)-(5), we find the nonlinear partial differential equation for $E = E(x,t)$:

$$(E \cdot \dot{E}'' - \dot{E} \cdot E'') + D_m^{-1} \cdot (\dot{E}^2 - E \cdot \ddot{E}) + \frac{a}{\nu} \cdot E^2 \cdot E'' = 0 . \tag{12}$$

Separating the variables by means of $E(x,t) = \psi(x) \cdot \phi(t)$, we find that this substitution is self-consistent provided that functions $\psi(x)$ and $\phi(t)$ satisfy the following conditions:

$$\psi_{x^2}'' = A = \text{const} . \tag{13}$$

and

$$\phi \cdot \ddot{\phi} - \dot{\phi}^2 = \frac{2}{\tau^2} \cdot \phi^3 , \tag{14}$$

where $\tau = (8\pi\nu\sigma/ac^2A)^{1/2}$ . The increasing disturbances $\dot{\phi} \geq 0$ are described by the solution:

$$\phi(t) = \frac{(B\tau)^2}{\sinh^2(\gamma - Bt)} ,$$

where $4B^2$ is the first integral of equation (4) and $\sinh \gamma = B\tau$ ( we assume that $\phi(0) = 1$). The expression for temperature and electrical field distribution have the following form (we are interested in the even solution of (13) with respect to the sample symmetry plane $x = 0$):

$$E(x,t) = \frac{(\frac{1}{2} Ax^2 + \tilde{E})}{\sinh^2(\gamma - Bt)} \cdot (B\tau)^2$$

$$T(x,t) = T_0 + \frac{1}{a} \cdot \left[ \sigma \cdot E(x,t) - 2 \frac{\sigma\nu}{a\tau} \cdot \frac{\sinh(Bt)}{\sinh(\gamma - Bt)} \right] \tag{16}$$

where $\tilde{E}$ is the integration constant, representing the intensity of the electrical field due to fluctuations within undisturbed region /4,15/. It is seen that the electrical field intensity as well as the sample temperature tend to become infinite for a finite time interval $\Delta t = \gamma/B$. Such a behaviour is typical for the so-called explosive instability development /16/, which has been predicted and observed in plasma /17-20/ and for hydrodynamic flow with a nonuniform velocity profile /21,22/. However, in superconductors the development of explosive instability is limited by the condition $T \leq T_c$. Therefore, the equation $T(x,t) = T_c$ determines the time-dependent position $\overline{x} = \overline{x}(t)$ of the boundary between normal and superconductive regions within the sample.

The obtained results enable us to give a more detailed description of the critical state instability development in type II superconductors. As it was shown earlier, the joint exponential growth of both the thermal and electromagnetic perturbations occurs if the magnetic field on the sample surface $H_e$ exceeds a certain value $H_j$ ($H_j$ being the flux jump field) /2-4/. This linear stage lasts for a time interval $\Delta t_i \sim$ $\sim t_j \ln(E^*/\overline{E})$, $t_j$ being the characteristic instability rise time /3/, $\overline{E} \cong E_0$ being an average magnitude of initial perturbation transferring the critical state into an unstable flux-flow regime /4/. As it follows from the results of Section 3, on the further stage of transport, the two current induced thermomagnetic shock waves (+ for $x \geqq 0$ and - for $x \leqq 0$) with amplitudes $E_{\pm} = E^*$ (correspondingly, $H_{\pm} = \mp H_j$) and velocities $v_{\pm} = \mp v$; $v = c \cdot E^*/H_j$, can propagate inside the superconductor in opposite directions. However, this picture represents only the intermediate-asymptotical regime of system evolution /23/, based on the solitary shock wave approach. These waves propagate without interference until they reach (at t = 0 for the present problem) the distance $\sim 2\Delta\xi$; $\Delta\xi$ being the width of shock wave front. The mutual nonlinear interaction occurring at $t \geqq 0$, form the explosive instability behaviour on the final stage of the magnetic flux jump.

## 6. DYNAMICS OF THE MAGNETIC FLUX PENETRATION

Using this qualitative picture and solutions (15)-(16) we can find the dependence $\overline{x}(t)$ for the case of the full magnetic flux jump. We consider a situation when shock wave propagation causes normal transition behind its front; hence $E^* \cong j_0/\sigma$ (to simplify estimations, we neglect below the difference between the normal and the flux-flow conductivities). We assume that at t = 0 : $\overline{x}_{\pm}(0) = \pm\Delta\xi$ and $\tilde{E} = 0$ (i.e. we neglect both the electromagnetic fluctuations of field intensity within undisturbed region and the waves overlapping at t = 0). With the help of condition $T(\Delta\xi,0) = T_c$ (or $E(\Delta\xi,0) =$ $= E^* = j_0/\sigma$) one finds the estimation for A: $A = 2E^*/(\Delta\xi)^2$ and, finally, the function $\overline{x}_-(t)$ ($\overline{x}_+ = -\overline{x}_-$):

$$\overline{x}_-(t) = -\frac{\Delta\xi}{\tau}\frac{\sinh(\gamma-Bt)}{B}\left(1+\frac{2\nu\sigma}{a^2(T_c-T_0)\tau}\frac{\sinh Bt}{\sinh(\gamma-Bt)}\right)^{1/2} . \tag{17}$$

In the limiting case when $B\tau \ll 1$, expression (17) may be simplified:

$$\overline{x}_-(t) = -\Delta\xi \cdot (1-t/\tau) \cdot \left(1+\frac{2D_m}{(\Delta\xi)^2} \cdot \frac{t}{(1-t/\tau)}\right)^{1/2} . \tag{18}$$

It is seen from expression (18) that magnetic flux penetration speed $u = d\overline{x}_-/dt$ has a square root singularity near the closure point: $u_t$

$$u|_{t\to\tau} \cong \left(\frac{D_m}{2(\tau-t)}\right)^{1/2} .$$

Note that closure speed $u|_{t\to\tau}$ does not depend on the joining waves amplitude $E^*$. With known value of A, the explosion time $\tau$ can be evaluated as:

$$\tau = \frac{2}{\pi} \cdot \frac{\Delta\xi\ell}{D_m} \cdot \left(\frac{\pi^3\nu(T_c-T_0)}{H_j^2}\right)$$

where $\ell = cH_j/(4\pi a(T_c-T_0))$ is the equilibrium macroscopic magnetic field penetration depth given by linear theory /3/. Taking into account the $H_j$ value for quasi-adiabatic flux jump $H_j^2 = \pi^3\nu(T_c-T_0)$ /2,3/, the velocity value of the shock wave with large amplitude is $v = cj_0/\sigma H_j \cong D_m/\ell$ and the wave front width estimation in the limit $\mu \ll 1$ is $\Delta\xi \cong 2D_m/v \cong 2\ell$ (9), we obtain finally:

$$\tau = \frac{4}{\pi} \cdot \frac{\ell^2}{D_m} \quad . \tag{19}$$

This time interval is much shorter than the linear instability rise time $t_j = 4t_m\pi^{-2}\mu^{-1/2}$ /3/ and (provided $b \geqq \ell$) it is comparable with the wave propagation time $t_w \cong b/v = bt_m/\ell$. Actually, $\tau$ coincides with the system magnetic diffusion time $t_m = \ell^2/D_m$ /2,3/.

## 7. CONCLUSIONS

Thus, nonlinear thermomagnetic wave is able to propagate inside the superconductor. We should emphasize that this result was obtained for an arbitrary temperature dependence of thermophysical parameters $\nu$ and $\kappa$ of superconducting material and for an arbitrary function $j_1(E)$. Moreover, since the system of equations (1)-(5) has a translational invariancy, the wave propagation conditions can be found for an arbitrary critical current density dependence on T and H. Note that for the samples of the finite thickness 2b the used boundary conditions are valid provided $\Delta\xi \ll b$. It is seen that the found nonlinear waves can be observed only for sufficiently thick samples. Possibly, such thermomagnetic waves have been observed by URBAN /25/, during his experiments on the limited instabilities in macroscopic superconducting samples.

As it was shown in Section 4, the nonlinear stationary wave is stable with respect to small variations of the wave profile. However, wave profile can be considerably modified because of the nonlinear interaction with another solitary wave. This situation can occur in the course of the evolution of the critical state instability in superconductors. For example, when magnetic flux jump is initiated by the current increase, mutual waves interaction can form an explosive critical state bahaviour on the final stage of instability development (Section 5). Note that in superconductors the explosive dynamics results from the interaction of "positive-energy" nonlinear waves. This mechanism of explosive instability formation is a new one and differs essentially from the three-waves-interaction mechanism responsible for such an instability in plasma /17/ or in a hydrodynamic boundary layer /16,22/. The dynamics of large ($E \gg E_0$) thermomagnetic disturbances of critical state in type II superconductors is characterized by the quadratic nonlinearity together with strong dispersion and intense dissipation. From the mathematical point of view the explosive critical state dynamics is connected with the explosive behaviour of the solution of the Burgers equation, reported in /24/ for media with negative viscosity. Indeed, regarding the asymptotics of (18) at $t \to \tau$:

$$|\bar{x}_{\pm}(t)| \cong (2D_m(\tau-t))^{1/2}, \quad t \to \tau \tag{20}$$

the magnetic flux confluence can be interpreted as a diffusion process with the negative diffusion coefficient $-2D_m$. It can be easily shown that magnetic field intensity has a square root singularity on the surface $x = \bar{x}_-(t)$:

$$H_-(\bar{x}_-(t),t) = \frac{16}{\pi^2} \cdot \frac{\Phi_i}{|(\bar{x}_-(t)|} . \tag{21}$$

Physically, such a sharpening of the magnetic field profile results from the initial magnetic flux ($\Phi_i = 0.5H_j$) conservation in the course of the wave evolution. Probably, such an effect was first observed by FLIPPEN /26/ who reported sharp voltage spikes on the final stage of the magnetic flux penetration inside the sample. From the spectral point of view the amplification of high-frequency modes, ensured by the energy pumping from the low-frequency harmonics, is responsible for the abrupt sharpening of the travelling wave profile in the systems with negative viscosity /16, 24/.

We should recognize that the exact analytical solution of system (1)-(5) for the explosive stage is not available in the general case. Physically, it is clear, however, that dependences $\nu = \nu(T)$, $j_c(H)$ of limited character cannot radically affect the explosive dynamics described above. For instance, temperature effect on heat capacity coefficient ($d\nu/dT \geq 0$) leads only to a certain instability retardation and results in the explosion time increase: $\tau \sim \sqrt{\nu}$; practically, value $\nu$ should be taken at $T = T_c$. Magnetic field effect on the explosion dynamics has to be investigated more thoroughly; for estimations one can relate critical current density to its value at $H = H_j$.

## REFERENCES


1. R.Hancox, Phys.Lett. 16 (1965) 208.
2. S.L.Wipf, Phys.Rev. 161 (1967) 404.
3. R.G.Mints, A.L.Rakhmanov, Rev.Mod.Phys. 53 (1981) 551.
4. R.G.Mints, A.L.Rakhmanov, Instabilities in Superconductors. Nauka, Moscow 1984 (in Russian).
5. I.L.Maksimov, R.G.Mints, J.Phys.D. 13 (1980) 1689.
6. J.Chicaba, Cryogenics 10 (1970) 306.
7. C.P.Bean, Rev.Mod.Phys. 36 (1964) 31.
8. R.G.Mints, A.L.Rakhmanov, J.Phys.D. 15 (1982) 2297.
9. N.E.Alexeevskij, I.Hlasnik, A.V.Dubrovin, Sov.Phys.JETP 27 (1968) 47.
10. D.Gentile, W.Hassenzahl, M.Polak, Cryogenics 20 (1980) 37.
11. V.I.Karpman. Nonlinear Waves in Dispersive Media. Nauka, Moscow 1973 (in Russian).

12. A.I.Larkin, Yu.N.Ovchinnikov, Sov.Phys.JETP 41 (1976) 960.

13. G.Whitham. Linear and Nonlinear Waves. Wiley, New York, 1974.

14. E.Whittaker, G.Watson. A Course of Modern Analysis. Cambridge University Press, New York, 1965.

15. M.Tinkham. Introduction to Superconductivity. McGraw-Hill, New York 1975.

16. M.I.Rabinovich, A.L.Fabrikant, Radiophys. and Quantum Electronics 19 (1976) 508.

17. V.M.Dikasov, L.I.Rudakov, D.D.Riutov, Sov.Phys.JETP 21 (1966) 608.

18. J.Fukai, S.Krishan, E.G.Harris, Phys.Rev.Lett. 23 (1969) 910.

19. C.T.Dum, R.N.Sudan, Phys.Rev.Lett. 23 (1969) 1149.

20. F.Homan, in: Tenth International Conference on Phenomena in Ionized Gases. Oxford, 1971, p.323.

21. P.S.Klebanoff, K.D.Tidstrom, L.M.Sargent, J.Fluid Mech. 12 (1962) 1.

22. A.D.D.Craik, J.Fluid Mech. 70 (1971) 437.

23. G.I.Barenblatt. Similarity, Self-Similarity and Intermediate Asymptotics. Van Dyke, New York 1979.

24. E.N.Pelinovskii, V.E.Fridman, Prikladnaya Matematika i Mekhanika 38 (1974) 991.

25. E.Urban, Cryogenics 10 (1970) 291.

26. R.B.Flippen, Phys.Lett. 17 (1965) 193.

# Nonlinear Dynamics of Electrons in Metals

*V.Ya. Demikhovsky*
Gorky State University, 23 Gagarin Ave., 603600 Gorky, USSR

The dynamics of electron conduction (further simply electrons) with a complex spectrum in the field of a longitudinal ultrasonic wave in a constant magnetic field is investigated. A nonlinear theory of magneto-acoustic oscillations of sound absorption in metals is presented. The absorption at isolated cyclotron resonances as well well as under the conditions of global electron stochastization at the Fermi surface is considered.

## 1. INTRODUCTION

The propagation of sound and electromagnetic excitations in metals placed in a magnetic field is accompanied with numerous resonance phenomena caused by the interaction of a wave with electrons (see, e.g. /1/). The wave resonates with the electrons the mean velocity of which along the magnetic field $\overline{v}_z$ and the cyclotron frequency $\omega_c$ meet the condition

$$k_{||}\overline{v}_z - \omega - \ell\omega_c = 0,$$

where $k_{||}$ is the component of the wave vector along the z-axis directed along the magnetic field $B_0$, $\omega$ is the wave frequency and $\ell = 0,\pm1,\pm2...$ . These electrons (below referred to as resonance electrons) determine the damping and the spectrum of sound waves, helicons, dopplerons, alfven and other electromagnetic excitations propagating in metals.

The resonance condition was formulated neglecting the wave effect on electron trajectories. Apparently, a finite-amplitude wave affects primarly the motion of resonance electrons, with the nonlinear motion in the wave and magnetic fields leading to nonlinear phenomena of the wave-particle type that are well known in plasma physics.

Such effects have been extensively studied in the last decade in solid state physics. Collisionless sound damping /2,3/, nonlinear cyclotron damping of dopplerons, nonlinear damping of helicons /4/, anomalous skin-effect /5/ and other effects were investigated theoretically and experimentally in pure metals and semiconductors.

The theoretical description of the nonlinear phenomena mentioned above was based as a rule, on simplified concepts of electron spectrum: the momentum dependence of energy was assumed to be quadratic and the Fermi surface - spherical or cylindrical. Apparently, such an approach gives only rough dynamic characteristics of conduction.

Research Reports in Physics Nonlinear Waves 3
Editors: A.V. Gaponov-Grekhov · M.I. Rabinovich · J. Engelbrecht

Below we shall analyse the relation of the electron spectrum $\varepsilon(\vec{p})$ and the Fermi surface geometry of a real metal to the dynamics of electrons.

We shall first consider the dynamics of electrons in the field of a wave propagating at an arbitrary angle to the direction of a constant magnetic field and establish possible types of resonances. It will be shown that ordinary resonances at which the motion of electrons is similar to the motion of a mathematical pendulum, as well as more complex resonances, are possible in metals with a complex spectrum. Then (i) the characteristic frequencies of oscillation of trapped particles and the resonance width will be determined and (ii) the constant of damping of the longitudinal sound in metal will be calculated, i.e. a theory of nonlinear magneto-acoustic oscillations will be built up.

According to the present concepts, the motion of resonance particles is regular until the neighbouring resonances overlap. Otherwise the motion becomes chaotic. We shall show that in metal samples with the mean free path of electrons of the order of 1 cm, the conditions may be realized under which cyclotron resonances overlap and the motion of electrons in the wave field becomes chaotic over a greater part of the Fermi surface. In conclusion we shall consider the damping of sound in the regime of global stochasticity.

Results presented in this paper were obtained in collaboration with V.A.Burdov /6,7/.

## 2. EQUATIONS OF MOTION IN ACTION-ANGLE VARIABLES. RESONANCES

Classical equations of motion of electrons in a magnetic field are known to have a form

$$\frac{d\vec{p}}{dt} = \frac{e}{c}\left[\vec{v},\vec{B}_0\right], \quad \frac{d\vec{r}}{dt} = \vec{v}(\vec{p}), \quad \vec{v} = \frac{\partial\varepsilon}{\partial\vec{p}} \tag{1}$$

where $\vec{p}$ is a kinematic momentum, $\vec{r}$ is a coordinate, $\vec{v}$ is a velocity, e is an electron charge and c is the velocity of light. Energy $\varepsilon$ and momentum component $p_z$ in the magnetic field of direction $B_0 \parallel 0z$ are the integrals of system (1), i.e. the electron moves along the intersection line of the constant-energy surface $\varepsilon(p)$ = const. and the plane $p_z$ = const. in momentum space. In coordinate space, the projection of the electron orbit into the plane perpendicular to $\vec{B}_0$ is similar to the cyclotron orbit in the $\vec{p}$-space and is turned by an angle $\pi/2$ with respect to it along the z-axis. In contrast to free electrons, the motion of conduction electrons along the amgnetic field is nonuniform, with a mean velocity in this direction

$$\bar{v}_z = \frac{\partial\varepsilon(S,p_z)}{\partial p_z} \quad ,$$

where the energy is a function of $p_z$ and S, i.e. the area bounded by the electron orbit in momentum space. The cyclotron frequency of the electron is determined as

$$\omega_c = \frac{2\pi eB_0}{c} \frac{\partial\varepsilon(S,p_z)}{\partial S} ,$$

and the distance covered by the electron along the magnetic field during the cyclotron period is

$$2\pi \frac{\bar{v}_z}{\omega_c} = \frac{c}{eB_0} \frac{\partial S(\varepsilon,p_z)}{\partial p_z} .$$

For the investigation of the dynamics of electrons with a complex spectrum $\varepsilon(\vec{p})$ in the fields of travelling sound (or electromagnetic) waves in the time-independent uniform magnetic field it is convenient to replace $\vec{p}$ and $\vec{r}$ by new canonical variables of the action-angle type. For this purpose, calibrating the vector potential as $\vec{A}(0,B_0x,0)$, the Hamiltonian in the magnetic field can be written in the form

$$H_0 = \varepsilon(p_x,x-x_0,p_z), \tag{2}$$

where $x_0 = cp_y/eB_0$ = const. is the coordinate of the Larmor orbit center. Instead of the canonical variables $p_x$, x and $p_z$, z, let us introduce the variables I and $p_z$ and $\nu$ and Z conjugate to them. The action I is conventionally defined:

$$I = \frac{1}{2\pi} \oint p_x(\varepsilon,p_z,x)dx = \frac{c}{2\pi eB_0} \Phi p_x dp_y = \frac{c}{2\pi eB_0} \cdot S . \tag{3}$$

All other variables are determined by means of a generating function that depends on new momenta and old coordinates:

$$F_2(I,\tilde{p}_z,x-x_0,z) = z\tilde{p}_z + \int p_x(I,\tilde{p}_x,x-x_0)d(x-x_0) .$$

According to /8/, we have

$$p_z = \tilde{p}_z, \quad Z = z + \int \frac{\partial p_x(I,p_z,x-x_0)}{\partial\tilde{p}_z} d(x-x_0) , \quad \nu = \int \frac{\partial p_x}{\partial I} d(x-x_0) .$$

Canonical equations of motion in new variables are

$$\dot{\tilde{p}}_z = 0, \quad \dot{Z} = \left.\frac{\partial H_0}{\partial\tilde{p}_z}\right|_I = \bar{v}_z = \text{const.} ,$$

$$\dot{I} = 0, \quad \dot{\nu} = \left.\frac{\partial H_0}{\partial I}\right|_{\tilde{p}_z} = \omega_c = \text{const.} , \tag{4}$$

where $\bar{v}_z(I,\tilde{p}_z)$ is the mean velocity in the z-direction and $\omega_c(I,\tilde{p}_z)$ is the cyclotron frequency. Since the variables $p_z$ and $\tilde{p}_z$ coincide below, we shall not distinguish one from another.

Our main problem is to investigate the dynamics of electrons in the field of a sound wave propagation at an arbitrary angle to the magnetic field direction. The corresponding Hamiltonian has the form

$$H = H_0 - \Phi(\vec{p})\cos(k_{||}z + k_{\perp}x - \omega t)$$

or in the action-angle variables

$$H = \varepsilon(I,p_z) - \Phi(I,p_z\nu)\cos\left[k_{||}(Z+\Delta z(I,p_z,\nu)) + k_{\perp}(x_0+\Delta x(I,p_z,\nu)) - \omega t\right]. \tag{5}$$

Here $\Phi(\vec{p}) = \Lambda_{ik}(\vec{p})u_{ik}$, $\Lambda_{ik}$ are the constants of the deformation potential, and $u_{ik}$ is the tensor of lattice deformation. The wave vector is in the x,z-plane.

Using the perturbation theory of resonances /9/, we shall analyse the system described by the Hamiltonian (5). A similar problem for free electrons ($\varepsilon=p^2/2m$) in the field of a longitudinal wave propagating at an angle to the magnetic field was solved by SMITH and KAUFFMAN /10/.

For convenience, further consideration will be carried out in the coordinate system moving along the magnetic field with a velocity $\omega/k_{||}$. For this purpose we shall use the generating function

$$F_2 = (k_{||}Z - \omega t)p + \nu I , \tag{6}$$

which introduces a new phase $\psi = k_{||}Z - \omega t$ and momentum $p = p_z/k_{||}$. Transforming (5) with the help of (6) and expanding the $\nu$-periodic function describing the perturbation into the Fourier series, we obtain

$$H = \varepsilon(I,k_{||}p) - \omega p - \Phi_0\sum_m v_m(I,p)\cos(\psi-m\nu+\nu_m), \tag{7}$$

where $v_m(I,p)$ are the coefficients of the Fourier transform

$$v_m(I,p) = \frac{1}{2\pi}\int\cos(\psi+k_{||}\Delta z(\nu) + k_{\perp}\Delta x(\nu))\cos m\nu d\nu , \tag{8}$$

and $\nu_m$ are the initial phases. From the condition of a constant phase of the m-th harmonic of perturbation in (7) $\psi - \ell\nu + \nu_m$ = const. follows the condition for electron resonance

$$k_{||}\bar{v}_z - \omega - \ell\omega_c = 0 . \tag{9}$$

Consider now the dynamics of particles at resonance. In accordance with the resonance theory of perturbations, let us introduce a slow $\alpha$ and a fast $\beta$ phase:

$$\alpha = \psi - \ell\nu + \nu_e$$
$$\beta = \nu \tag{9a}$$

and conjugate momenta $p_\alpha = p$ and $p_\beta = I + \ell p_\alpha$. Now the term with $m = \ell$ in the sum in (7) depends only on the slow phase $\alpha$ while all other terms depend also on $\beta$. Averaging (7) with respect to the fast phase, we shall obtain an effective Hamiltonian at the $\ell$-th resonance:

$$H = \varepsilon(p_\alpha,p_\beta-\ell p_\alpha) - \omega p_\alpha - \Phi_0 v_\ell(p_\alpha,p_\beta)\cos\alpha, \tag{10}$$

which does not depend on $\beta$. Taking into account that $p_\beta$ = const., let us introduce into (10) the expansion in $(p_\alpha - p_\ell)$ where $p_\ell$ is the resonance momentum:

$$H = \varepsilon(p_\ell, p_\beta - p_\ell) - \omega p_\ell + \frac{\partial H_0}{\partial p_\alpha}(p_\alpha - p_\ell) + \frac{1}{2!}\frac{\partial^2 H_0}{\partial p_\alpha^2}(p_\alpha - p_\ell)^2 +$$

$$+ \frac{1}{3!}\frac{\partial^3 H_0}{\partial p_\alpha^3}(p_\alpha - p_\ell)^3 - \Phi_0 v_\ell \cos\alpha \ . \qquad (11)$$

Here

$$H_0 = \varepsilon(p_\alpha, p_\beta - \ell p_\alpha) - \omega p_\alpha \ .$$

The first derivative in (11) turns to zero when at resonance:

$$\frac{\partial H_0}{\partial p_\alpha} = \left.\frac{\partial H_0}{\partial p_\alpha}\right|_I + \left.\frac{\partial H_0}{\partial I}\right|_{p_\alpha} \cdot \left.\frac{\partial I}{\partial p_\alpha}\right|_{p_\beta} = k_{||}\bar{v}_z - \omega - \ell\omega_0 = 0 \ .$$

Omitting the constant term and calculating the second and the third derivatives, the Hamiltonian (11) can be written as

$$H = \frac{G}{2}(p_\alpha - p_\ell)^2 + \frac{R}{3}(p_\alpha - p_\ell)^3 - \Phi_0 v_\ell \cos\alpha \qquad (12)$$

where

$$G = -\frac{1}{2\pi m_c} \cdot \frac{\partial^2 S}{\partial p_z^2} \cdot k_{||}^2 \ , \qquad R = -\frac{1}{2\pi m_c} \cdot \frac{\partial^3 S}{\partial p_z^3} \cdot k_{||}^3 \ .$$

In the case of an ordinary resonance the terms of the order of $(p - p\ )^3$ can be neglected and (12) is reduced to a pendulum Hamiltonian with the characteristic frequency

$$\tilde{\omega}_\ell = k_{||}\left(\frac{\Phi_0 v_\ell}{2\pi m_c} \cdot \frac{\partial^2 S}{\partial p_z^2}\right)^{1/2} \qquad (13)$$

The maximal deviation of the momentum $p_\alpha$ from the resonant value is

$$\Delta p = 2\tilde{\omega}_\ell / G \ . \qquad (14)$$

When the resonance momentum $p_\ell$ is at the extreme point of the function $\partial S/\partial p_z$, coefficient G proportional to $\partial^2 S/\partial p_z^2$ vanishes and the Hamiltonian at the first resonance has the form

$$H = \frac{R}{3}(p_\alpha - p_\ell)^3 - \Phi_0 v_\ell \cos\alpha . \qquad (15)$$

Further we will consider in more detail the nature of motion at this resonance. The canonical equations which follow from (15) now have the form

$$\dot{p}_\alpha = -\Phi_0 v_\ell \sin\alpha \ ,$$

$$\dot{\alpha} = R(p_\alpha - p_\ell)^2 \ . \qquad (16)$$

The corresponding phase portrait is shown in Fig.1. The equilibrium points $p_\alpha = p_\ell$ and $\alpha = \pi n$ can be considered as the result of merging of the singular points of the

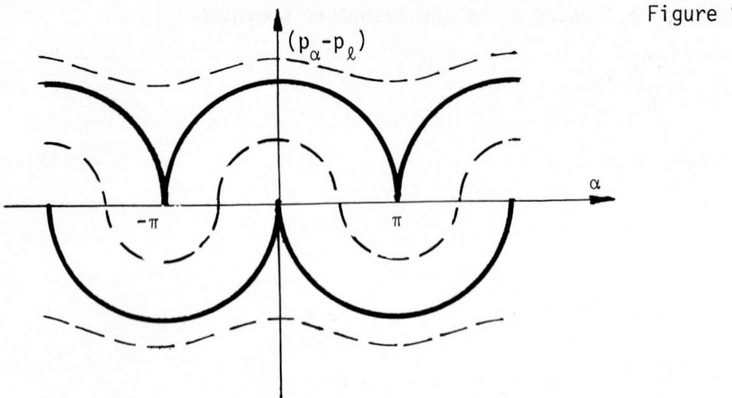


Figure 1.

"center" and the "saddle" types. The characteristic oscillation frequency is equal to

$$\tilde{\omega}_\ell^{\,ext} = (\Phi_0^2 v_\ell^2 R)^{1/3}$$

and the resonance width is $\Delta p_\ell^{\,ext} = (6\Phi_0 v_\ell R^{-1})^{1/3}$.

Let us consider, in particular, the case when the sound wave propagates perpendicularly to $B_0$. If the x-axis is directed along the wave vector, the Hamiltonian has the form

$$H = \varepsilon(I,p_z) - \Lambda_{xx}(\vec{p})k_\perp u_0 \sum_m v_m(I,p_z)\cos(m\nu - \omega t + \nu_m) \ . \tag{17}$$

In contrast to free electrons, a nondegenerate case is usually realized in metals, i.e. the cyclotron frequency depends on action (as well as on $p_z$) and $\partial^2\varepsilon/\partial I^2 \neq 0$. Therefore, the motion at resonance $\omega = \ell\omega_c$ in this case is similar to the motion of a mathematical pendulum with the Hamiltonian

$$H = \frac{G}{2}(I - I_\ell)^2 - \Phi_0 v_\ell \cos\alpha \ , \tag{18}$$

where $G = \ell^2(\partial^2\varepsilon/\partial I^2)$ and $\alpha = \ell\omega - \omega t$ is the slow phase. The trapped particle frequency is then

$$\tilde{\omega}_\ell = \left(\ell^2 \left.\frac{\partial^2\varepsilon}{\partial I^2}\right|_I \cdot \Phi_0 v_\ell\right)^{1/2} . \tag{18a}$$

Resonances with the cubic dependence of the Hamiltonian on action seem to be unlikely in this geometry.

## 3. DISTRIBUTION FUNCTION

The distribution function of electrons in the field of an ultrasonic wave may be found from a kinetic equation which in the action-angle variables has the form

$$\frac{\partial f}{\partial t} + \dot{p}\frac{\partial f}{\partial p} + \dot{I}\frac{\partial f}{\partial I} + \dot{\psi}\frac{\partial f}{\partial \psi} + \dot{\nu}\frac{\partial f}{\partial \nu} = -\frac{f - f_0}{\tau_p} \ . \tag{19}$$

The derivatives of the canonical variables $\dot{I}$, $\dot{p}$, $\dot{\nu}$ and $\dot{\psi}$ are determined by means of (7). The collision integral is written in the relaxation time ($\tau_p$) approximation and $f_0(\varepsilon+v)$ is a local equilibrium distribution function. As is common in the problem of sound interaction with an electron system of a conductor, we shall seek a solution of (19) in the form

$$f = f_0 + g , \tag{20}$$

where the nonequilibrium correction meets the equation

$$\frac{\partial g}{\partial t} + \dot{p}\frac{\partial g}{\partial p} + \dot{I}\frac{\partial g}{\partial I} + \dot{\psi}\frac{\partial g}{\partial \psi} + \dot{\nu}\frac{\partial g}{\partial \nu} + g/\tau_p = -\frac{df_0}{dt} . \tag{21}$$

The kinetic equation (21) is found in the neighbourhood of an isolated resonance assuming that the resonances are rather far apart from each other. For this purpose the characteristic equation coinciding with the equations of motion must be considered in the resonance variables $p_\alpha$, $p_\beta$, $\alpha$ and $\beta$ determined by (9a). After averaging with respect to the fast phase $\beta$, we obtain

$$(k_{||}\bar{v}_z-\omega-\ell\omega_c)\frac{\partial g}{\partial \alpha} - \Phi_0 v_\ell \sin\alpha \frac{\partial g}{\partial p_\alpha} + \frac{g}{\tau_p} = \omega\Phi_0 v_\ell f_0'\sin\alpha . \tag{22}$$

If the resonance momentum $p_\ell$ does not coincide with the extreme point of the function, $\partial S/\partial p_z$, we have $k_{||}\bar{v}_z-\omega-\ell\omega_c = G(p_\alpha-p_\ell)$ near the resonance. Then Eq. (22) in dimensionless variables will be written as

$$S\frac{\partial g}{\partial \alpha} - \sin\alpha\frac{\partial g}{\partial S} + a_\ell g = \omega\tilde{p}_\ell \cdot f_0'\sin\alpha . \tag{23}$$

We have introduced the dimensionless velocity $S = (p_\alpha-p_\ell)/\tilde{p}_\ell$ , $\tilde{p}_\ell = (\Phi_0 v_\ell G^{-1})^{1/2}$ and the nonlinearity parameter at the first resonance is $a_\ell = (\bar{\omega}_\ell\tau_p)^{-1}$ where $\tilde{\omega}_\ell =$ $= (\Phi_0 v_\ell G)^{1/2}$ is the oscillation frequency of trapped particles. The solution of (23) for trapped and untrapped particles is

$$\begin{aligned} g_t &= \omega\tilde{p}_\ell f_0'\left[a_\ell\alpha - S\right], \\ g_{ut} &= \omega\tilde{p}_\ell f_0'\left[a_\ell(\alpha-\bar{\alpha}) - (S-\bar{S})\right] . \end{aligned} \tag{24}$$

Here the mean velocity of the untrapped particles is $\bar{S} = \pi/(\kappa K(\kappa))$, $\kappa^2 = \frac{1}{2}(1+H)$, $\bar{\alpha} = F(\alpha/2,\kappa)/K(\kappa)$ and K and F are, respectively, the complete and the incomplete elliptic integrals of the first kind. When the resonance coincides with the extremum $\partial S/\partial p_z$, we have $k_{||}\bar{v}_z-\omega-\ell\omega_c = R(p_\alpha-p_\ell)^3$ and the kinetic equation is

$$s^2\frac{\partial g}{\partial \alpha} - \sin\alpha\frac{\partial g}{\partial s} + a_\ell^{ext}g = \omega\tilde{p}_\ell^{ext}f_0'\sin\alpha . \tag{25}$$

Here the dimensionless velocity and time are determined as

$$s = (p_\alpha-p_\ell)/\tilde{p}_\ell^{ext},\ \tilde{\omega}_\ell^{ext} = (\Phi_0^2 v_\ell^2 R)^{1/3},\ \tilde{p}_\ell^{ext} = (\Phi_0 v_\ell R^{-1})^{1/3},\ a_\ell^{ext} = (\tilde{\omega}_\ell^{ext}\tau_p)^{-1}.$$

The solution of (25) has the form similar to (24):

$$g = \omega\tilde{p}_\ell^{ext} f_0'(\varepsilon)\left[a_\ell^{ext}(\xi-\overline{\xi}) - (s-\overline{s})\right]. \tag{26}$$

The mean velocity is determined conventionally $\overline{s} = \frac{1}{T}\int_0^T s(\tau)d\tau$ but now it is not related to energy by the elliptic integral of the first kind. Variable $\xi$ does not coincide with the $\alpha$-coordinate and $\overline{\xi} = \overline{s}\tau$ where $\tau = \overline{\omega}_\ell^{ext}t$.

## 4. MAGNETO-ACOUSTIC OSCILLATIONS OF SOUND ABSORPTION IN THE REGIME OF ISOLATED RESONANCES

The coefficient of damping of a sound propagating in a metal placed in a magnetic field is known to oscillate magneto-acoustically with varying $|\vec{B}_0|$. When the sound propagates strictly perpendicularly to $\vec{B}_0$, cyclotron resonance and geometric absorption oscillations may occur. These resonance phenomena are related to resonance electrons.

Nonlinear damping, intense generation of harmonics and other nonlinear phenomena related to the trapping of particles at resonances are apt to occur in high-intensity fields.

Here we consider nonlinear magneto-acoustic oscillations. Let us find the work dome by the wave particles:

$$A = \left\langle\frac{\partial v}{\partial t} \cdot n_{res}\right\rangle , \tag{27}$$

where

$$v = -\Phi_0 \sum_m v_m \cos(\psi - m\nu), \qquad n_{res} = 2\int g \frac{d\vec{p}}{(2\pi\hbar)^3} ,$$

which is the resonance particle density. The damping coefficient is determined from the condition of energy balance in the wave - particle system:

$$\frac{dS}{d\eta} = -A = -2\Gamma S , \tag{28}$$

where S is the sound energy flux, $\Gamma$ is the damping coefficient and $\eta$ is the coordinate.

It is known that magneto-acoustic oscillations may be related to the extreme point $\partial S/\partial p_z$ /1/. The oscillations will be the strongest because the electron state density in such cross-sections is large. The contribution of the corresponding group of electrons to $\Gamma$ can be calculated using the distribution function (26) as well as (27) and (28). Then the contribution of the first resonance to sound damping is

$$\Gamma_\ell = \Gamma_L \cdot C(a_\ell^{ext})^{2/3} , \tag{29}$$

where $\Gamma_L$ is the linear damping coefficient, C is the dimensionless constant of order unity and $a_\ell^{ext} = (\tilde{\omega}_\ell^{ext}\tau_p)^{-1}$ is the nonlinearity parameter.

Similar calculations at resonances described by the standard Hamiltonian give the total damping coeffieient

$$\Gamma = \frac{|\vec{k}|}{2\rho w\hbar^3 \cos\phi} \cdot \sum_\ell \frac{v_\ell^2 m_c}{\partial^2 S/\partial p_z^2\Big|_{p_\ell}} \cdot 2a_\ell , \tag{30}$$

where w is the sound velocity, $\rho$ is the density of metal, $\phi$ is the angle between $\vec{k}$ and $\vec{B}_0$ and $a_\ell = (\tilde{\omega}_\ell \tau_p)$ is the nonlinearity parameter at the first resonance.

According to (29) and (30), the amplitude oscillations of damping coefficient in the nonlinear regime can be expected to decrease, being $\Phi_0^{-1}$ in the first case and $\Phi_0^{-1/2}$ in the second case.

## 5. CHAOS AT THE FERMI SURFACE

The particle motion is regular at isolated resonances only until the neighbouring resonances overlap. As the wave amplitude grows, the invariant surfaces in the phase space that are related to resonance are destroyed and the motion becomes chaotic. It is easy to see that according to (13) and (14) the condition under which ordinary resonances overlap (the Chirikov criterion) has the form

$$K = \frac{2(\tilde{\omega}_{\ell+1} + \tilde{\omega}_\ell)}{\omega_c} \gtrsim 1 \;, \qquad (31)$$

while at resonances described by the Hamiltonian (15) we have

$$K = \frac{6^{2/3}\tilde{\omega}_\ell^{ext}}{\omega_c} \gtrsim 1 \;. \qquad (32)$$

Estimations show that the condition (31) can be met in a typical metal over the greater part of the Fermi surface when the sound intensities are of order $10^2\,W/cm^2$. Criterion (32) is met beginning with higher intensities of a sound wave. For the calculation of sound damping under the conditions of global stochasticity it is, again, necessary to calculate the distribution function of electrons. At the Fermi surface with overlapped resonances and random electron motion the main part of the distribution function satisfies the diffusion equation of the form /11/

$$\frac{\partial}{\partial p}\left(D \frac{\partial f^{(0)}}{\partial p}\right) = - \frac{f^{(0)} - f_0(\varepsilon)}{\tau_p} \qquad (33)$$

with the diffusion constant

$$D = \frac{\pi}{2} \Phi_0^2 \left\langle \frac{v_\ell^2}{\omega_c} \right\rangle \;. \qquad (34)$$

The flows at the boundaries of the stochasticity region $p_1$ and $p_2$ are equal to zero.

$$D \frac{\partial f^{(0)}}{\partial p}\bigg|_{p_1} = D \frac{\partial f^{(0)}}{\partial p}\bigg|_{p_2} = 0 \;.$$

The calculation of damping needs, besides $f^{(0)}$, the correction $f^{(1)}$ which is linear with respect to $v_\ell$.

Such a program for metals with an isotropic spectrum was realized in /7/. It follows from /7/ that the value $b = (p_1 - p_2)/p_0$, i.e. the relation of the stochasticity interval to the diffusion impulse $p_0 = \sqrt{D\tau_p}$, is the nonlinearity parameter in stochastization.

When conditions (33) and (34) are met and the resonances overlap in weak magnetic fields, magneto-acoustic oscillations of the sound damping coefficient $\Gamma$ can be expected to be weak, with the absorption coefficient being smaller than the linear one, $\Gamma_L$. The difference $\Gamma_L - \Gamma$ must increase with the wave intensity and electron relaxation time $\tau_p$ because they both lead to the increase of the parameter b. The oscillations $\Gamma$ related to isolated resonances must evidently be observed in a stronger magnetic field, when conditions (33) and (34) do not hold.

It should be noted in conclusion that the approach presented in this paper can also be used for the investigation of nonlinear dynamics of electrons and of kinetics in the field of electromagnetic waves propagating in metals.

## REFERENCES


1. A.A.Abrikosov. Fundamentals of the Theory of Metals. Nauka, Moscow, 1987 (in Russian).
2. Yu.M.Galperin, V.D.Kagan, V.I.Kozub. Sov.Phys.-JETP, 1972, 62, 4, 1521.
3. V.D.Frill, V.I.Denisenko, P.A.Bezuglyi. Fizika Nizkikh Temperatur, 1975, 1, 9, 1217-1219 (in Russian).
4. V.F.Voloshin, G.A.Vugalter, V.Ya.Demikhovsky, L.M.Fisher, Y.A.Yudin. Sov.Phys.-JETP, 1977, 72, 1, 257;
   G.A.Vugalter, V.Ya.Demikhovsky. Sov.Phys.-JETP, 1976, 70, 4, 1419.
5. O.I.Lyubimov, N.M.Makarov, V.A.Yapol'sky. Sov.Phys.-JETP, 1983, 85, 2, 614.
6. V.A.Burdov, V.Ya.Demikhovsky. Sov.Phys.-JETP, 1988, 94, 5, 150.
7. V.A.Burdov, V.Ya.Demikhovsky (to be published).
8. L.D.Landau, E.M.Lifshits. Mechanics. Nauka, Moscow, 1988 (in Russian).
9. A.Lichtenberg, M.Lieberman. Regular and Stochastic Motion. Springer-Verlag, New York, Heidelberg, Berlin, 1983.
10. G.R.Smith, A.N.Kauffman. Phys.Fluids, 1978, 21, 2230.
11. G.M.Zaslavsky. The Stochasticity of Dynamic Systems. Nauka, Moscow, 1984 (in Russian).

# The Outlook for Solid-State Lasers in the Laser Market

*I.G. Zubarev, G.A. Pasmanik, V.G. Sidorovich, and E.I. Shklovsky*

Institute of Physics, USSR Academy of Sciences, Lenin Ave. 53, 117924 Moscow, USSR

The paper deals with a brief characterization of solid-state lasers and the outlook for their application in various fields including scientific research, medicine, material processing, remote measurements, microelectronics, interference holography and space geodesy.

## 1. INTRODUCTION

This paper is seeking to give a brief characterization of solid-state lasers (SSL) which are widely used in practical applications. Here we have attempted to formulate some recommendations on developments in the involved areas of quantum electronics and to outline apparent priorities in laser research. We shall consider optical systems of pulsed, pulse-frequency and CW glass-crystal Nd lasers, i.e. primarily YAG lasers, which show relatively high effeciency and divergence nearing a diffraction limit.

The 1988 solid-state laser share in total laser sales throughout the world was 18% (about $112-115 million out of nearly $630 million), and this is not likely to be changed substantially with an increase of total laser production over the coming years. However, the proportions within the entire variety of currently available SSL are expected to change. It appears that the role of laser-diode-pumped SSL will be growing. The share of Nd:YAG-based devices remains at present the largest among all other solid-state lasers, but it tends to decline with the emergence of new types of laser crystals (GSGG, YSGG, etc). Today commercially available lasers with phase-conjugation devices do not exist. Such lasers, characterized by higher average power and a diffraction divergence of radiation are expected to enter the market in the next few years.

## 2. SSL APPLICATIONS

Judging by the current situation in the world laser market and its prospects, solid state lasers have advanced and may see continuing advances in such applications as scientific research, medicine, material processing, remote measurements, microelectronics, holographic interferometry and space geodesy.

Research Reports in Physics **Nonlinear Waves 3**
Editors: A.V. Gaponov-Grekhov · M.I. Rabinovich · J. Engelbrecht

Scientific research rates first in the SSL sales volume. In 1988, sales of lasers for scientific R&D approached $33 million. The major applications are spectroscopy, studies of ultrafast processes, impurity diagnostics, laser plasma production and investigations into high-power laser radiation interaction with matter.

Medcine is the second-largest SSL market: in 1988 sales of lasers for medical applications amounted to nearly $30 million. However, the trend here is towards a slight decline in the use of SSL which are being ousted, partly, by excimer, dye and ion lasers.

Material processing comes third in significance of all application areas for SSL. The number of lasers sold for these needs in 1988 was 1200 units with a value of nearly $30 million. The forecast calls for further rise up to about 5000 units in 1995 (with an annual unit rise of 20%). At this rate material processing is likely to become a leading area among other SSL applications.

SSL for remote measurements of environmental characteristics are still in the early stage of commercialization. With a growing interest in ecology, one can predict a continuous rise in the demand for SSL in this area to tens of millions of dollars over the next few years. Solution of a wide variety of practical problems involved in remote diagnosis (i.e. measurements of minor impurities, wind velocities, atmospheric turbulence, oscillations of the Earth's crust) requires optical systems (lidars) that vary a great deal in components and scheme configuration. Unique lidars may cost very much.

In microelectronics solid-state lasers are mainly used for trimming of film-type resistors and condensers, for brazing electronic components, for scraping semiconductor and for hole-drilling in ceramics. When used for photolithography, a solid-state laser has to operate at the 4th or 5th harmonic (wavelength of about 0.2-0.25μm) which nearly always yields a coherent radiation. However, lithographic imaging in a coherent light has certain disadvantages related to the radiation diffraction at sharp edges of templates or masks and to its scattering on dust, defects of optical elements, etc. Besides, the intensity instability of higher harmonics is much larger than the instability of the first harmonic. For these reasons it is now difficult to predict the market for SSL in this area, though the trend will certainly continue in the future. There are grounds for such an optimism, because, in particular, there are possibilities to correct (under coherent radiation) the aberrations of optical elements (including large-sized ones) incorporated in a projection lithographic system.

The potential of SSL as a point source of hard X-ray and $\gamma$-radiation as well as of an ion flow may also attract microelectronics industry. The major consumers here are branches specializing in X-ray photography and epitaxy of thin films. The current research in this field shows hopeful signs for pulse-frequency SSL with an average power of over 100W which are expected to make rather a big share in the market of SSL-based devices for these applications. Multimode lasers with a power reaching hundreds of watts are available today. The next move is towards the development of single-mode lasers with powers exceeding 100W.

Holographic interferometry, though well-established in physical terms, has not found extensive application yet. This is associated with the problems of fabricating reliable and convenient sources capable of producing highly coherent high-energy pulses required for the illumination of diffusely scattering large-sized objects. One can expect that the needs of machine manufactoring industries in nondestructive test equipment for detection of hidden defects in various structures will encourage the development of mobile holographic registering systems which are fit for in-shop and field operation.

Laser illumination seems also to be a promising trend. Unlike optical systems of non-coherent image formation, narrow bands and high directivity of laser radiation permit to illuminate objects and receive radiation in the intensive glow from other sources, for example, in fire. The major factor to push the development of this trend is the laser efficiency and the ease of use.

Another aspect of the same problem is using solid-state lasers for sequential hologram recording, or for holographic filming. Illuminating a 10 x 10 $m^2$ stage will require a SSL producing pulsed-periodic coherent radiation with an average power reaching several tens of watts.

Space geodesy is one more application area of SSL where lasers provide a high-precision distance control to satellites and from satellites to any spot on the Earth. High-precision data about satellite coordinates is relevant, for example, in the case when satellite-borne narrow-band radiofrequency generators serve to guide ground-based objects in determining their own coordinates by Doppler shifts with respect to their standard generators. Here it is important to use pulse-frequency single-mode lasers with a short (subnanosecond) pulse duration and rather high (~1J) pulse energy at the second frequency of the second harmonic, which is a region for the most sensitive photon counters. Subnanosecond-pulse lasers are being currently introduced at the artificial satellite tracking stations worldwide.

## 3. SCIENTIFIC RESEARCH

A number of factors are important for the use of SSL in scientific research. First, the capability to excite a 4.06 mcm dye lasers by the 2nd or 3rd-harmonic radiation. This accounts also for new types of broad-band SSL ($Ti:Al_2O_3$-lasers, for example), which, in their turn, are employed in spectral investigations and studies on generation of ultrashort picosecond and femtosecond pulses. These studies are underway in many R&D centres now. On the one hand, major emphasis is paid to the problems of producing narrow-band frequency-tunable radiation for measurements of spectral characteristics in atoms, ions and nuclei, and, in case of a high-power radiation, for selecting and isolating them from other particles. On the other hand, partially synchronized SSL pulses, used for dye pumping, together with an additional absorber are capable of generating ultrashort picosecond pulses. Their subsequent nonlinear self-modulation in optical fibres allows effective compression of these pulses using additional elements with a rather strong material (prisms) or intermode (diffraction gratings or waveguides) dispersion.

Secondly, a SSL higher-harmonic (the 3rd, 4th and 5th) operation enables one to study new peculiarities of a radiation-matter interaction, for example the high-temperature plasma production and the face investigations of solids. The diagnostics of fast pulse processes is of particular interest. For example, investigations of a light scattering on ions require high-power sounding pulses. Since plasma can have its own glow, the sounding radiation has to be spatially coherent and have a narrow spectral line to be able to discern the scattered light against the plasma glow. A similar technique is used for interferometric measurements of spatial distribution of densities in incandescent gases.

IR spectroscopy of condensed media is one more scientific SSL application. A solid-state dye laser pumped by the second-harmonic radiation is used here and SRS yields tunable radiation in the near IR region. The transmission coefficient of this radiation measured in various media (including excited ones) provides versatile spectroscopic data. Other spectroscopy versions are based on a SSL-pumped parametric light oscillator generating in the near and medium IR region. For higher average power of radiation of the parametric oscillator operating in a pulse-frequency pulsed-periodic mode, one should lengthen the SSL pulses to 100...200 ns. To keep up the high efficiency of a nonlinear conversion in crystals with quadratic nonlinearity and to avoid optical breakdown in this case, it is advisable to generate a sequence of 10...20 subnanosecond pulses instead of a single pulse. The intervals between the pulses are determined by the cavity length.

## 4. MEDICAL LASERS

Among the great variety of lasers currently in use for medical needs, the two major types which appear to be the best to meet the requirements, are solid-state Nd and Er lasers ($\lambda$ = 1.64 mcm and $\lambda$ = 2.94 mcm, respectively). There is a good reason for that. A Nd:YAG laser is capable of every possible generation mode including ultrashort pulses (from femtosecond to picosecond), Q-switched pulses (1 - 100 ns), free generation (pulse duration ranging from tens of microseconds to tens of milli-seconds) and CW generation. When a laser operates in the first two modes, the radiation may be converted to other spectral regions using the techniques of non-linear optics. In all generation modes described above, Nd:YAG lasers provide pulse powers and energies required for different medical applications. Generation versa-tility is called forth by wide applications, since each area has its optimal radiation parameters affecting specific biological objects. Thus, ultrashort pulses absorbed in an eyeball may be used to control intraocular pressure; Q-switched pulses excite acous-tic waves in internal organs which can be used for shattering stones in these organs. Free generation is fit for cutting tissues and bones, elimination of false neural canals which cause arhythmia, and other surgery operations. Free generation delivered by a light guide proves very effective for moving atherosclerotic plaque in blood vessels and various deposits. CW radiation may be used to stop bleeding and for cauterization.

The Er laser, though not a CW generator, shows a unique performance. Its radiation wavelength hits the maximum of a water absorption band with the absorption coefficient $\alpha = (30 - 60)\ 10^3\ cm^{-1}$ (average $45000\ cm^{-1}$). Therefore, the radiation is absorbed in a thin surface layer of biological tissues, which allows for tissue removal without thermal effects in the nearby areas, i.e. ablation. The Er laser radiation proves to be most effective for removing necrotic tissues. As soon as the necrotic layer is gone, the radiation affects living tissue causing it to bleed, which is a sure sign of a clean surface. Such surface-cleaning operations yield positive results in healing wounds. Er lasers are also used to remove plaque: the radiation is introduced into a blood vessel by a $ZrF_4$-based light guide and the entire procedure is controlled with a low-intensity X-ray equipment.

Ophthalmic applications of SSL radiation are still more impressive. These include procedure for recovering lacrimal canals, welding-on detached retina, for correcting nearsightedness by cutting into the corneal edges and many other operations. This versatility of SSL generation makes them indispensable in ophthalmology. Excimer lasers belong to another laser type currently used in this area of medicine. The radiation is also absorbed in a thin surface layer of tissue. However, according to some researchers, the UV-radiation impact on tissue may lead to adhesions in DNA molecules, which, in its turn, can result in a mutagenesis and in cancerogenic effects.

Finally, a few words must be said in favour of therapeutic effects of laser radiation (healing of burns, ulcers, etc.). They were discovered after the introduction of He-Ne laser into medical applications. The reason for this applicability is thought to be an excitation of singlet levels of molecular oxygen followed by chemical reactions.

Figure 1 is a schematic representation of the molecular oxygen levels (the space between oscillatory transitions $v = 0.1, \ldots$ , of the oxygen molecules $\Delta\nu = 1550\ cm^{-1}$ and of the two first transitions in the electron spectrum). These electron transitions are forbidden, but become allowed in collisions. Hence the absorption at possible transitions is increasing along with pressure in a gas phase and is strong in liquids dissolving oxygen. The absorption cross-section is maximum at the main transition: $v = 0 \rightarrow v' = 0$ ($\lambda = 1.27$ mcm). The radiation with 1.06 mcm wavelength leads to the transition $v = 0 \rightarrow v' = 1$. In liquids this and other transitions corresponding to shorter wavelengths have nearly equal absorption cross-sections.

Chemical reactions involving excited singlet oxygen are unknown, but the nearly equal cross-sections of transitions at different wavelengths (except for $\lambda = 1.27$ mcm) may account for the weak sensitivity of this effect to the wavelengths of the employed radiation.

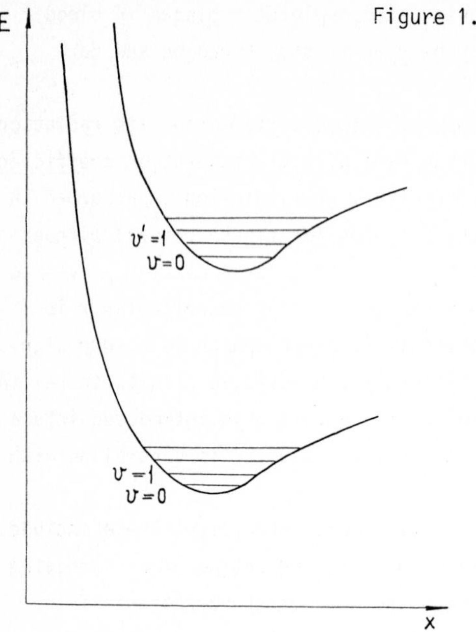


5. MATERIAL PROCESSING

Material processing traditionally includes welding, cutting and drilling, thermal treatment, and alloying. The two processes that find major application in microelectronics are marking and scribing.

Smooth (spike-free) millisecond pulsing has the biggest potential for welding. For connection of two bars, for example, an optical radiation is used to melt off their ends spaced at a fraction of a millimeter. The diameter of the focused laser beam should be larger than this gap. Welding ordinarily requires power densities to about $10^5$ W/cm² and argon supplies to protect the seam surface from oxidation. The characteristic laser power is 100...200 W. The rate of a continuous welding for a steel sheet sheet of 1...2 mm thick is about 1 m/min. Generally, a single-mode radiation is not compulsory here, but for welding in rather large sealed volumes (for example, large-sized cylinders) it is more convenient. Single-mode laser welding is used in microelectronics (linking wire to a contact) when the typical dimensions of a light spot do not exceed 15...20 mcm. Average power of a laser in these conditions may not be high (~ 1 W).

Laser cutting and drilling, especially on refractory metals and diamonds requires much higher power densities (up to $10^8...10^9$ W/cm²). This is associated with the necessity to vaporize the biggest part of the removed material. Such densities are achieved though modulation of several-millisecond pulses by short spikes of several hundreds nanoseconds. Spikes to tens of nanoseconds and shorter generate a surface-shielding plasma, while those longer than a microsecond cause thermal stresses and cracks especially in cutting and drilling of brittle materials.

Theoretically, the wall-reflection factor being neglected, a cut depth may reach the length of a focal waist which for a single-mode beam equals $2ka^2$, where a is the minimal radius of the beam in the focal waist region, k is the wave number. The ratio of the beam width to the cut depth is 1/ka. For a multimode radiation this ratio increases proportionally to the beam divergence increase over the diffraction limit.

Let us consider an example. Suppose we have to cut a metal sheet, which is 1 cm thick, using a single-mode beam with a 1.06 mcm wavelength. It follows from the formula above that the radius of a light beam in the focal waist should be $a \cong 30$ mcm. Note for comparison that for cutting the same depth with a single-mode radiation of a 10.6 mcm laser will require a beam with a radius $a \cong 100$ mcm. The example shows that $CO_2$ lasers in this application produce material waste several times larger as compared to Nd lasers, which reduce productivity of a $CO_2$-based technological setup (all other conditions being equal). Besides, the 1.06 mcm radiation lowers heating efficiency due to the beam's reflection from the processed surface. Multimode radiation used for cutting requires thicker weld, which leads to still larger waste in material.

In this context, shorter wavelengths in the visible and UV regions seem to hold more promise, in particular due to harmonic generation using KTP and BBO crystals.

For effective removal of the evaporating material, a gas blow coaxial with the beam is used in this process (oxygen in particular, as it sustains combustion and facilitates surface processing).

In the above example one 300...400 ns spike with the power density of $3 \cdot 10^8$ W/cm² is used to remove a material layer of nearly 50 mcm in thickness. If a several-millisecond pulse ("flash") generates 100 spikes, one such a "flash" is enough to remove a 5 mm-thick layer. For the 100 m J - energy of one spike (10 J in one "flash"), the area of the illuminated surface will be $10^{-4}$ cm² or 50 x 200 mcm². It allows one to estimate the average cutting rate for 1 cm-thick metal: a 200 mcm cut in two "flashes" with a 10 Hz frequency (average power 100 W) corresponds to the cutting rate of 1 mm/s. This rate is characteristic for both routine and circuit cutting, and also for drilling.

It is convenient to deliver radiation to the processed surface by some "flexible" technique, for example, by a light guide of 1...2 mm in diameter. A double-pass scheme may be used to achieve single-mode radiation at the output of this light guide. Single-mode radiation from a fibre or compact portable laser is supplied to one end of a light guide by which it is transmitted to an optical system to be amplified, phase-conjugated, re-amplified and, finally, transmitted by the same light guide in the reverse direction. However, fabrication of double-pass systems with an average power of 100 W requires new optical components including nonreciprocal elements based on Faraday cells with permanent magnets, PC-mirrors of highly purified liquids and gases, laser amplifiers providing a multifold (up to hundreds of times) increase of radiation in two or four passages with a rather high efficiency. Though the use of these elements will nearly double the cost of a laser, it is a worthy project for these lasers will be able to replace a wide class of precision cutting machines and perform efficient high-precision (due to a single-mode radiation) metal processing on large-sized stationary works.

Thermal hardening is actually a process of laser radiation heating of a surface to the depth of 100...200 mcm followed by immediate cooling. Heating is performed either with a CW radiation or with several-millisecond pulses free of spikes. Unlike with $CO_2$-lasers, SSL does not require a special absorbing coating on a processed surface because a 1 mcm-wavelength radiation is generally eagerly absorbed by material. Various alloying additions are introduced onto the surfaces before heating, which then are remelted. These operations do not call for a single-mode radiation, and the currently available lasers with a 15-20 mrad divergence capability meet the consumer's demand.

Processing of semiconductor and other materials in microelectronics consists mainly of scribing (i.e., cutting of substrates, marking and hole-drilling). A competitive technique to laser scribing methods is the so-called disc cutting which is more reliable in preserving the structure of a surface. However, by optimizing a laser generation for scribing applications, a defectless cutting of surface structures was achieved recently. Cleavage appearing in a substrate depth has no noticeable effect on the IC operation. This scribing technique is equally good for silicon substrates ad for ceramic ones, especially when only a small part of them has to be processed. A typical laser generation here is 100 ns-pulsing with the repetition frequency of 10...20 kHz and the average power of several watts. Laser scribing is also useful for processing glass masks. In this application CW radiation is more convenient as the beam "pierces" glass due to thermal stress generating a crack which "traces" the light beam path. CW SSL are applicable here.

## 6. REMOTE MEASUREMENTS

These involve detection of the atmospheric and water pollution, chemical analysis of impurities in these media, determining air- and water-flow velocities, density and temperature distributions, detection of acoustical waves in air and water as well as determination of space-time distribution of turbulence in the atmosphere. An average power of radiation used for remote measurements is always relevant, as an increase in this power has an immediate increasing effect on the power of a recorded signal. Frequency-tunable lasers have the required potential for lidars which control the atmospheric pollution. In creating tunable sources in the visible and near IR regions, the emphasis is on achieveing high efficiencies of SSL and high coefficients of conversion to tunable radiation. Today an average power of tunable radiation generally does not exceed 1 W. It is enough for laboratory use, but much too insufficient for remote spectroscopic investigations. Therefore, increasing the average power of a frequency-tunable radiation to tens of watts is highly recommended. Lasers with such power are certain to be in demand.

High average powers can be achieved, mainly, by pulse-frequency generation. It is accounted for by the fact that nonlinear processes of frequency conversion are not very effective, generally, under the low-power CW radiation. The increased CW radiation power, however, does not always lead to higher excitation efficiency of frequency-shifted components, since it simultaneously encourages side parasitic effects hindering nonlinear conversion processes.

The currently underway SSL-based lidars provide tuning capabilities in the 1...4 mcm region and, with the use of the second harmonic, in the visible regions. When required, these lidars can produce a two-frequency radiation consisting of two harmonics either with a small or large frequency shift (in the latter case the frequency of one harmonic generally coincides with the SSL frequency). Such lidars are used for measurements of the atmospheric and water pollution with organic and inorganic particles, and for diagnostics of chemical composition of gas impurities.

Coherent detection of radiation, in particular, dynamic hologram recording between the reference wave and the weak signal in atmosphere, finds increasing application now. A matrix quantum receiver has more advantages here as it is capable of recording microstructure of a scattered light, operating in a photon-counting mode. Note that sensitivity of the thermal-scattering detection method increases in this case not with the shortening of wavelength $\lambda$ (as one would expect proceeding from the fact that the scattered light intensity is proportional to the 4th power of frequency), but with an increase in the wavelength. This is caused by the fact that the number of scattered photons in one transverse and one longitudinal mode (per space-time resolution element) is proportional to the probability of induced transition on the beam path (which depends on energy, but not on the wavelength of a light pulse), multiplied by the number of thermal phonons in the mode. The thermal phonon number $\bar{n}$ in the mode, in its turn, depends on the ratio of the energy yielded by thermal motion of molecules to the phonon energy $\hbar\Omega$ , where $\Omega$ is the frequency of the back-scattering sound. An increase in the wavelength leads to decreased $\Omega$, which results in larger $\bar{n}$. Therefore, the number of scattered quanta grows proportionally to the increase of the wavelength, i.e. IR-band is a better source for coherent detection of thermal scattering light than UV-band. The principle of coherent detection is used in lidars intended for wind velocity measurements. Coherence is essential here for detecting a slight (Doppler) frequency shift of the back-scattered light.

The situation is somewhat different with aerosol scattering. In this case the number of photons scattered in one transverse mode for time $\sim\lambda/v$, sufficient for the Doppler shift detection does not depend on the wavelength $\lambda$.

Development of lidars for remote measurements of temperature and water flow velocities seems to be promising.

The current laser technology has the capability of developing a new generation of lidars based on nonlinear effects excited in the air by a rather powerful laser pulse with a small angular divergence. These involve thermal effects, i.e. producing a thermal lens by the resonance-absorbed radiation in order to determine absorption coefficient and, consequently, concentration of absorbing impurities, and the SMBS effect for driving a returning pulse precisely onto the lens, which largely improves the sensitivity of the receiving equipment.

Let us consider the thermal effects. We assume the resonance-absorbing radiation to be focused into the air layer we want to investigate. If the energy of pulsed radiation is higher than the critical energy of thermal self-defocusing which depends on the absorption coefficient, the focal waist of this radiation is broader than the diffraction limit. A thermal lens formed in the air in these conditions can be identified by recording a spatial structure of the probing pulse back-scattered by molecules and aerosols present in the thermal space in the focal waist region. The fact is, the focal waist broadening leads to a change in transverse speckle-structure of the back-scattered light. By recording these changes, one can estimate the value of the absorption coefficient. Estimates show that even very small values of the absorption coefficient, such that cause a nearly neglible integral path attenuation of a light beam, can be measured in this way. We mean the absorption coefficients which cannot be measured by standard methods using two different-frequency pulses, and the comparison of their back-scattering attenuation.

## 7. MICROELECTRONICS

Thin-film microtechnology is used in versatile device fabrication, mostly in intergrated-circuit (IC) processing. This processing is based on planar technology, i.e. a step-by-step formation of layers with the given pattern, which are placed one over another and made of materials with different electric properties. These litographic layers can be produced by changing the properties of a substrate under the action of some external force. Pattern formation at present is mainly a photolithographic process. The screen pattern is projected on a substrate surface coated with a photoresist (generally, a polymer). Exposed to light, a photoresist is capable of either dissolving in a certain group of solvents or, on the contrary, resisting their action (the light either breaks molecular bonds or binds them). Having been developed in a photoresist solvent, the projected pattern remains on the substrate surface, and the unshielded spots undergo ion, plasma or chemical treatment.

In photolithography spatial coherence of light beams has to be taken into account, since even weak scattered light under strong coherent radiation can cause an appreciable distortion of an image. It is impossible to diminish coherence of the radiation with a pulselengh from nanoseconds to tens of nanoseconds using simple methods. Therefore one has to process microcircuits in which spatial coherence will have no impairment on the quality of lithographic images.

It is necessary to produce a radiation with a plane or spherical wavefront, which allows either contact or projection printing to be used. Major disadvantage of contact lithography is in that accidental touching of substrate and mask results in defects. In many IC with a resolution exceeding a few microns, a clearance is allowed between substrate and mask to prevent contact. Minimal linewidth on an IC surface pattern for such noncoherent printing when a mask is exposed to the plane-wavefront radiation, is determined by the diffraction at the clearance depth $\ell$ and is equal to $\sqrt{\lambda\ell}$.

Optical projection systems are more suitable for imaging applications. They provide capabilities for reducing an image with respect to a mask size. Since coherent properties of radiation impair the image quality due to the diffraction at sharp edges of a semi-transparent photomask, it is a good idea to use "soft"-edges (semi-transparent) photomasks with allowance for a limited light-gathering power of optical projection system. In this case one can use both plane- and spherical-wavefront radiation and the uniform amplitude distribution over the beam cross-section. Here it is important that a resist should have a threshold response and weak diffraction lobes are not shown in the image.

For all the shortcomings, coherence of radiation has one major advantage as compared to noncoherent light in this application. It is the coherent radiation that is most effective when used to correct aberrations of the optical elements of the projection system. This correction is essential when a resolution less than 1 mcm is required and projection lenses operate under the diffraction limitations. Fabrication of aberration-free lenses with an operational field of the order of 10 cm poses quite a serious problem for such resolution requirements, and imaging is therefore preformed stage-by-stage (a mask of a substrate, or a substrate alone keeps moving past the lens, but stops during the pulse exposition time). Phase-conjugation (PC) can also be used for correcting aberrations of optical elements. This process is feasible, in fact, only under coherent radiation. PC is believed to be the major tool for correcting aberrations of fast lenses.

An essential difference between laser and thermal-source radiation is the capability of producing high average and high peak intensities of radiation by a laser. These effects enhance nonlinear processes in a photoresist. Currently the photoresists are under developments which allow high contrast range for the energy densities $w \cong 0.5$ J/cm$^2$ used for dry treatment (ablation). This is sufficient to increase resolution capability to 0.3 mcm is a short-wave-length UV illumination radiation is used. Submicron resolution can be feasible with diffraction optics free of chromatic aberrations. For diffraction optics, the characteristic linewidth of excimer lasers reaching tens of inverse centimeters have no effect on a diffraction resolution realized on the IC surface. Since fabrication of achromatic lenses in the UV band is not an easy problem, one may either resort to reflection optics, or change a laser linewidth using intracavity selection techniques. In practice, however, the use of reflection optics reduces the field of vision, while narrowing of the linewidth due to the intracavity selectors may lead to power losses for excimer lasers. This is why the narrow-band SSL harmonic radiation competes with the excimer laser radiation.

Another feature of coherent radiation is the capability of a laser pulse compression from tens of nanoseconds to one nanosecond by stimulated Mandelstam-Brillouin scattering. Decrease in the pulse-length down to 1 nanosecond is advisable when a submicron resolution is required. It is associated with the fact that in ablation, when the surface is exposed to action of long ($\tau \geqq 10$ ns) laser pulses, the transverse radiation scattering on optical inhomogeneities at the surface layer of a resist, created by the light beam itself, becomes very important. If the pulselength does not exceed 1 ns, ablation takes place after the pulse effect, and optical inhomogeneities near the resist surface, which induce light scattering, have no time to be formed. In this respect, realization of a single-pulse exposure with laser pulselength shorter or of the order of 1 ns seems very attractive. The stability requirements to a projection system are less stringent in this case. However, the energy of UV radiation in such short pulse must reach tens of Joules. To ensure reasonable efficiency of photolithography equipment, the repetition frequency has to exceed 1 Hz, i.e. an average power of radiation must be $p \gtrsim 100$ W.

SSL in microelectronics are used for IC control. It involves control of spatial arrangement of elements on the integrated-circuit surface by imaging illuminated surface of a photoresist. Coherent holographic recording and comparing the obtained data with standard results is also possible.

Most common microelectronics problems calling for sources of hard quanta and particles, are the X-ray lithography and the laser-plasma thin-film epitaxy. The X-ray generation requires power densities on the target surface up to nearly $10^{13}...10^{14}$ W/cm², and for a thin-film epitaxy $I \cong 10^{11}$ W/cm². Let us consider the first problem in detail.

At present contact lithography is the only feasible way of using X-rays for microlithography, since the field size and resolution capabilities of X-ray lenses are still inadequate for projection imaging on a sample. The relatively soft X-ray radiation ($\lambda \cong 0.4$-05 nm) is used for this application so that a resist would absorb sufficient energy. The diffraction effects at these wavelengths are neglibly small up to a linewidth on the pattern surface, which approaches the order of 2 mcm for attainable 50 mcm spacing between a substrate and a mask. The photomask is fabricated not from a quarz plate like for lithography applications, but on thin (under 10 mcm) silicon or silicon-carbide membrane (quarz is not used in X-ray lithography, as it absorbs soft X-ray radiation). Absorbers here are metal films with large atomic mass, applied by means of electron-beam lithography.

X-ray lithography provides higher resolution capacity compared to any optical projection camera, and lower cost that electron-beam lithography.

X-ray radiation sources with laser-excited plasma may be an alternative to a rather expensive synchrotron, though the now available radiation output in them is tens of times less.

X-ray radiation of these sources is due to a presence of high-order ions in plasma, which induce generation of linear and continuous spectrum with the quantum energies approaching units keV. By target and laser-parameter selection one can have the peak of the X-ray radiation corresponding to the required wavelength range. The pulse length of a laser-plasma X-ray ranges from units to tens of nanoseconds, the size of the radiation source being 100 mcm and less. The employed laser type is responsible for producing either high-peak or high-average power of the X-ray radiation.

The conversion efficiency of laser energy to the "soft" X-ray radiation ($\hbar w \cong$ $\cong 0.1$-$0.7$ keV) is slightly affected by the change in the target flow from $10^{12}$ W/cm² to $10^{14}$ W/cm², remaining 8-10% for $\lambda$ = 1 mcm and 35-40% for $\lambda$ = 0.25 mcm. This dependence is somewhat different for copper and gold targets, which is accounted for by different contributions of the corresponding shells of the copper and gold ions into the indicated spectral bands. As for the "hard" X-ray radiation ($\hbar w \cong$ 1-2 keV) for $I \lesssim 5\ 10^{12} - 10^{13}$ W/cm², a sharp decrease of the conversion value is observed, with a special distinction on copper for both the "soft" and "hard" X-ray radiation. At the same time a gold target is not good for effective laser energy conversion to the "hard" X-ray radiation. The conclusion to be drawn here is as follows. For effective radiation conversion the minimal plasma temperature $T_{p\ell}$ has to be maintained in the region of several keV near the target with a corresponding atomic number, which should match the decreased ion emissivity. The light flows on target should provide $I \gtrsim 10^{13}$ W/cm². Very short light pulses with $\tau \lesssim$ 0.4 ns fail to produce high conversion efficiency because bulk plasma emitting X-rays diminished with shorter wavelengths (all other conditions being equal). Laser pulsing (100 ns) is most productive of effective conversion for large values of light flow on target. It follows from the above that an acceptable length range of a laser pulse covers 0.5-10 ns which can be broadened up to 25 ns.

Analysis shows that for an average power of 200 W in SSL operating at wavelength $\lambda$ = 1.06 mcm the average power of 16 W (in $2\pi$ ster.) can be realized in the "hard" X-ray, which corresponds to 10 mW/cm² (spaced 20 cm from the source) In the "soft" X-ray region an average power density at a distance of 20 cm form the source will make 30 mW/cm² at least. Pulse-periodic KrF-laser with the average power of 200 W provides 25 mW/cm² in the "hard" X-ray and 62 mW/cm² in the "soft" X-ray band. These values largely exceed the average-power flows of the X-ray radiation which can be produced by electron-beam X-ray sources. This fact is relevant for lithographic applications of X-ray sources.

As it has been mentioned above, laser-plasma sources have rather broad radiation spectra: $0.1 \leq \hbar w \leq 5$ keV. Note that the "hard" radiation component decreases abruptly at $\hbar w \geq 2$ keV. This has its advantage, since electrons appearing in a resist at $\hbar w \geq 3$ keV scatter within the resist bulk thus reducing the resolution and the image quality. Another advantage of a laser-plasma source is its small size (less than 100 mcm), which reduces shadows (the penumbra effect) in contact printing. For example, a 0.1 mcm resolution can be achieved if the proximity between a mask and a substrate is 20 mcm, and the radiation source is spaced at 10 cm from the resist. These parameters as well as high efficiency of X-ray conversion make laser-plasma radiation sources attractive for X-ray lithography. With these sources, the exposure can be performed in one or several laser flashes (in the "soft" X-ray region sensitivity changes from 1 to 10 mJ/cm², in the "hard" X-ray from 10 to 100 mJ/cm²). Higher sensitivity of an X-ray resist in the "soft" region is attributed to the high absorption coefficient of the resist at $\hbar w \leqq 1$ keV.

These relations and the requirements on the laser pulse length (the sensitivity of a resist is taken as a given parameter) can be used to determine characteristics of a laser system adequate for an effective X-ray lithographyic source. Thus, SSL or KrF lasers with an average power of 20 W (w = 1 J, f = 20 Hz) can prove effective for an exposure in the region $\hbar w \sim 1$ keV with an X-ray resist sensitivity lower than 10 mJ/cm². Operation with an X-ray source in the region $\hbar w \cong 1$ keV imposes more stringent requirements to the laser. The sensitivity of the resist in this region is no better than 10 mJ/cm², and the light flux on target should be no less than $10^{13}$ W/cm². The sensitivity of 25 mJ/cm², which is feasible, will require an average laser power of no less than 100 W.

Such a laser can be based on various approaches. For one, it can be a laser with a low pulse energy (1 J, for example), but operating at high frequencies (f = 100 Hz). The pulse length of this laser has to be $\tau$ = 1 ns, and the beam quality high enough to ensure a small-size focal spot (100 mcm). Such a laser can be based on active elements of the "slab" type or on ordinary cylindrical ones. The major advantage of the "high-frequency"-source is that the mask is not heated much due to low energy of each X-ray pulse.

Another approach implies the development of laser systems having high pulse energies (for example, 20 J) and relatively small repetition frequency (f = 5 Hz). With KrF lasers due to UV radiation ($\lambda$ = 0.249 mcm) one can succeed in the X-ray conversion to 35%. The required efficiency of a lithographic process can be ensured here by a KrF laser having 3 times less power compared to SSL. However, the development of KrF lasers with the necessary beam parameters will involve considerable effort because of additional requirements for the value of the light intensity of target ($\sim 10^{13}$ W/cm²).

Finally, it should be noted that sources of X-ray radiation described above can find application in medicine, primarly, for probing local parts of body on small surface areas.

## 8. INTERFERENCE HOLOGRAPHY

The principal idea of a double-exposure holography is superposition of optical fields scattered by the investigated object at different moments of time. This object (vehicle, building, vessel or just a honeycomb sheating structure) vibrates under the action of applied vibration generator causing displacement of each element on the object surface. The larger the displacement, the weaker the dependence of an element on other elements in the structure. It is relatively small in a rigid defectless structure, and the field of displacements forms a geometrical figure of some particular, usually regular, form. If a link of an element with other elements of the structure breaks off, for example, due to defects in a weld, this element is still greater displaced. In this case an "ejection" is observed in the displacement field where the defects of the structure are present.

Coherent light fields carry all the information about small (of the order of a wavelength) displacements of the illuminated onjects. For example, in the object image plane we have

$$|E(t,\vec{r}_\perp)+E(t+\Delta t,\vec{r}_\perp)|^2 = |E(t,\vec{r}_\perp)|^2 + |E(t+\Delta t,\vec{r}_\perp)|^2 + E(t,\vec{r}_\perp)E^*(t+\Delta t,\vec{r}_\perp) + \text{c.c.}$$

Depending on the sign of a phase in the interference component, the intensity sum of both the fields increases or decreases. It means that for rather large values of $\Delta t$ a moirê is superimposed on the object image, its contrast being stronger with a greater deviation from one or another surface element.

Illumination of large-scale objects (bridges, other large-sized structures) requires fairly large pulse energies. Estimates show that with a highly sensitive receiving system a diffusely scattering object has to be illuminated by a coherent pulse with the energy density no less than $10^{-6}$ J/cm². Most suitable lasers for such an illumination are SSL (ruby or neodymium). The energy of a laser pulse must be ~1 J in order to illuminate the surface area of nearly $10^6$ cm². For industrial measurements the required frequency of the pulse repetition may reach either tens of hundreds fractions of one Hz, or tens of Hz. In the first case the coherent radiation with the intensity proportional to $|E(\vec{r}_\perp,t) + E(\vec{r}_\perp,t+\Delta t)|^2$ is directly recorded, generally. In processing of the corresponding images the moirê has to be singled out against the speckle-inhomogeneous background typical of diffusely reflecting objects. In the second case one can average the images of the illuminated objects, recorded at different moments at different angles. It is easier to identify the moirê in these conditions, since it is imposed on a half-tone image. The average power of the illumination source appears to depend on the repetition frequency and changes, respectively, from fractions to tens of watts. To achieve sensitivity capable of receiving light from a diffusely scattering object with the surface energy density equal to $10^{-6}$ J/cm², a matrix photoreceiver with a high quantum output is required. In practice this receiver should be linked up to a narrow-band filter necessary for work in illuminated areas, especially when a holographic interferogram has to be recorded through incandescent gas, for example, through a burning jet escaping an engine's nozzle.

For the real-time image recording one has to use media with rather inertial optical memories. For this purpose, the photorefractive crystals are used, in which, at first, holograms of the pulses scattered with the first and second brightening are recorded, and then read by the third pulse which forms a double-exposure image of the object. Such an image recording is preferable to direct entry of interferograms, recorded for each brightening, into a computer to be processed and compared. It is quite understandable, since the information involved in the process is extremely large, which makes discrimination of a moirê nearly impossible in a real-time image-recording. A receiver-transmitter system capable of realizing the indicated parameters is quite expensive.

The interference holography involves the problem of measuring small displacements at great distances, in particular, control of seismic waves on the Earth's solid and water surface, earthquake control, etc. The major problem here is to identify the effect under study against a variety of background spurious factors. The principles of interference holography may be used in optical systems for high-precision measurements of velocities and angular rotation of remote objects.

It seems that the development of a moving interference system for the solution of the above problems is quite a realistic idea for both the laboratory conditions and technology manufactoring, as well as for full-scale measurements at large distances.

A related problem is the observation of objects under laser illumination. Here the single-exposure registration of an image is used. This approach makes possible the fabrication of a coherent analog of a conventional searchlight with superior parameters because of the high directivity of laser illumination pulses allowing for object imaging in the conditions of parasitic background and back-scattering on the beam path.

The current key problem mentioned above in connection with a holographic cine-filming area is no more associated with producing a light beam required for a reliable hologram recording, but with a realization of conditions for recording a great number of holograms in three-dimensional media, for example, in photorefractive crystals. In this case a set of these crystals could function as a film in conventional cinematography. Since every crystal contains information about a large number of stills, their total number may not be large. In the coming years holographic cinematography may generate interest for scientific investigations of kinetic processes in hot gases and plasma, and of fast processes in continuous media.

## 9. SPACE GEODESY

Location of spacecraft and the Moon enables one to solve two groups of problems. The first involves problems like determination of the path and manner of motion of a spacecraft itself, mainly in order to particularize their location using the data obtained by ground-based objects. The second group deals with problems like determination of certain geophysical characteristics, such as the level of the ocean surface or polar caps, continent drift parameters, coordinates of control geodetic points, etc. by known parameters of a spacecraft motion.

The location of the Moon (and of all natural space objects) is technically accessible. The motion of the Moon has been thoroughly studied, while numerical methods are used to define it more exactly. The laser radars of the fourth generation which are currently under development, must have pulse lengths $\tau \leq 100$ ps. When the range of metering precision is upgraded to less than 1 mm, it will be possible to investigate relativistic effects in the character of the Moon motion. This will provide possibilities for experimental verification of different relativistic theories of gravity. The location of the Moon calls for optical systems generating picosecond pulses of high average power. Obviously, SSL is the best laser type to suit this application.

Besides the Moon, artificial satellites of the American "Lageous" or of Soviet "Etalon" type also have to be located for various geodetic data. These satellites are massive baloons; their geometric centres coincide with the centre of gravity to a high degree of precision. "Lageos" is about 0.5 m in diameter, the average distance from the Earth to this satellite is nearly 6000 km. The corresponding parameters for "Etalon" are about 1.5 m and 2000 km, respectively.

The problems (here referred to as the second group) can be solved using two approaches which have a principal difference. By the first approach, transmitter and receiver are ground-based, while a spacecraft carries a retroreflector which transmits the radiation back to the Earth. By the other approach, vice versa, the transmitter-receiver device is spacecraft-borne and the retroreflectors are based on the ground.

The Moon, as a space object used for geodetic data acquisition, has practically lost its significance compared to artificial satellites. It is much farther from the Earth, and its location requires technological installations of a considerably improved design. Therefore, only four radars in the world (2 American, 1 French and 1 Soviet) are intended for the locations of the Moon, while some forty radars are engaged in satellite location.

The precision of parameter measurements of various ground-based targets, as for example in setting up a geodetic network, depends particularly on a number of points where simultaneous measurements are taken. It is not profitable to produce a large number of laser radars because of their high cost. Therefore, satellite-borne laser transmitters based on SSL have become available for these applications. They are the so-called altimeters providing vertical measurements of altitudes from satellite to the Earth. They were used to detect local lowering of ocean level, and are currently employed for observations of the Earth's polar caps, etc. However, the solution of some urgent geodetic problems, detection and measurements of the Earth' surface deformations among them, requires the development of a more perfect satellite locator. In particular, a satellite ranger are capable of locating a few hundreds (~300) of targets in a single pass from the corner reflectors fixed on special places. Every target will be ~20 cm in size and is projected to consist of 15 corner reflectors of 3 cm in diameter. A very effective guidance system must be developed for such a satellite, since the search and location of one target takes just about 3 sec.

For this and other above-mentioned applications SSL are the most suitable, especially those with divergence close to the diffraction limit.

# Autowave Mechanism in Low-Temperature Chemistry of Solids

*V.V. Barelko, I.M. Barkalov, V.I. Goldanskii, A.M. Zanin, and D.P. Kiryukhin*

Institute of Chemical Physics, USSR Academy of Sciences,
Chernogolovka, Moscow Region, 142432, USSR

The results of the study of new phenomena in low-temperature chemistry of solids of the travelling wave type observed for various solid-state chemical reactions at nitrogen and helium temperatures are summarized. The studied processes are characterized by anomalously high conversion rates for these temperature regions, comparable with the rates of the fastest reactions in classical chemistry - those of high-temperature combustion. A mechanism with an autocatalytic link of mechanochemical nature is suggested, which allows us to give a qualitative explanation of the observed phenomena. The above mechanism is assumed to be widely spread in cosmochemistry and the processes of chemical evolution of the universe.

## 1. INTRODUCTION

This review reports the phenomenon which was hardly believed to exist a few years ago. We speak of solid-state chemical reactions proceeding at very low temperatures (as low as the helium ones) with velocities that are comparable with or even exceed the values characteristic of the fastest chemical reactions, those of high-temperature combustion. Even the regimes of the interaction are like combustion processes - they are defined as a narrow reaction front propagating spontaneously over a solid sample and described by the autowave dynamics. Naturally, this likeness was only a seeming one, and the classical self-activation mechanisms with a thermal feedback inherent in combustion have no connection with the discovered phenomena. An analysis of the results has helped to formulate a hypothesis based on the leading role in the matrix activation of the process of layer-by-layer autodispersion of the initial solid solutions of reagents. In other words, to interpret these phenomena, a positive feedback mechanism was introduced, the essence of which is in the interaction of the stage of initial inert matrix destruction (the stage providing chemically active newly formed fracture surfaces) with the stage of conversion itself (the stage initiating subsequent dispersion of a solid sample). A sequence of transitions between different forms of energy realized in the above mentioned mechanism can be represented as follows: chemical reaction energy evolving in the conversion front is changed to potential mechanical energy of a solid matrix under elastic deformation and is accumulated in an unreacted sample layer adjacent to the front region; when an overthreshold value is reached in this accumulation process, a dispersion of a subsequent layer takes place, which results in switching on of the reaction followed by heat release, etc., until the process is completed through the entire sample volume /1/.



Research Reports in Physics **Nonlinear Waves 3**
Editors: A.V. Gaponov-Grekhov · M.I. Rabinovich · J. Engelbrecht

## 2. OBJECTS OF THE STUDY

Autowave regions of solid-state conversion were discovered for a large number of reaction systems different both in the mechanism of chemical conversion and in the state of a solid matrix. In a series of the systems investigated, there are processes of hydrocarbon chlorination, olefin hydrobromination and polymerization. Among the studied reactions, there are examples of the processes taking place both in vitreous and polycrystalline matrices.

To carry out an autowave conversion, some systems required to be previously radiolyzed or photolyzed (accumulation of stabilized active centres). There were some systems that did not require any preliminary energy pumping-up in the matrix (for instance, solid co-condensates of chlorine and ethylene).

A brief characterization of the data corresponding to the chlorination reaction in vitreous solid-state solution of $Cl_2$ (mole ratio 1:3) in methylcyclohexane (MCH) and butyl chloride (BC) previously irradiated by $^{60}Co$ $\gamma$-rays is given below.

## 3. EXPERIMENTAL EVIDENCE FOR AN ACTIVATING ROLE OF BRITTLE FRACTURE IN REACTION INITIATION

In order to produce brittle fracture in a sample, we initially used the method of thermoelastic stresses, ensuring that the rate of temperature change was large enough to cause the sample to crack. For recording this rate the thermal signals accompanying the process, a differential scanning calorimeter was employed (for more details, see /2/).

Slow heating of the sample from 4.2 to 77 K did not produce any cracking and the calorimeter registered no thermal effects (Fig.1a,b; solid lines). Fast heating led to to the fracture of the sample, and the concomitant thermal effect was registered by the calorimeter (Fig.1c,d; solid lines ). After determining the rate of heating which

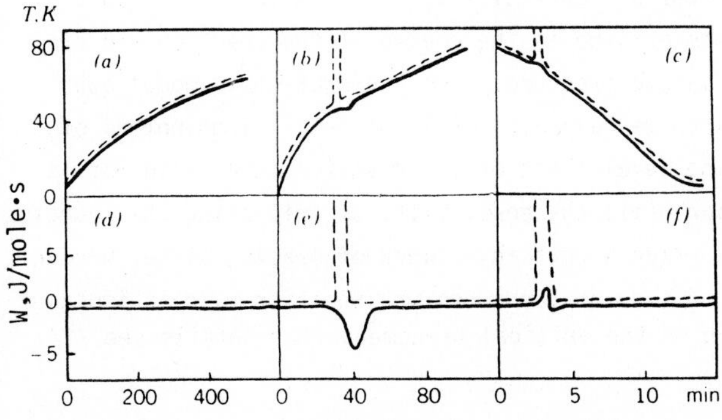


Figure 1. (a,b,c) - time dependence of the temperature of the sample ($Cl_2$+MCH, molar ratio 1:3) and (d,e,f) - thermal effects. Solid lines: nonirradiated samples; dashed lines: samples irradiated by $\gamma$-rays from $^{60}Co$ at 77 K, dose 27kG.

results in brittle fracture of the sample, analogous experiments were performed with samples preirradiated with a certain dose of $^{60}Co$ $\gamma$-rays to produce stabilized active centres in the system. In the course of slow heating of the irradiated sample, unable to cause any fracture, no heat production due to the chemical reaction could be determined (Fig.1a,b; dashed lines). On the contrary, at the instant of brittle fracture during fast heating, the reaction did take place (Fig.1c,d; dashed lines) /3/.

Brittle fracture of the sample could also be produced during fast cooling. Figure 1e,f presents data for a sample which was cooled rapidly enough to cause cracking. As expected, the fracture of the irradiated sample was accompanied by the fast chemical conversion, whereas no such conversion was observed in the nonirradiated sample.

To prove the decisive role of fracturing in the initiation of chemical conversion in a sample, a direct experiment was carried out in which the sample was subjected to nonthermal fracture. The brittle fracture was accompanied at a constant thermostat temperature of 4.2 K by turning a frozen-in thin metallic rod. At the instant of the disturbance of the sample containing stabilized active centres, a rapid (explosive) chemical conversion occurred. Being initiated locally, the reaction then spread over the sample - the dark colour due to $\gamma$-radiolysis disappeared.

These results led us to two conclusions of principal importance. First, in the studied systems the brittle fracture of the sample produces a burst of chemical conversion. Second, the process initiated by the fracture is self-accelerated and spreads over the entire sample. Therefore, the formation of a primary crack acts like a trigger switching on a certain positive feedback mechanism, providing a self-sustaining reaction regime. To shed light on this mechanism, an experimental investigation on the dynamics of the reaction response to an external distrubance was performed.

These features of the reaction burst dynamics evidence convincingly that the classical thermal mechanism of feedback does not play any decisive role in this phenomenon. The discovered crucial significance of newly formed surfaces in fracturing the sample suggests the following nonthermal mechanisms of self-acceleration in the studied system: the chemical reaction occurring on a newly formed surface and in its adjacent regions in its turn generates discontinuities, for example, as a result of the temperature or density gradients arising in the course of the reaction and producing stresses which lead to the sample fracture. Such a branch-chain model qualitatively describes the observed experimental facts: (i) the critical phenomena of the generation and self-accelerating development of the reaction in a solid matrix followed by its intensive dispersion; (ii) the possibility of initiating the reaction burst by creating at the initial instant a certain network of cracks, either mechanically or thermally (in the absence of such disturbances the reaction rate is immeasurably small); (iii) the degeneration of the critical phenomena (for details see /1/ and references therein).

## 4. EXPERIMENTAL RECORDING OF AUTOWAVE CONVERSION REGIMES

It is known that the principal feature of systems described by nonlinear models of the branched-chain type is their ability to generate autowave phenomena. The autowave regimes of conversion must occur in samples of sufficient extent in response to a local disturbance. These considerations have determined the experimental problem at this stage of research, namely, searching for autowave phenomena, ascertaining the fact of their existence and studying the dynamic characteristics and properties of the chemical-reaction wave front propagating over a solid sample.

Already in the first experiments it was possible to initiate spontaneous propagation of a reaction wave by a local disturbance produced either mechanically or thermally (pulse heater) in a limited area of the solid matrix. Below we describe the typical results obtained in experiments with cylindrical samples of frozen reactant mixture 0.5-1.0 cm in diameter and 5-20 cm in length. As the propagation of the reaction over a sample was accompanied by the change in its colour, the method of cinegram could be used to measure the velocity of front propagation. The front structure was studied thermographically with the help of thermocouples embedded in frozed samples. As seen from the cinegram, a flat front was rapidly formed after local mechanical destruction, which propagated parallel to itself along the reaction-tube axis with constant velocity equal to 1-2 cm/s.

To study the wave-front structure (its temperature profile), the autowave process was registered thermographically in a series of experiments. The propagation velocity was measured by the time required for the wave to travel between two thermocouples (Fig.2).

Figure 2 presents typical profiles of the travelling temperature wave front displayed both in time and in space at initial temperatures of 77 and 4.2 K. The front

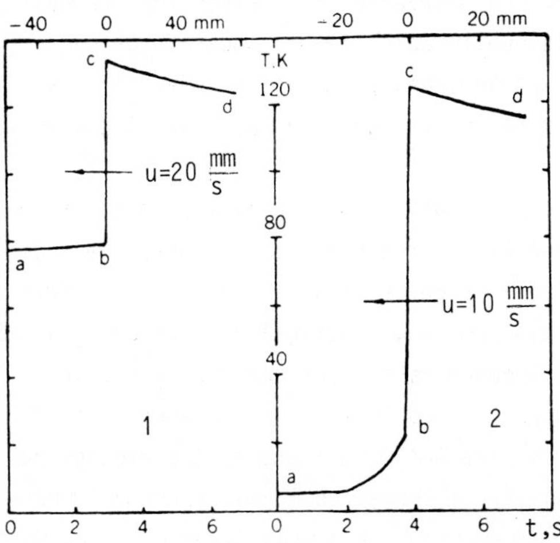


Figure 2. Temporal and spatial display of typical temperature profiles of a $BC+Cl_2$ reaction wave at (1) 77 K and (2) 4.2 K. Preirradiation dose 27 kGy.

profile remains qualitatively the same on changing the temperature of the thermostat. It includes three stages: ab, inert heating (at 77 K this stage, being practically isothermic, was not registered); bc, the jumplike onset and ignition of the reaction (the characteristic break point b); cd, the stage of cooling. Stage bc is a rapid one. Its duration was found oscillographically to be 0.1 s at 77 K and 0.3 at 4.2 K, so that at the wave velocity U = 2 cm/s the width of the zone of the travelling wave front was $\delta \cong 0.2$ cm, and at U = 1 cm/s, $\delta \cong 0.3$ cm.

This structure of the travelling front of a low-temperature reaction exhibits features utterly atypical of classical thermal self-propagation. They are: (i) a weak or nonexistent stage of inert preburst heating; (ii) a jumplike switching on and off of the reaction-characteristic break points b and c, the temperature where the reaction switches on at point b being far below that for the initiation of a thermoactivated reaction ($T_g \cong 100$ K); (iii) similarity of the dynamical patterns of the process development in the case of spontaneous initiation of a reaction by a travelling wave and in the case of forced reaction ignition by the accelerated heating; (iv) weak alteration in the wave velocity on changing the temperature in the thermostat from 77 to 4.2 K, accompanied by strong alteration in the heat release.

The next important step in the elucidiation of the role of the thermal factor in the mechanism of the phenomena in question was studying the effect of the sample size on the characteristics of the autowave process. The question was if the self-sustained wave regime of convection could be made possible by intensification of heat release at the expense of a decrease in the diameter of a cylindrical sample containing the reactant mixtures.

It has been established, in experiments performed in capillaries of 0.5-1.0 mm diameter (which corresponds to more than a tenfold increase in the parameter characterizing the intensity of heat release), that none of the studied systems displays a generation of the autowave process. Moreover, the characteristic values for the self-propagation of the reaction do not undergo any noticeable changes under these conditions. Figure 3 shows a cinegram of the reaction front propagation ($BC+Cl_2$) in a capillary of 1-mm diameter immersed in liquid helium (irradiation dose 45 kGy, wavefront propagation velocity $\cong$2.5 cm/s).

The next step was to perform a series of experiments with thin-film samples of reactants. The experiments were carried out with films frozen on a flat substrate and immersed in liquid nitrogen, i.e. in the absence of any heat screen between the free film surface and the pool of cooling agents. The autowave chemical conversion in a film was initiated by a local mechanical disturbance in the form of a puncture or a scratch. The changeover to such objects, characterized by the most intense heat absorption, allowed the realization of quasi-isothermal conditions of the process development and thus favoured the manifestation of the above mentioned isothermal mechanism of wave excitation, which involves autodispersing the sample layer by layer due to the density difference between the initial and final reaction products. The new conditions did not only suppress the phenomenon, but made it even more pronounced.

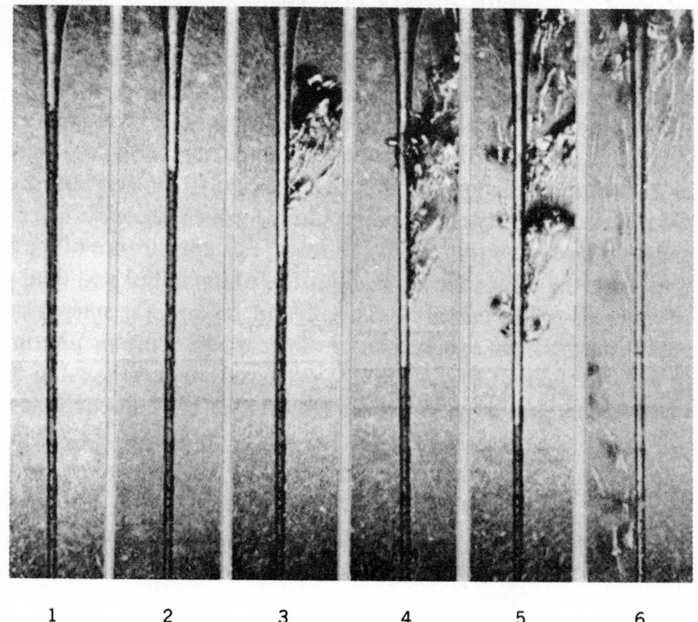


Figure 3. Cinegram of reaction front propagation in a capillary, diameter ~1 mm, at 4.2 K. Time intervals between frames 1 and 2: 0.14 s; 2 and 3: 0.06 s; 3 and 4: 0.14 s; 4 and 5: 0.14 s; 5 and 6: 0.72 s. Dose 45 kGy.

## 5. THEORETICAL ANALYSIS OF AUTOWAVE PROCESSES IN SOLID-STATE CRYOCHEMICAL CONVERSIONS

The experimental results have made it possible to further develop the hypothesis of the mechanochemical nature of the self-activation of solid-state conversion and to describe qualitatively the possible pattern of the autowave process of chemical interaction in the considered systems. It can be pictured in the form of a narrow region propagating over a solid sample and cut all over with an intricate network of newly formed cracks. The chemical reaction occurs exactly in this region, on the surface of or near the cracks, and in its turn creates the conditions for continuation of the process of fracture, which plays an activating role, in the neighbouring layer of the solid matrix. This dispersing of the sample, which proceeds layer by layer, is caused by the propagating field of stresses. The appearance of the field may be due to, as has been already said, the difference in the densities of the initial and final products (isothermal mechanism) or to the steep temperature gradients resulting from the exothermicity of the reaction (nonisothermal mechanism). Thw two factors can, naturally, act simultaneously as well. For that reason, we restrict ourselves at this stage of research to the analysis of the nonisothermal mechanism under the conditions

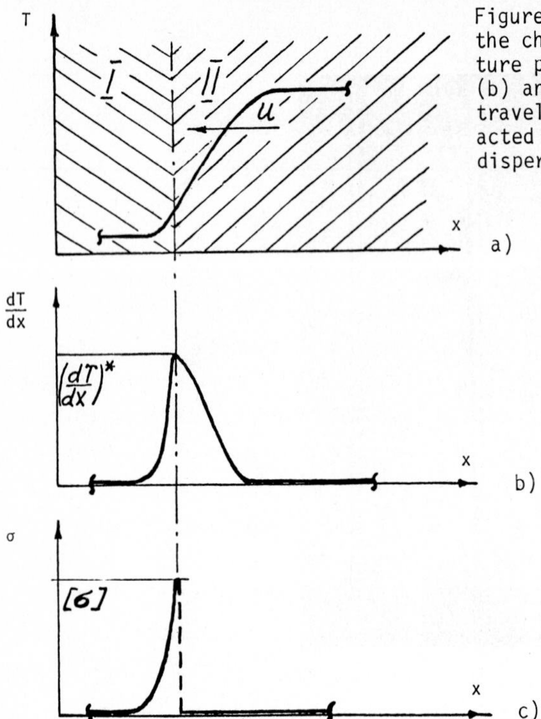


Figure 4. A qualitative illustration of the characteristic form of the temperature profile (a), temperature gradient (b) and stress field (c) in the travelling wave front. I is the unreacted region of the sample; II is the dispersed region of the sample

realized in massive cylindrical samples, in which, as shown above, heating due to the reaction is significant and the role of thermal stresses in the sample dispersion must be most pronounced.

The model proposed in /4/ describes a sample of infinite length without regarding the heat exchange with the environment, i.e. it is assumed that the characteristic time of the reaction heat release is much below of that of the heat transfer. This assumption makes it possible to perform the analysis in terms of an one-dimensional model. In accordance with the hypothesis, we believe that switching on of the reaction at a given cross section of the sample occurs when that cross section suffers a stress exceeding the ultimate strength of the solid matrix. Since we speak about thermal stresses and are concerned with elastic deformation and brittle fracture, we can easily change over, in the framework of our task, from stresses to temperature gradients, assuming a single-valued dependence between them (Fig.4). Such a substitution of variables greatly simplifies the model and allows the elimination of the mechanical equations and the reduction of the analysis to only one equation of thermal balance, in which we introduce a certain critical value of the temperature gradient $dT/dx = (dT/dx)^*$ as a parameter.

With the assumptions given above, the equation describing the autowave process in systems in question takes a form which looks similar to the fundamental combustion equation /5/:

$$\lambda \frac{d^2T}{dx^2} - Uc\rho \frac{T}{dx} + Q = 0 \ . \tag{1}$$

Here T is the temperature; x is the coordinate; $\lambda$ is the thermal conductivity; c and $\rho$ are the heat capacity and the density of the solid mixture of reactants, respectively; Q is the rate of the reaction heat release, and U is the propagation velocity of the temperature wave front.

The prinicpal difference between the autowave and combustion processes is that in the former case Q is a function of the temperature gradient but not the temperature itself. Let Q change jumpwise from 0 to $Q^*$ at $dT/dx = (dT/dx)^*$, and let it retain this value for a certain time interval. This implies physically that the heat release in a reaction is switched on only in response to a brittle fracture of the sample produced by thermal stresses equal to the ultimate strength of the material. The employment of the reaction time $\tau$ as a parameter reflects that part of the hypothesis according to which the time period of chemical activity in the course of fracture formation is limited by the deactivation process (for instance, by recombination of active particles on fracture surfaces). In therms of the stationary model (1) considered here, parameter $\tau$ contains information on the reaction zone (zone of dispersivity) at the propagating front, the size of which is equal to $U\tau$. At each point within the zone the reaction proceeds at a rate $Q^*$, and outside it $Q = 0$.

The boundary conditions can be written as:

$$T(x=-\infty) = T_0, \quad T(x=+\infty) = T_0 + Q\tau/c\rho \equiv T_m \ ,$$

where $T_0$ is the initial temperature of the sample and $T_m$ is the maximum temperature of its adiabatic heating due to the reaction heat. The model is considered to be quasiharmonic, i.e it is assumed that the characteristic size of a grain in the zone of dispersion is much less than the width of the temperature wave front.

Equation (1) with the heat source as a function of the type indicated and with the above mentioned boundary conditions is integrated analytically and has a solution in the form of a travelling wave. The principal attention will be paid to the analysis of the dependence of the wave propagation velocity on the parameters. This dependence, shown in Fig.5, has the form

$$g(u) \equiv \{1 - \exp(-u^2)\}u^{-1} = G \ . \tag{2}$$

Here $u = U(\tau/a)^{0.5}$ is the dimensional velocity of a propagating wave (the parameter to be determined), $G = (dT/dx)^*(a\tau)^{0.5}/(T_m-T_0)$ is the dimensionless critical temperature gradient; $a = \lambda/c\rho$ is the coefficient of thermal conductivity.

As seen from Fig.5, the autowave solution of Eq.(1) exists only at $G < G_0 \cong 0.64$. Physically, this implies that in a system described by this model the autowave mode of the reaction propagation over a solid reactant mixture becomes impossible at a definite increase in the strength of the sample, decrease in the thermal effect and reaction velocity, and increase in the thermal conductivity.

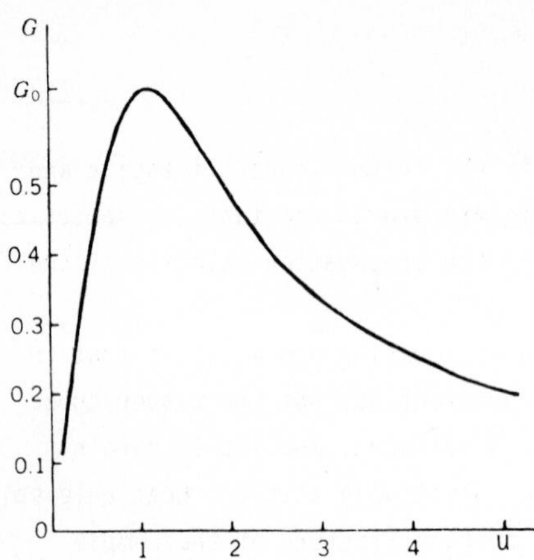


Figure 5. Critical temperature gradient as a function of reaction-wave propagation velocity.

The essential feature of Eq.(1) is the nonuniqueness of its solution. In the range $G < G_0$, two different values of the steady-state velocity of wave propagation correspond to the same vale of G. Let us compare the characteristics of the two autowave regimes. For the sake of simplicity, we shall consider range $G << G_0$ in which Eq.(2) is solvable for u. The expression for u corresponding to the smaller velocity mode has the form

$$u = a(dT/dx)^*/(T_m - T_0) \quad \text{or} \quad u = a(dT/dx)^*/(Q^*\tau/c\rho) \;, \tag{3}$$

and that for the greater velocity mode

$$u = (T_m - T_0)/(dT/dx)^*\tau \quad \text{or} \quad u = (Q^*/c\rho)/(dT/dx)^* \;. \tag{4}$$

The slower autowave process is similar in some respect to the classical combustion, despite the difference in their physical nature. The wave velocity shows the same dependence on thermal conductivity as in the case of flame propagation. Analogously to combustion, the reaction zone is near the maximum temperature $T_m$ (it is near $T_m$ that the critical gradient $(dT/dx)^*$ switching on the reaction is realized), whereas the greater part of the front temperature profile corresponds to inert heating of the sample (the "Michelson zone" in the combustion theory).

The faster process is significantly different. The most conspicuous feature of (4) is that the velocity does not depend on the thermal conductivity. This result, unexpected for a problem with conductive heat transfer, is connected with the peculiarities of the structure of the faster wave front. The temperature profile of the wave front is considerably steeper than that of the slower wave, and the coordinate of the switching on of the reaction is in its fore part, near $T_0$. This explains the independence of the wave process of parameter a, because the temperature gradients are high, approaching a critical value even in the unheated part of the front.

Qualitatively the regime of a faster wave is closer to the experimental conditions.

To test the fitness of the theoretical mechanism to the experimentally observed phenomena, it seemed principally important to try to realize experimentally the

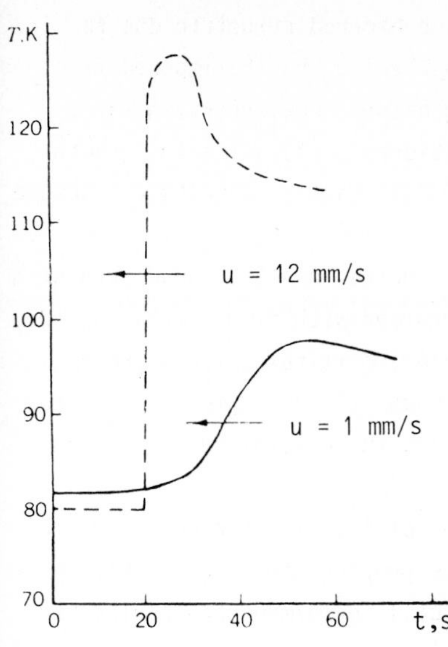


Figure 6. Time dependence of the temperature profiles of $BC+Cl_2$ reaction wave. The system was placed in the vapour of liquid nitrogen. The solid curve resulted from initiation by slow heating, and the dashed curve by rapid heating (discharge of capacitor through heater).

second mode of the autowave process. Its initiation was performed, in accordance with the theory, not by pulse heating but with the help of a heater the temperature of which could be raised slowly.

In specially designed experiments the sample was placed in a cryostat in the vapour of boiling nitrogen. In that case we could observe, along with the faster wave (having the same velocity as that described earlier), also the slower wave (Fig.6). As seen, the velocity of the slower wave is about one order of magnitude less than that of the faster wave, which coincides with the calculations by (3) and (4).

It should be noted that the realization of the slower wave involved certain difficulties. Not in every experiment could it be initiated by the slow local heating. Frequently the faster wave was initiated as well. There were cases when the well-developed slower wave transformed into the faster one in the course of its propagation. This resulted from the fact that the slower regime was unstable.

## 6.AUTOWAVE CONVERSION UNDER CONDITIONS OF UNIFORM COMPRESSION OF THE SAMPLE

From the theoretical analysis it follows that acting on the strength characteristics of the solid matrix of a frozen reactant mixture may be an effective means for testing the developed concepts. This was an impetus for a study of the effect of high pressures on the dynamics characteristics of the autowave regimes of chemical conversion /6/ and the role of loading dynamics and regimes of sample fracture on the mechanical initiation of the reaction /7/.

As predicted by the theory, the increase in the pressure was accompanied by a decrease in the reaction wave-front velocity. At $6x10^8$ Pa the velocity is significantly reduced - approximately by a factor of 2.5-3 below the uncompressed sample. Thus, the achievement by the matrix of a more homogeneous and ordered structure due to the compression, together with its simultaneous strengthening and the impeded fracture formation, slows down the propagation of a travelling dispersion-conversion wave. This can naturally be connected, on the basis of relationship (4), with the growth of the critical value of the temperature gradient $(dT/dx)^*$ causing a brittle fracture of the sample.

The experiments on the investigation of the role of mechanical loading determining a brittle or plastic character of a fracture were performed with film samples 50-100μ thick, which were prepared from a chlorine + butyl chloride mixture. The microphographs demonstrate that an impact with a needle (pressure $\cong 10^8$ Pa, time $\cong 0.1$ s) produces a network of cracks. Such a brittle fracture leads to an autowave initiation in the preirradiated film.

A local disturbance with a needle (diameter 0.6 mm) of the film subjected to the same dose of γ-radiation in the regime of the autowave propagation of chemical reaction, left a distinct trace as a pressed-in, plastically deformed area having the form of the cross section of the needle.

Therefore, neither the appreciable plastic deformation (both in the case of uniform compression and of local fracture) of the studied solid reaction systems nor their static state of high stress is a factor conditioning the critical phenomena and autowave processes observed during the chemical conversion in the systems. In other words, this series of experiments has provided another telling argument for the decisive role of the brittle fracture in the mechanism of the considered phenomena (see /1/ and references therein).

## 7. CONCLUSION

This review sums up the results of the first cycle of investigations that are new for low-temperature chemical kinetics of the solid-state conversion autowave mechanism. This cycle was developed mainly in the framework of a phenomenological approach. The next stage of the investigations is naturally faced with a number of significant problems connected with elucidating the elementary steps of conversion, defining the detailed schemes of conversion of mechanical forms of energy, accumulated in a solid matrix, into chemical ones to establish the nature of chemically active states arising in the travelling wave front and to determine their life times.

Nevertheless, summing up the material, it is necessary to stress that the considered stage, in spite of some phenomenological narrowness, is of basic importance for the creation and development of the autowave ideology in cryochemistry of solids. The role of the performed study is not only in the fact of discovery itself or in the development of simple macrokinetic models of the phenomenon. The most important stage is also a conclusion of wide spreading of the autowave conversion mechanism in the low-temperature solid-state chemistry. The notions of the nature of autoactivation of frozen reagents solutions will naturally deepen with new experimental evidence. We should mention, in this connection, some recent experimental results.

The reaction of autowave polymerization of solid acetaldehyde seems to be of great interest /8/. The system is considered particularly interesting in three respects. First, its initial reagent is one-component. Second, the thermal effect of the reaction (comparable with crystallization heat) is very small as compared with those studied earlier. Third, the phenomenon of travelling wave conversion can be realized in the same reaction for samples with different starting states - both in vitrous and in polycrystalline matrices.

A marked acceleration of the wave propagation process was recorded for another polymerizing system - that of cyclopentadiene with butyl chloride /9/. This feature is associated with the factor of formation and accumulation, under $\gamma$-irradiation, of poorly condensing radiolysis products (hydrogen, carbon, oxide, methane) in some matrices. Gases desorbing from the matrix fractures during self-dispersion in the travelling wave front, facilitate the process of further destroying the sample, which is the cause of the increasing velocity of the front propagation.

References /10/ and /11/ produce additional considerations supporting a special role role of the potential mechanical energy, accumulated in the radiolyzed samples and converted subsequently into the chemical form of energy in the autodispersion - conversion wave front.

The reactions in co-condensates of chlorine with ethylene hydrocarbons as well as the earlier studied chlorination reactions in the transition from matrix activation induced by $\gamma$-irradiation to photolysis /12/ stand out from the ordinary series of investigations for their dynamics of autowave conversion. In the above mentioned processes, the velocities of reaction wave propagation over the solid sample measured by magnitudes of the order of meters and dozens of meters a second were registered. These systems are assumed to acquire a new mechanism of matrix autodispersion, which is close, in its essence, to the detonation mechanism. However, here we speak of the usual scheme of initiation of a detonation wave known in the classical combustion theory and of a special type of detonation which may be called gasless.

The extensively studied (especially during the recent years) transitions of solids from the metastable amporphous state to the polycrystalline state /13/ are of autowave character and resemble very much the above described regimes of solid-state cryochemical reactions. The action of autodispersion, which facilitates phase transition by allowing it to proceed on the surface of a fracture instead of in the glass volume, cannot be excluded in the case of those processes either. Actually, the two classes of processes are similar in their physical nature: both are connected with rearrangement of the solid matrix and are of exothermic character, differing only in the extent of the thermal effect.

The same mechanism may describe the most interesting process /14/ of rearrangement of the high pressure metastable crystalline state in the alloy Zn - Sb to the more stable amorphous state.

## 8. LAST REMARKS

The developed concepts can rightfully be called also an account for the "cold" evolution of matter in the universe. In particular, one can imagine the formation, from the frozen mixture of elements, of compounds such as ammonia and methane that are found in appreciable amounts as solids on the cold planets of the solar system. Daily temperature variations and the concomitant thermal stresses in the solid cover of a planet might serve as a mechanism of its continuous destruction, resulting in the chemical binding of the components dispersed in the sun mill.

## REFERENCES


1. V.V.Barelko, I.M.Barkalov, V.I.Goldanskii, A.M.Zanin, D.P.Kiryukhin. Uspekhi khimii, 1989 (in press); Advances in Chem.Phys., 1988, 74, 339-384; Preprint OIKhF, USSR Acad.Sci., Chernogolovka, 1988, 48p.
2. I.M.Barkalov, D.P.Kiryukhin. Vysokomolek.soyed., 1980, 22A, 4, 723-737.
3. A.M.Zanin, D.P.Kiryukhin, I.M.Barkalov, V.I.Goldanskii. Pis'ma v JETP, 1981, 33, 6, 336-339.
4. V.V.Barelko, I.M.Barkalov, D.A.Vaganov, A.M.Zanin, D.P.Kiryukhin. Khimich.fizika, 1983, 2, 7, 980-984.
5. Ya.B.Zeldovich, G.I.Barenblatt, V.B.Librovich, G.M.Makhviladze. Mathematical Theory of Combustion and Blast. Nauka, Moscow, 1980, 478 p.
6. A.M.Zanin, D.P.Kiryukhin, V.S.Nikolskii, I.M.Barkalov, V.I.Goldanskii. Izv. Akad. Nauk SSSR, Khimia, 1983, 6, 1228-1231.
7. D.P.Kiryukhin, A.M.Zanin, V.V.Barelko, I.M.Barkalov, V.I.Goldanskii. Sov.Phys.-Dokl., 1986, 288, 2, 406-409.
8. G.A.Kichigina, D.P.Kiryukhin, A.M.Zanin, I.M.Barkalov, V.I.Goldanskii. Khimich. fizika, 1988, 7, 4, 543-547.
9. G.A.Kichigina, D.P.Kiyukhin, A.M.Zanin, I.M.Barkalov. Vysokomolek.soyed., 1989 (in press).
10. D.P.Kiryukhin, V.V.Barelko, I.M.Barkalov. Sov.Phys.-Dokl., 1989 (in press).
11. V.V.Barelko, D.P.Kiryukhin, A.M.Zanin, I.M.Barkalov. Khim.vysokikh energii, 1989 (in press).
12. I.M.Barkalov, V.I.Goldanskii, A.M.Zanin, D.P.Kiryukhin. Sov.Phys.-Dokl., 1987, 296, 4, 891-894.
13. V.A.Shklovskii. Sov.Phys.-JETP, 1982, 82, 2, 536-547.
14. O.I.Barkalov, I.T.Belash, A.I.Bolshakov, E.G.Ponyatovskii. Sov.Phys.-Tech.Phys., 1988, 30, 9, 2724-2729.

# Coherent Amplification of Weak Pulses

*S.P. Bondarev*

Institute of Physics, Belorussian Academy of Sciences,
220602 Minsk, USSR

The structure of a pulse at the two-level amplifier output depending on the weak input pulse shape has been determined. It is shown that at large linear amplifications, as well as in the case of nonlinear amplifications, pulses close to the amplifier response to the δ-pulse are formed. The parameters of amplified pulses are associated with the spectrum of the weak input pulse.

## 1. INTRODUCTION

Coherent amplification of weak pulses in a medium consisting of two-level atoms with a non-uniformly broadened shape of the spectral line is considered. It is shown that large amplifications of weak pulses lead to the formation of pulses close to the amplifier response to the δ-pulse. The distinction of the input pulse from the δ-pulse leads to the appearance of a slowly varied function defined by the input pulse spectrum (see the expression for the amplified pulse envelope).

The case when the effect of non-uniform broadening is not essential was discussed in /1/. The conclusion in /1/ that the amplified pulse is close to the self-similar one agrees with the result of this paper, because in the absence of non-uniform broadening the self-similar solution coincides with the response to the δ-pulse.

## 2. BASIC EQUATIONS

Equations describing the pulse propagation in a two-level amplifying medium with a non-uniformly broadened spectral shape are of the form

$$\partial_z U = \alpha_0 \int d\Delta g(\Delta) R(\Delta), \tag{1}$$

$$\partial_\tau R = -i\Delta R + nU, \tag{2}$$

$$\partial_\tau n = -\frac{1}{2}(UR^* + U^*R), \tag{3}$$

where $U = \mu E/\hbar$, $R = ip/\mu N_0$, $E$ and $p$ are slowly varying amplitudes of the electric field and polarization, respectively, $\mu$ is the transition dipole moment, $N_0$ is the initial density of inverted atoms, $n$ is the difference between the populations of the upper and ground levels, $\alpha = t - Z/C$ is the decay time, $C$ is the light velocity, $\alpha_0 = 2\pi N_0 \omega_0 \mu^2/\hbar C$, $\omega_0$ is the carrier frequency of the input pulse. Function $g(\Delta)$ describes the non-uniform broadening shape. It will be assumed below that $g(\Delta)$ is an

Research Reports in Physics Nonlinear Waves 3
Editors: A.V. Gaponov-Grekhov · M.I. Rabinovich · J. Engelbrecht

even function. Frequency $\omega_0$ coincides with the maximum position of the non-uniform broadening shape.

It is assumed that the amplifying medium occupies the half-space $Z > 0$ and on the boundary at $Z = 0$ a weak input pulse $U(Z=0,t) = U_0(\tau)$ is given such that $U_0(\tau) = 0$ at $\tau < 0$. The input pulse duration $\tau_p$ is less than the relaxation time $T_2{}^*$, conditioned by the non-uniform broadening.

At the initial stage of amplification when the pulse intensity is not very large and the inversion variation can be neglected, i.e. $n = 1$ in Eqs. (1)-(3) can be assumed, a linear amplification regime is realized. In this case the pulse envelope can be represented as the integral

$$U(Z,\tau) = \frac{1}{2\pi} \int d\nu \varepsilon(\nu) \exp\left[iZ\kappa(\nu) - i\nu\tau\right] , \tag{4}$$

where

$$\kappa(\nu) = \alpha_0 \int \frac{g(\Delta)}{\nu - \Delta} d\Delta \tag{5}$$

and $\varepsilon(\nu)$ is input pulse spectrum.

Pulses formed at large linear amplifications exhibit characteristic properties, which will be considered below.

## 3. AMPLIFICATION

Large amplifications are realized at a sufficientlylarge value of amplification length $Z$ and at sufficiently large times $\tau$ after the arrival of the front edge of the pulse. In this case integral (4) comprises a rapidly varying function $\exp\left[iZ\kappa(\nu) - i\nu\tau\right]$. The saddle-point method is used for its approximate analysis.

Having calculated integral (4) by this method, we obtain the following approximate expression for the envelope:

$$U(Z,\tau) = A(Z,\tau) \exp\left[\psi(Z,\tau)\right] , \tag{6}$$

where

$$A = \varepsilon(\nu_1)/(-2\pi i Z \partial_\nu{}^2 \kappa(\nu_1))^{1/2} , \tag{7}$$

$$\psi = i\left[Z\kappa(\nu_1) - \tau\nu_1\right] . \tag{8}$$

When $g(\Delta)$ is even, function $\psi$ takes real positive values. We shall assume here that the main contribution to (4) is made by one saddle point which can be determined from the equation

$$Z\partial_\nu \kappa - \tau = 0 . \tag{9}$$

When the contribution of several saddle points is taken into account, the envelope is represented as a sum of expressions analogous to (6).

If the amplified pulse is observed at $\tau << T_2{}^*$, the relaxation effect is neglible. Assuming $g(\Delta) = \delta(\Delta)$ in (5), we find for this case $\kappa(\nu) = \alpha_0/\nu$ . In this case (6) takes the form

$$U(Z,\tau) = \frac{\varepsilon(i\sqrt{\alpha_0 Z/\tau})}{2\sqrt{\pi}(\tau^3/\alpha_0 Z)^{1/4}} \exp(2\sqrt{\alpha_0 Z\tau}) . \quad (10)$$

From the applicability conditions of (6) it follows that A is a slowly varying function and $\psi$ is a rapidly varying function of time. Extracting the factor $\varepsilon(\nu_1)$ in (6) we can write

$$U(Z,\tau) = \varepsilon(\nu_1)\theta(Z,\tau) . \quad (11)$$

The function

$$\theta(Z,\tau) = (-2\pi i Z\partial_\nu^2\kappa(\nu_1))^{-1/2} \exp\left[iZ\kappa(\nu_1) - i\tau\nu_1\right] \quad (12)$$

describes the amplifier response to the input $\delta$-pulse. In the case of input pulse $U_0(\tau) = U^0\delta(\tau)$ we have $\varepsilon(\nu_1) = U^0 = \text{const}$. The distinction of the input pulse from the $\delta$-pulse leads to the appearance of the slowly varying function $\varepsilon(\nu_1)$ in the expression for the envelope. This function is defined by the spectrum of the input pulse with frequency $\nu_1 = \nu_1(Z,\tau)$, that can be found from (9). Since the rapidly varying function $\exp\left[\psi\right]$ is defined by the amplifier response to the $\delta$-pulse, the envelope of the pulse generated at large linear amplifications behaves locally as the envelope of the pulse formed in the case of the $\delta$-shaped input pulse.

We shall show now that the property of the amplified pulse being close to the amplifier response to the $\delta$-pulse and its relation to the spectrum of a weak input pulse exists also at the non-linear amplification stage.

## 4. AMPLIFICATION OF WEAK SIGNALS

To analyse the pulse formed at the nonlinear stage of weak signal amplification, we shall use the inverse scattering transform. The standard details of this method are given in /1,2/. Solution of the problem is reduced then to the investigation of the system of singular integral equations. The integrals are calculated using the saddle-point method and the following approximate expression for the envelope is obtained

$$U(Z,\tau) = 2\partial_\tau\psi \operatorname{sech}(\psi - \ln\left|\frac{4\partial_\tau\psi}{A}\right|) \exp\left[i \arg(A/\partial_\tau\psi)\right] , \quad (13)$$

where A and $\psi$ are defined by expressions (7) and (8). At $|A/4\partial_\tau\psi|^2 < 1$ expression (13) goes into expression (6) describing the pulse in the case of large linear amplifications. If the non-uniform broadening is absent, then expression (13) takes the form

$$U(Z,\tau) = 2(\alpha_0 Z/\tau)^{1/2}\operatorname{sech}(2\sqrt{\alpha_0 Z\tau} - \ln\left|\frac{8(\pi^2\alpha_0 Z\tau)^{1/4}}{\varepsilon(i\sqrt{\alpha_0 Z/\tau})}\right|)\exp\left[i \arg(\varepsilon(i\sqrt{\alpha_0 Z/\tau}))\right]. \quad (14)$$

The pulse formed at the nonlinear stage of amplification has an oscillating structure. Expressions(13), (14) describe this pulse in the region of the first main part.

As in the case of large linear amplifications, it follows from the applicability condition of expression (13) that A is a slowly varying , but $\psi$ is a rapidly varying function. Since the input-pulse-dependent function is slowly varying, the local behaviour of the envelope of the amplified pulse is determined by the nonlinear amplifier response to the weak $\delta$-pulse. The difference between the input and the $\delta$-shaped pulse leads to the appearance of the slowly varying function $\varepsilon(\nu_1)$ in the expression for the envelope. This function is determined by the input signal spectrum at frequency $\nu_1 = \nu_1(Z,\tau)$ which can be found from expression (9).

The pulse formed at large linear amplifications is close to the response to the $\delta$-pulse. This property takes place independently of whether the inverse scattering transform is applicable or not for the pulse analysis at the nonlinear stage of amplification. The system of equations (1)-(3) which can be integrated by the method of the inverse scattering transform provides the possibility of substantiating the assumption that the property of the amplified pulse being close to the $\delta$-pulse, realizable at large linear amplifications, is preserved at the nonlinear stage of amplification.

The solution of nonlinear equations corresponding to the weak input signal can be constructed proceeding from the solutions corresponding to the input $\sigma$-pulse. Such an approach is of interest in the cases where the inverse scattering transform fails. If the expression for the amplifier response $G(U^0,Z,\tau)$ to the weak $\delta$-pulse $U_0(\tau) = U^0\delta(\tau)$ is known, the solution corresponding to the weak input pulse is obtained under the assumption that parameter $U^0$ is a slowly varying function of $\{Z,\tau\}$ . We shall establish the connection between $U^0$ and the input signal by matching the solution of the amplification problem obtained in a linear approximation and the expression for $G(U^0,Z,\tau)$ in the region of their overlapping. From expression (11) defining the envelope at large linear amplifications it follows that the difference of the input pulse from the $\delta$-shaped one will lead to the replacement of $U^{00}$ by $\varepsilon(\nu_1)$. Accordingly, at the nonlinear stage of amplification the pulse caused by a weak input signal will be described by the expression

$$U(Z,\tau) = G(\varepsilon(\nu_1),Z,\tau). \tag{15}$$

If the effect of non-uniform broadening is insignificant, the system of equations (1)-(3) is equivalent to the Sine-Gordon equation. In this case, as shown in /1/, the pulses close to the self-similar pulse are formed at large amplification of weak signals. This conclusion agrees with the results of the present paper, since in the absence of non-uniform broadening the self-similar solution corresponds to the input $\delta$-pulse. The models of coherent interaction of pulses with a nonlinear medium when the pulses close to the self-similar solution are formed at the amplification of weak signals, are considered in /3/.

In the absence of non-uniform broadening the amplified pulse is described by the expression

$$U(Z,\tau) = \sqrt{\alpha_0 Z/\tau}\ s(s_0,\xi) \tag{16}$$

where $\xi = \sqrt{\alpha_0 Z\tau}$ is the self-similar variable. Parameter $s_0$ is a slowly varying function of $\{Z,\tau\}$ determined by the input pulse spectrum $s_0 = \varepsilon(\nu_1)$.

In /1/ it ia stated that $s_0$ is determined by the behaviour of the input pulse envelope in the vicinity of the front edge of this pulse. This conclusion is valid in the case of a long amplifier at large values of $|\nu_1| = \sqrt{\alpha_0 Z/\tau}$ when the condition

$$\sqrt{\alpha_0 Z/\tau} >> \tau_p^{-1} \tag{17}$$

holds. In this case, due to the action of the cutting function $\exp(-\eta\sqrt{\alpha_0 Z/\tau})$ the integral

$$\varepsilon(\nu_1) = \int_0^\infty d\eta U_0(\eta)\exp(-\eta\sqrt{\alpha_0 Z/\tau}) \tag{18}$$

will depend on the behaviour of $U_0(\eta)$ at $\eta \to 0$. Another regime of amplification when $s_0$ is determined by the input signal area has been studied theoretically and experimentally in /4/. It follows from (18) that this regime is realized for short input pulses at moderate amplifier lengths when $|\nu_1|$ is small and the condition inverse to (17) is fulfilled.

REFERENCES


1. S.V.Mankov. Propagation of an ultrashort optical pulse in a two-level laser amplifier. Sov.Phys.-JETP, 1982, v.83, No.1, 37-52.
2. I.R.Gabitov, V.E.Zakharov, A.V.Mikhailov. Maxwell-Bloch equation and inverse scattering method. Theor.and Math.Phys., 1985, v.63, No.1, 328-343.
3. O.P.Varnavskii, V.V.Golovlev, A.N.Kirkin et al. Coherent propagation of small area pulses in activated crystals. Sov.Phys.-JETP, 1986, v.63, No.5, 937-944.
4. A.A.Zabolotsky. Multiwave coherent interaction and nonlinear frequency shift. Zh. Eksp.Teor.Fiz., 1989, v.94, No.11, 33-45.

# Nonlinear Dynamics and Kinetics of Magnons

*V.S. L'vov*

Institute of Automation and Electrometry,
Siberian Branch of USSR Academy of Sciences,
630090 Novosibirsk, USSR

A review on nonlinear spin waves (magnons) is presented. The corresponding physical models in magnetic dielectrics are analyzed. The parametric dynamics and kinetics of magnons are described in detail.

## 1. NONLINEAR MAGNONS IN MAGNETIC DIELECTRICS

### 1.1. Introduction

During the recent two decades, there has been a steady growth of interest in the highly nonequilibrium systems. It concerns above all studies on the behaviour of substances at high power levels of external action: of dielectrics in the powerful laser wave field, magnets in strong microwave fields, plasma heated to thermonuclear temperatures, etc. In fact, a new physical discipline has emerged - *the physics of nonlinear waves*. Its purpose is to study, possibly from a single viewpoint, the phenomena and processes arising upon the excitation, propagation and interaction of limited-amplitude waves in various media. These studies have revealed that some phenomena, such as formation of the "tenth wave" in a stormy sea, self-focusing of light and Langmuir wave collapse in plasma have a common physical reason.

In physics of nonlinear wave processes, dealt with by the Gorky schools, the section on spin waves (magnons) is one of the most advanced ones. This is partly due to the fact that experiments on spin waves in magnetic dielectrics are much simpler than the similar experiments on sound in crystals, on Langmuir and other types of waves in plasma, in nonlinear optics, etc. They may often be carried out at room temperature, in a customary frequency range and on top-quality monocrystals. Spin waves are easily excited by the microwave magnetic field to an essentially nonlinear level, when their behaviour is completely determined by their interaction with each other. There is a great amount of experimental material, as well as well-developed, highly advanced theoretical models. They may certainly prove useful in elaborating other sections of nonlinear physics. Regretfully, these results are almost unknown. The aim of this paper is to draw the attention of researchers engaged in the theoretical and experimental studies on plasma physics, nonlinear optics, hydrodynamics, nonlinear acoustics etc. to a related field: nonlinear spin waves. We suppose that reduction of a barrier separating these sections of nonlinear physics from each other will be mutually fruitful. For that purpose the author of the present paper has written a book "Nonlinear



Research Reports in Physics Nonlinear Waves 3
Editors: A.V. Gaponov-Grekhov · M.I. Rabinovich · J. Engelbrecht

Spin Waves" /1/, which will be published in a considerably revised and updated form by Springer in 1990, under the title "Nonlinear Dynamics and Kinetics of Magnons".

## 1.2. Magnetically-Ordered Dielectrics

Today, a lot of magnetically-structures materials are known (dielectrics, semiconductors and metals, both crystalline and amorphous). Their structure includes paramagnetic atoms (ions) with uncompensated electron spin magnetic moments $\vec{\mu} = \mu_B \vec{S}$ ($\mu_B$ is the Bohr magneton, $1/2 < S < 7/2$ is the atom spin). At low temperatures, these moments are oriented with respect to each other in a definite fashion. In *ferromagnets*, which present the simplest case, magnetic moments of all atoms are parallel. This results in the macroscopic magnetic moment, equal to their sum. The physical reason which causes magnetic ordering is the *exchange interaction*. It has an electrostatic nature and is associated with the Pauli principle, forbidding the existence of two electrons in one quantum-mechanical state. The Hamiltonian of exchange interaction

$$H_{ex} = -J\vec{S}_1\vec{S}_2 \tag{1}$$

is called the *Geisenberg* Hamiltonian, J is the *exchange integral*. In ferromagnets, $J > 0$. In order of magnitude, $S(S+1)J/3 \cong T_C$ is the *Curie temperature*. At $T > T_C$, magnetic ordering disappears. For different substances $T_C$ varies in the range 1-1000K.

## 1.3. Spin Waves and Equations of Motion

In the excited state, the *magnetic moment density* $\vec{M}(\vec{r},t)$ depends on coordinate $\vec{r}$ and time t. This dependence may be expanded into the series in plane waves, which at $T << T_C$ are weakly interacting. These are called *spin waves*, or *magnons*. Similarly, sound waves are called phonons, when one wants to emphasize their quantum-mechanical properties.

The exchange interaction determines the magnitude $M(\vec{r},t)$ with high accuracy at $T << T_C$. Thus, spin waves represent the magnetic moment precession waves. Magnitude $M(r,t)$ obeys the phenomenological *Bloch equation* /1/,

$$\partial M/\partial t = g\left[\delta W/\delta M x M\right] , \tag{2}$$

which defines this precession, conserving both the full energy of a magnet W and value $|\vec{M}|$. Here W is the functional of $M(r,t)$; $g \cong \mu_B/\hbar$ is the magnetomechanical electron factor.

## 1.4. Classical Hamiltonian Formalism and Spin Waves in Magnets

We see that the Bloch equation (2) has a specific form which is very distinctive, e.g. from the Maxwell equations in a nonlinear dielectric. The latter equations radically differ from the Euler equations for compressible fluid. As a matter of fact, spin, electromagnetic and sound waves are, first of all, waves, i.e. medium oscillations transferred in a relay fashion from one point to another. If we are interested only in some propagation characteristics of the small (but limited) amplitude waves, such as diffraction or self-focusing, then it is absolutely unnecessary to know what is it that oscillates, whether it is the magnetic moment, electric field, or density. All information on the type of a medium, which is necessary and sufficient for investigating the propagation of noninteracting waves in this medium, is given by their dispersion law $\omega(\vec{k})$. Likewise, the existence of other universal functions which describe some of the properties of the medium may be assumed, the knowledge of which is sufficient to describe wave interactions. Such functions appear on passing to the Hamiltonian method of describing motion. This method is applicable to a wide class of weakly interacting and weakly dissipative wave systems. It reveals their common properties.

1.4.1. Transition to the Canonical Description of Magnons. Let us write the Bloch equation (2) in the Hamiltonian form /1/:

$$i\partial a/\partial t = \delta H/\delta a^* . \tag{3}$$

Here H is the *Hamiltonian function* representing in this case the energy W defined through the *canonical variables* $a(\vec{r},t)$ and $a^*(\vec{r},t)$. These are connected with magnetization by the ansatz found in 1969 by ZAKHAROV, L'VOV and STAROBINETS /1/:

$$M_x + iM_y = a\left[2gM_0(1-gaa^*/2M_0)\right]^{1/2} , \qquad M_x - iM_y = a^*\left[2gM_0(1-gaa^*/2M_0)\right]^{1/2}, \quad M_z = M_0 - gaa^* . \tag{4}$$

This is a classical analogy of the Godstein-Primakov ansatz, defining the spin operators $\hat{S}$ through the Bose operators $\hat{a}$ and $\hat{a}^+$. The canonical variables a and a* represent the classical limit $\hat{a}$ and $\hat{a}^+$.

1.4.2. The Hamiltonian Function at Small Nonlinearity. At a small nonlinearity, when $gaa^* << M$, the Hamiltonian function H may be expanded into the series in $a(\vec{r},t)$, $a^*(\vec{r},t)$. In the k-representation the expansion is as follows:

$$H = H_0 + H_1 + H_{int} , \quad H_{int} = H_3 + H_4 + \dots , \qquad H_0 = \text{const} , \quad H_2 = \sum_{\vec{k}} \omega_{\vec{k}} a_{\vec{k}} a_{\vec{k}}^* . \tag{5}$$

Constant H , which is independent of a, a*, does not arise in equation (3) and may be omitted. The first-order terms in a, a*, $H_1$ are absent, because we have assumed that in the absence of waves, when a = 0, a medium is in equilibrium. Among the second-order terms in a, a*, the expansion contains no terms of $a_k a_{-k}$ and $a_k^* a_{-k}^*$. If they do appear, they may be eliminated using the linear canonical (U-V) transformation

$$b_{\vec{k}} = U_{\vec{k}} a_{\vec{k}} + V_{\vec{k}} a_{-\vec{k}}^* \tag{6}$$

with an appropriate choice of U and V. From now on, we shall consider this transformation fulfilled, with Hamiltonian H having the form (5). Then in variables a, a* the linearized equations of motion become trivial:

$$\partial a_{\vec{k}}/\partial t + i\omega_{\vec{k}} a_{\vec{k}} = 0 . \qquad (7)$$

They describe the propagation of noninteracting spin waves having the dispersion law $\omega(\vec{k})$. All information on wave interactions is given by the coefficient of $H_{int}$ expansion to the series of a, a*:

$$H_3 = (1/2) \sum_{1,23} \left[ V(1,23) a_1{}^* a_2 a_3 + H.c. \right] \delta(1-2-3) +$$

$$+ (1/6) \sum_{123} \left[ U(123) a_1 a_2 a_3 + H.c. \right] \delta(1+2+3) , \qquad (8)$$

$$H_4 = (1/2) \sum_{12,34} T(12,34) a_1{}^* a_2{}^* a_3 a_4 + \dots .$$

Here and below $a_j = a(k_j)$, $V(1,23) = V(k_1,k_2k_3)$, etc., $\sum_{1,23} = \sum_{k_1k_2k_3} \delta(\vec{k}_1-\vec{k}_2-\vec{k}_3)$, etc. The physical sense of $H_3$, $H_4$ could be easily understood by analogy with quantum mechanics: $H_3$ describes three-magnon processes of the type $1 \leftrightarrow 2$ and $0 \leftrightarrow 3$ (transformation of one wave into two, and vice versa, generation of three waves from vacuum and vice versa). The terms of $H_4$-expansion describe the four-magnon scattering processes of the type $2 \to 2$. In the presence of external magnetic pumping field, there appears an additional term in the Hamiltonian $H_{int}$

$$H_p = (1/2) \sum_k \left[ h \exp(-i\omega_p t) V_k a_k{}^* a_{-k}{}^* + H.c. \right] , \qquad (9)$$

describing the *parametric excitation* of spin waves, i.e. the induced process of decay of a photon (with frequency $\omega_p$ and zero wavevector) into two magnons with wavevectors k, -k and frequencies $\omega_k = \omega_p/2$.

1.4.3. Equations of Motion. We must note that equations (5) conserve the energy of magnon system H. As a matter of fact, in reality the interactions such as magnon-phonon interactions, interactions of magnons with lattice defects and those of other types always exist, which leads to a small dissipation of their energy. This process may be taken into consideration phenomenologically: by adding the imaginary part $\gamma_k$ to frequency $\omega_k$. As a result we obtain

$$\partial a_k/\partial t + \gamma_k a_k + i\omega_k a_k = -i\delta H_{int}/\delta a_k{}^* . \qquad (10)$$

The interaction Hamiltonian (5) - (9) together with the equations of motion (10), give the canonical formulation of the problem of nonlinear behaviour of magnons, understandable for the physicists unfamiliar with magnetism. The whole specific character of magnetics is given by function $\omega(k)$ and coefficients of the interaction Hamiltonian.

## 2. PARAMETRIC DYNAMICS OF MAGNONS

A wide class of nonlinear wave processes is described by the dynamic equations of motion for the complex wave amplitudes (10), where phase correlations play an essential role. It would therefore be reasonable to call them *dynamic processes* as distinct from the processes of another class which are described by the kinetic equations for occupation numbers, and which may be called *kinetic processes*, one can mention confluence of two waves into one, generation of the second harmonic, decay of one wave into two waves, and various types of four-magnon processes, including self-focusing and collapse /1-3/. We shall consider here only one dynamic process, viz. the parametric excitation of magnons. This phenomenon was discovered in 1957 by SUHL /4/ (*transverse pumping*, when $\vec{h} \perp \vec{M}$) and in 1960 by MOGRENTHALER /5/ (*parallel pumping*, when $\vec{h} \parallel \vec{M}$).

### 2.1. Introduction into the S-Theory

2.1.1. Parametric Instability. In order to calculate the threshold of parallel pumping, we shall substitute $H_{int} = H_p$ into equation (10) from formula (9). As a result, we have linear equations of motion for *slow amplitudes* $b_k(t)$:

$$b_{\vec{k}}(t) = a_{\vec{k}}(t)\exp(i\omega_p t/2), \quad b^*_{\vec{k}} = a^*_{-\vec{k}}(t)\exp(-i\omega_p t/2) \ , \tag{11}$$

$$\partial b_{\vec{k}}/\partial t + \gamma_{\vec{k}} + i(\omega_{\vec{k}}-\omega_p/2)\, b_{\vec{k}} + ihV_{\vec{k}} b^*_{-\vec{k}} = 0 \ . \tag{12}$$

A solution to this equation will be as follows:

$$b_k(t) = |b_k(0)|\exp(\nu_k t - i\phi_k) \ , \qquad b_{-k}{}^*(t) = |b_{-k}(0)|\exp(\nu_k t + i\phi_{-k}) \ . \tag{13}$$

Then, for the increment of parametric instability $\nu_k$, we have:

$$\nu_k = -\gamma_k \pm \left[|hV_k|^2 - (\omega_k-\omega_p/2)^2\right]^{1/2} \ . \tag{14}$$

The minimal threshold of excitation $h_1$ (corresponding to max $\nu_k = 0$) is determined from the condition

$$h_1 = \min(\gamma_k/|V_k|) \ . \tag{15}$$

At $h \quad h_1$, the exponential increase (13) of the amplitude of pairs with increment (14) begins. It follows from (12) that

$$\cos(\phi_k-\phi_{-k}-\phi_p) = (\omega_k-\omega_p/2)/|hV_k| \ . \tag{16}$$

This means that at the linear stage of parametric instability, a definite correlation between the phases of waves in a pair is reached. The phase correlator of waves with equal and oppositely directed wavevectors may be called, by analogy with superconductitivity, *pairing*.

2.1.2. Diagonal Hamiltonian of the S-Theory. The growth of wave amplitude continues until the interactions of waves with each other become essential. As all the parametrically excited waves have almost equal frequencies (close to $\omega_p/2$), the free magnon interaction proves to be nonresonant. In the Hamiltonian $H_4$ (8), the following terms are most essential:

$$H_S = \sum_{\vec{k},\vec{k}'} T_{\vec{k}\vec{k}'}|a_{\vec{k}}|^2|a_{\vec{k}'}|^2 + (1/2)\sum_{\vec{k},\vec{k}'} S_{\vec{k}\vec{k}'}a_{\vec{k}}^* a_{-\vec{k}}^* a_{\vec{k}'}a_{-\vec{k}'} ,$$
$$T_{\vec{k}\vec{k}'} = T_{\vec{k}\vec{k}',\vec{k}\vec{k}'} , \quad S_{\vec{k}\vec{k}'} = T_{\vec{k},-\vec{k},\vec{k}',-\vec{k}'} . \tag{17}$$

They either do not depend on wave phases at all, or depend only on the sum of phases $\phi_{\vec{k}} = \phi_{\vec{k}} + \phi_{-\vec{k}}$ in pairs. All other terms in $H_4$ become zero averaging over chaotic wave phases and make contribution to the equation of motion only in the second order of the perturbation theory in $H_4$. The reduction of $H_4$ to the form $H_S$ (17), which is the diagonal in the pairs of waves, was suggested by ZAKHAROV, L'VOV and STAROBINETS /6/. This resulted in the creation of a simple and efficient "S-theory", which in 1970-74 promoted the studies on the above-threshold behaviour of magnons. In particular, it allowed to give qualitative explanation to many experimentally observed effects and to obtain in most cases a good qualitative agreement with the experiment /7/. Later on, the S-theory was adopted by researchers (mostly the Soviet ones), who obtained interesting and important experimental results on the nonlinear behaviour of parametric magnons, associated with a new insight into the physical sense of the phenomena. The most comprehensive description of the S-theory and the pertinent experiments are given in monograph /1/ and reviews /7-9/. We are sure that the significance of these results goes beyond the limits of the physics of magnetic dielectrics. They have played, and will certainly keep on playing, an important role in the development of the physics of nonlinear waves in other media. A brief account of the fundamentals of the S-theory will be given here.

2.1.3. Basic Equations of the S-Theory. Introducing the interaction Hamiltonian into equation (10), we obtain the basic equations for the S-theory:

$$\partial b_{\vec{k}}/\partial t + \left[\gamma_{\vec{k}} + i(\omega_{NL}(\vec{k})-\omega_p/2)\right]b_{\vec{k}} + iP_{\vec{k}}b_{-\vec{k}}^* = 0 ,$$
$$\partial b_{-\vec{k}}^*/\partial t + \left[\gamma_{\vec{k}} - i(\omega_{NL}(\vec{k})-\omega_p/2)\right]b_{-\vec{k}}^* - iP_{\vec{k}}b_{\vec{k}} = 0 . \tag{18}$$

They differ from the linear equations (12), describing the parametric instability, only by renormalization of the frequency $\omega(\vec{k}) \to \omega_{NL}(\vec{k})$ and of the pumping $hV(\vec{k}) \to P(\vec{k})$ due to the first and second term in equation (17), respectively:

$$\omega_{NL}(\vec{k}) = \omega(\vec{k}) + 2\sum_{\vec{k}'} T_{\vec{k}\vec{k}'}|b_{\vec{k}'}{}^2| , \quad P_{\vec{k}} = hV_{\vec{k}} + \sum_{\vec{k}'} S_{\vec{k}\vec{k}'}b_{\vec{k}'}b_{-\vec{k}'} . \tag{19}$$

It is evident that the approximation of the diagonal Hamiltonian (17) is essentially the *approximation of the self-consistent field*. Classical examples of this approximation are the Curie-Weiss theory of molecular field, the Landau theory of second-order phase transitions, the Landau theory of weakly supercritical flows in hydrodynamics, and the BCS theory of superconductivity.

## 2.2. The Ground State in the S-Theory

2.2.1. Stability of the Ground State. Assuming in (18) $\partial b/\partial t = 0$, let us consider the stationary solutions of this equation. It is readily evident that in the points of the k-space where $b_k = 0$, the determinant of this system is equal to zero:

$$|P_{\vec{k}}|^2 = \frac{2}{k} + \left[\omega_{NL}(k) - \omega_p/2\right]^2 . \tag{20}$$

This is the equation of two surfaces, and on their arbitrary part we may assume $b_k = 0$. Thus we have a great number of stationary states. The requirement of their stability with respect to the growth of waves in the points where in the stationary state $b_k = 0$ strongly reduces the class of possible stationary states. Firstly, the solution is stable if two surfaces (20) coalesce into one

$$\omega_{NL}(\vec{k}) = \omega_p/2 . \tag{21}$$

Secondly, in the points of *resonance surface* (21) where $b_k \neq 0$

$$|P_{\vec{k}}| = \gamma_{\vec{k}}, \quad i\gamma_{\vec{k}} = P_{\vec{k}} \exp(i\psi_{\vec{k}}) . \tag{22}$$

In the residual parts of this surface

$$|P_{\vec{k}}| < \gamma_{\vec{k}} . \tag{23}$$

It should be noted that the ambiguity of solutions of stationary equations and elimination (complete or partial) of this arbitrariness using the stability conditions is not an inherent property of the S-theory. This is a general feature of the approximation of self-consistent field in the theory of nonlinear waves.

2.2.2. The Simplest Solution to the S-Theory. In the simplest, isotropic case, when $V_k = V$, $\gamma_k = \gamma$, and $S_{kk'} = S$, there must obviously be isotropic solution $N(\Omega) = N/4\pi$ ($\Omega = \theta, \phi$, are the polar vectorial and azimuthal angles). In this case, it follows from (19) and (22) that

$$N = \sum_{\vec{k}} |b_k|^2 = \left[(hV)^2 - \gamma^2\right]^{1/2}/|S| , \quad hV \sin\psi = \gamma . \tag{24}$$

The second equation represents a condition of energy balance: $W_+ = W_-$, where $W_+ = \omega_p hVN \sin\psi$ is the energy influx from pumping, and $W_- = \gamma(\omega_k + \omega_{-k})N$ is the energy consumed. The limitation of amplitude N is achieved due to the fact that the phases of pairs $\psi_k$ differ from the optimal value $\pi/2$, at which the energy flux into the system is maximal. Equation (24b) is fundamentally important in the S-theory. It was verified in the direct experiments involving of the pair phase in the parametric excitation of magnons in the ferromagnet $Y_3Fe_5O_{12}$ /10,11/. It was shown that, within experimental measurements error, the points lie on the bisector of the quadrantal angle, in complete agreement with (24b). This indicates that the S-theory correctly describes the essential features of the above-threshold behaviour of parametric magnons.

2.2.3. Reshaping the Distribution Function with Increase in Supercriticality. It is reasonable to give the geometric interpretation of the stability condition (23) $|P(\Omega)| < \gamma(\Omega)$, which defines the angular distribution function $N(\Omega)$ of the parametric magnons: surface $|P(\Omega)|$ lies entirely inside surface $\gamma(\Omega)$ and touches it in the points where $N(\Omega) \neq 0$. In the case of axial symmetry, characteristic for many ferromagnets (including $Y_3Fe_5O_{12}$), this touching occurs along the lines $\Theta$ = const, and $0 < \phi < \pi$ (along the resonant surface parallels). The first touching occurs on the equator ($\Theta = \pi/2$), and the state with one group of pairs $N(\Theta,\phi) = N_1\delta(\Theta-\pi/2)$ exists in a wide range of supercriticalities $h_1 < h < h_2$. For $Y_3Fe_5O_{12}$, $h_2 \approx (3\text{-}4)h_1$. For magnetic fields $H > 1500$ Oe at $h = h_2$, the second touching is on the parallel $\Theta \approx 50°$ and the second group of pairs is generated. At higher h the third group is generated, etc. At $H < 1500$ Oe in $Y_3Fe_5O_{12}$ the distribution function evolution takes place in quite a different manner at $h > h_2$ (depending on H), surface $|P(\Omega)|$ coalesces with surface $\gamma(\Omega)$ on the band $|\Theta-\pi/2| < \delta$, the width of which is $\delta \sim (h-h_2)$. As a result, there appears a continuous distribution of pairs near equator with the width $2\delta$. The theoretical conclusions described here are in a quantitative agreement with the result of the experiment on $Y_3Fe_5O_{12}$ /12/.

## 2.3. Nonstationary Self-Consistent Dynamics of Parametric Magnons

2.3.1. The Spectrum of Collective Oscillations of Parametric Magnons and Methods of Their Excitation. Linearizing the nonstationary equations of the S-theory (18) with respect to the ground state of the system and assuming the amplitudes of excitation $\alpha_k$, $\alpha_k^* \sim \exp(-i\Omega t)$, we obtain a system of algebraic equations homogeneous with respect to $\alpha$ and $\alpha^*$. The condition for their solvability determines the frequency Re $\Omega$ and damping Im $\Omega$ of collective oscillations. In the case of axial symmetry, one can obtain /1/:

$$\Omega_m = -i\gamma \pm \left[4S_m(2T_m+S_m)N_1^2 - \gamma^2\right] . \qquad (25)$$

Here m is the number of axial mode, $T_m$ and $S_m$ are the corresponding axial-angle Fourier harmonics of function (17) $T_{kk'}$ and $S_{kk'}$, respectively. For the sake of simplicity, it is assumed that $T_m = T_{-m}$ and $S_m = S_{-m}$. It should be noted that collective oscillations may be spatially nonuniform. In this case, $\Omega_m$ depends on their wavefactor $\vec{\kappa}$, with $\Omega_m$ in Eq.(25) being $\Omega_m(0)$. For the simplest case the dispersion law $\Omega_m(\vec{\kappa})$ has been given in /1/.

The above-described collective oscillations of a system of parametric magnons may be excited (similarly to oscillations of any other nature) in various ways: with the aid of external resonant influence, parametrically or by hit. All these methods of their excitation have been used in the experiments on ferro- and antiferromagnets. It is simple to devise a theory and to interpret experiments for the resonance method of excitation of collective oscillations. It was exactly by this method that these oscillations were experimentally discovered and studied by ZAUTKIN, L'VOV and STAROBINETS in 1972 /19/. In addition to parallel pumping, they applied to the $Y_3Fe_5O_{12}$ sample another microwave signal, the frequency of which differed from the pump frequency $\omega_p$ by the frequency of collective oscillations $\Omega_0$. The role of external resonant force was played in this case by the beats between two microwave signals. Later on, in 1975 /14/, OREL and STAROBINETS implemented the direct resonance method of excitation using the alternate magnetic field with a middle-wave frequency (of the order of 1 MHz). In principle, collective oscillations may also be excited by the sound the frequency and wavevector of which coincide with those of the oscillations. Closely related to the resonance methods of exciting collective oscillations is a simple and pleasant method of hit excitation by means of a drastic change of the pump frequency or phase, suggested in 1974 by PROZOROVA and SMIRNOV /10/. Their data on eigenfrequency of collective oscillations in antiferromagnet MnCO as well as the data presented in /13/ for $Y_3Fe_5O_{12}$ are in a quantitative agreement with the formula

$$\Omega_0{}^2 = 4\gamma^2(2T_0+S_0)S_0{}^{-1}\left[(h^2/h_1{}^2) - 1\right]^{1/2} , \tag{26}$$

which follows from the theory (see (24) and (25) at $\gamma << hV$). This indicates that the S-theory adequately describes the parametric excitation of magnons in ferro- and antiferromagnets.

Apart from the above-considered linear interaction ($H_m\alpha^*$ +H.c.) of the collective oscillations of parametric magnons with the external radiofrequency field $H_m$, which leads to resonance at the frequency $\Omega = \Omega_0$, the S-theory predicts the nonlinear interaction of the type ($H_m\beta^*\beta^*$+H.c.). It should lead to the parametric resonance of collective modes $\beta$ in the radiofrequency field with a frequency $\Omega = 2\Omega_0$. It is evident that the same effect also involves the instability of the original mode $\alpha$ with a frequency $\Omega$, excited with the radiofrequency field, with respect to its decay into two modes $\beta$ with the frequency $\Omega/2$. The action of both mechanisms above a certain critical amplitude of the radiofrequency field H give rise to the *double parametric resonance of magnons* - i.e. the parametric excitation of collective oscillations in the system of parametrically excited magnons. This phenomenon was discovered and experimentally studied by ZAUTKIN et al. in 1977 /15/.

2.3.2. Self-Oscillations of Magnetization in the Parametric Excitation of Magnons. In the very first experiments on the parametric excitation of magnons in 1961, HARTWICK, PERESSINI and WEISS /16/ found that the steady-state condition is often established and magnetization performs complex self-oscillation around an average value. Since then, physical origin of self-oscillations has been one of the principal problems of parametric excitation. Various hypotheses were put forward (see, for example, /17/) but all of them are of purely historical interest. In terms of the S-theory, self-oscillations were explained as the result of the generated instability of collective oscillations considered above. Indeed, if

$$S_m(2T_m+S_m) > 0 \; , \tag{27}$$

then it follows from (25) that Im $\Omega_m > 0$. As evident from the same formula, in this case Re $\Omega_m = 0$, i.e. the ground-state instability is purely periodical. Therefore, the interactions of different modes of self-oscillations are extremely strong, and the problem of self-oscillation behaviour at the nonlinear stage requires computer analysis. This was given in /18,19/. An essential result of these numerical experiments was the proof of the fact that at supercriticalities $p = h/h_1$ smaller than $p_c \cong (1 \div 1,5)$dB, the system of parametric magnons sets to a stable limiting cycle, the region of attraction of which is the whole phase space. At $p > p_c$ the trajectories close to this cycle become exponentially unstable, with the medium-cycle increment of divergence (the Lyapunov factor) growing in proportion to $p - p_c$. At small values of $(p-p_c)$ a narrow layer filled with exponentially unstable trajectories is formed near the limiting cycle and a transition to chaotic self-oscillations occurs. These and other features of self-oscillations in the numerical experiment with the nonstationary equations of the S-theory (18) are in a qualitative agreement with the result of laboratory studies /18,19/. As far as the conditions for the generation of self-oscillations are concerned, the detailed experimental studies carried out by ZAUTKIN and STAROBINETS on the ferromagnet $Y_3Fe_5O_{12}$ in a wide temperature and field range, using samples of different forms, showed a good qualitative agreement with the instability condition (27).

## 3. NONLINEAR KINETICS OF MAGNONS

### 3.1. The Kinetic Equation, Thermodynamic Equilibrium and Relaxation

As stated above, the interaction of magnon packets which are wide in the k-space may be described with the kinetic equation for the occupation numbers $n_k(t) = \langle |b_k|^2 \rangle / \hbar$

$$dn_k/2dt = St_k\{n_{k'}\} \; . \tag{28}$$

The collision term St is a functional of occupation numbers in a definite region of the k-space. If the principal interaction is the three-magnon interaction $H_3$(8), then

$$St_k^{(3)}\{n_{k'}\} = (\pi/\hbar^2) \sum_{\vec{k}=\vec{1}+\vec{2}} |V_{\vec{k},\vec{1}\vec{2}}|^2 \, *$$

$$* \left[ n_1 n_2 (n_k+1) - n_k (n_1+1)(n_2+1) \right] \delta(\omega_k-\omega_1-\omega_2) \delta(\vec{k}-\vec{k}_1-\vec{k}_2) \, +$$

$$+ \, (2\pi/\hbar^2) \sum_{2=k+1} |V_{2,k1}|^2 \left[ n_2 (n_k+1)(n_1+1) - n_k n_1 (n_2+1) \right] *$$

$$* \, \delta(\omega_2-\omega_k-\omega_1) \delta(\vec{k}_2-\vec{k}-\vec{k}_1) \; . \tag{29}$$

If the three-magnon interaction is absent or forbidden by the laws of conservation of energy and momentum, the leading interaction is the four-magnon one, and

$$St_k = St_k^{(4)}\{n_{k'}\} = (2\pi/\hbar^2) \sum_{k+1=2+3} |T_{k1,23}|^2 *$$
$$* \left[(n_k+1)(n_1+1)n_2n_3 - n_kn_1(n_2+1)(n_3+1)\right] *$$
$$* \delta(\omega_k+\omega_1-\omega_2-\omega_3)\delta(\vec{k}+\vec{k}_1-\vec{k}_2-\vec{k}_3) . \qquad (30)$$

Equations (29) and (30) are easily obtained with the help of the "golden rule of quantum mechanics" in the perturbation theory, and are given in the text-books on theoretical physics, nonlinear acoustics, etc.

A stationary solution of the kinetic equations (29)-(30) is the Bose-Einstein distribution:

$$n_k = n_0(k) = \left[\exp(\hbar\omega_k/T) - i\right]^{-1} , \qquad (31)$$

which describes the thermodynamic equilibrium with temperature T. If at some $k = k_0$, the value $n(k_0)$ is made slightly deviating from the equilibrium (31), then

$$\left[n(k_0) - n_0(k_0)\right] \sim \exp(-2\gamma_0(k_0)t) . \qquad (32)$$

Here $\gamma_0(k)$ is the magnon *damping decrement*, i.e. the factor of proportionality at n(k) in the collision term. For the three-magnon decay processes,

$$\gamma_d(\vec{k}) = (\pi/\hbar^2) \sum_{\vec{k}=\vec{1}+\vec{2}} |V_{\vec{k},\vec{1}\vec{2}}|^2 \left[n_1+n_2+1\right] \delta(\omega_{\vec{k}}-\omega_{\vec{1}}-\omega_{\vec{2}}) . \qquad (33a)$$

For the three-magnon confluence processes,

$$\gamma_{sp}(\vec{k}) = (2/\hbar^2) \sum_{\vec{2}=\vec{k}+\vec{1}} |V_{\vec{2}.\vec{k}\vec{1}}|^2 \left[n_1-n_2\right] \delta(\omega_2-\omega_k-\omega_1) . \qquad (33b)$$

For the four-magnon scattering processes,

$$\gamma_{sc}(\vec{k}) = (2\pi/\hbar^2) \sum_{\vec{k}+\vec{1}=\vec{2}+\vec{3}} |T_{\vec{k}\vec{1},\vec{2}\vec{3}}|^2 * \left[n_1(n_2+n_3+1)-n_2n_3\right] \delta(\omega_k+\omega_1-\omega_2-\omega_3) . \qquad (33c)$$

Near the equilibrium we can substitute the Bose-Einstein distribution (31) for $n_0(k_0)$. Definitely, of greatest interest are the kinetic effects, which are far away from the thermodynamic equation and which just represent the subject of nonlinear kinetics. Of the great variety of nonlinear kinetic effects in magnets, we shall consider only two effects which seem to be instructive.

### 3.2. Damping of the Monochromatic Wave in the Nonlinear Medium /21/

In the above calculation of the magnon damping decrement $\gamma(k_0)$, we regarded all the residual reservoir of magnons (with $k = k_0$) to be in the thermodynamic equilibrium. This may be so if the number of magnons in the packet N under study is far smaller than the equilibrium number of thermal magnons $N_T$. However, the energy and momentum conservation laws allow only small part of the entire reservoir of thermal waves $\Delta N_T$ to participate in the relaxation of a narrow ($\Delta\omega_k << \omega_k$) packet. Therefore, at a relatively low intensity of the narrow packet N, comparable to $N_T$, the energy dissipated by it may lead to a substantial deviation of the occupation numbers of thermal magnons in this region from the equilibrium values. For this reason, the relaxation time of the packet will depend on it amplitude, i.e. the relaxation becomes nonlinear.

This effect must be the strongest in the relaxation of the monochromatic wave, which can interact, in accordance with the kinetic equation, with the waves with vectors $\vec{k}$ lying on a definite surface. For example, in decaying the wave with $\vec{k} = \vec{k}_0$, this surface is defined as

$$\omega(\vec{k}_0) = \omega(\vec{k}) + \omega(\vec{k}_0-\vec{k}) \ . \tag{34}$$

Formally, the number of waves $\Delta N_T$ on this surface is zero. If, however, we take into account that in reality the conservation law (34) is implemented with an accuracy of damping $\gamma(\vec{k})$, then

$$\Delta N_T{}^0 \cong 4\pi n_0(k)k^2\gamma(k)/(\partial\omega/\partial k) \ . \tag{35}$$

For this packet, we can schematically write the kinetic equation as follows

$$d\Delta N_T/dt = -\gamma(k)(\Delta N_T-\Delta N_T{}^0) + \Phi \ , \tag{36}$$

where $\Phi \cong \gamma(k_0)|A|$ is the in-term, which shows that in every relaxation of the monochromatic wave with amplitude A the number of waves in the packet increases by 1. Therefore the in-term of Eq.(36) should coincide (except for alternating the sign) with the out-term for the monochromatic packet. At $\Phi = 0$, Eq.(36) describes the relaxation of the number of waves $N_T$ to the thermodynamically equilibrium value $N_T{}^0$ (35). From Eqs.(35) and (36), an estimate for the relative variation of the number of waves in a packet follows:

$$a = \frac{\delta n(k)}{n_0(k)} \cong \frac{\Delta N_T-\Delta N_T{}^0}{N_T{}^0} \cong \frac{\gamma(k_0)\,|A|^2}{\gamma^2(k)} \cdot \frac{\partial\omega/\partial k}{n_0(k)k^2} \ .$$

If we substitute here the estimate for

$$\gamma(k_0) \cong |V|^2k^2(k_0)/(\partial\omega/\partial k)$$

following from (29), then

$$a \cong |V|^2|A|^2/\gamma^2(k) \ . \tag{37}$$

It is evident that at $a << 1$

$$\gamma(k_0,|A|^2) - \gamma(k_0,0) \cong a\gamma(k_0,0) \ . \tag{38}$$

But if $a \cong 1$, the damping of the monochromatic wave should substantially differ from that of the equilibrium. Accurate calculation /21/ gives for the decay processes,

$$\gamma_d(\vec{k},|A|^2) = \frac{\pi}{\hbar^2} \sum_{\vec{k}=\vec{1}+\vec{2}} \frac{|V_{\vec{k},\vec{1}\vec{2}}|^2 (n_1+n_2+1)\delta(\omega_k-\omega_1-\omega_2)}{\left[1-|V_{\vec{k},\vec{1}\vec{2}}|^2|A|^2/\gamma_1\gamma_2\right]^{1/2}} \quad , \tag{39a}$$

and for the confluence processes

$$\gamma_{sp}(\vec{k},|A|^2) = \frac{2\pi}{\hbar^2} \sum_{\vec{2}=\vec{k}+\vec{1}} \frac{|V_{\vec{2},\vec{k}\vec{1}}|^2 (n_1-n_2)\delta(\omega_2-\omega_{\vec{k}}-\omega_1)}{\left[1+|V_{2,k1}|^2|A|^2/\gamma_1\gamma_2\right]^{1/2}} \quad . \tag{39b}$$

At A = 0 these equations coincide with Eqs.(33). At small A values estimate (38) is confirmed. It is important that in the confluence process the damping decreases with the increase in A (the effect of medium clarification) and in the decay processes it increases. At A = A, where

$$|V_{k,12}|^2 A_1^2 = \gamma_1\gamma_2 \quad , \tag{40}$$

the damping (39a) formally becomes infinite. Recall that formula (40) defines the decay instability threshold of the monochromatic wave. At $A > A_1$ secondary waves with $k = k_1$ and $k = k_2$ grow exponentially with time, and formula (39a) is inapplicable.

### 3.3. The Kinetic Instability of Magnons.

Let us assume that magnons are parametrically excited in a magnet, on the resonance surface $2\omega(\vec{k}_p) = \omega_p$. Function $n_p(\vec{k})$ is their distribution function. In the isotropic case,

$$n_p(k) = N_p k_p^2 \delta(\omega_k-\omega_p/2) 2\pi^2 \partial\omega/\partial k \quad . \tag{41}$$

The general distribution function n(k) includes the thermodynamically equilibrium term $n_p(k)$

$$n(k) = n_0(k) + n_p(k) \quad . \tag{42}$$

The divergence of the distribution function (42) form the equilibrium alters wave damping in the whole k-space. Let us consider first the case when the decay processes are allowed for the parametric magnons

$$\omega_p/2 = \omega(\vec{k}_p) = \omega(\vec{k}_1) + \omega(\vec{k}_2) \quad , \qquad \vec{k}_p = \vec{k}_1 + \vec{k}_2 \quad . \tag{43}$$

Now let us consider the damping of magnons with $\vec{k} = \vec{k}_1$. For them, (43) are the confluence processes. According to (33b) and (42):

$$\gamma_{sp}(\vec{k}_1) = \gamma_{sp}{}^0(\vec{k}_1) - (2\pi/\hbar^2) \sum_{\vec{k}_p=\vec{k}_1+\vec{k}_2} |V(\vec{k}_p,\vec{k}_1\vec{k}_2)|^2 \ast$$

$$\ast\ n_p(\vec{k}_p)\delta\left[(\omega_p/2) - \omega_1 - \omega_2\right] \quad . \tag{44}$$

The first term in (44) arises from the equilibrium part of the distribution function

(42), and the second term from the parametric magnons. It is important that this term is negative, and at a reasonably high $n_p$ the general damping $\gamma_S(k_1)$ may also become negative. As a consequence, the number of magnons with $k = k_1$ will exponentially grow. This phenomenon may be called the *first-order kinetic instability*. The kinetic instability threshold may be estimated from (44) by introducing distribution (41) there. As a result, we have:

$$\gamma_{sp}(\vec{k}_1) - \gamma_{sp}{}^0(\vec{k}_1) \cong |V|^2 N_p/k_p(\partial\omega/\partial k) \ , \tag{45a}$$

$$|V|^2 N_{cr} \cong k(\partial\omega/\partial k)\gamma_k \cong \omega_k\gamma_k \ . \tag{45b}$$

Comparing this estimate with formula (40), we see that the first-order kinetic instability threshold (in the number of waves N in a packet) is $\omega_k/\gamma_k$ times higher than the threshold of decay instability (of the monochromatic wave), which is dynamic in its nature.

The theory of kinetic instability resulting from three-magnon processes has been most comprehensively developed for the antiferromagnets, in which the decay of parametric magnons to magnons and sound is allowed /22/. Experimental evidence of this instability has been found in the antiferromagnet FeBO /23/. Quite recently, the first-order instability has been discovered in $Y_3Fe_5O_{12}$ /27/.

In the non-decay region of the spectrum, the first-order instability is impossible. However, with the decreased divergence of magnon distribution (42) from the equilibrium (with increase in $N_p$), there arises the *second-order kinetic instability*, in which *two* parametric magnons confluence to give two secondary magnons. By contrast with (43), the conservation laws have the form

$$\omega_p = \omega(\vec{k}_{p1}) + \omega(\vec{k}_{p2}) = \omega(\vec{k}_1) + \omega(\vec{k}_2) \ , \quad \vec{k}_{p1} + \vec{k}_{p2} = \vec{k}_1 + \vec{k}_2 \ . \tag{46}$$

Substituting the distribution function (42) into formula (33c), we obtain the following equation for the damping of secondary magnons

$$\gamma_{sc}(\vec{k}_1) = \gamma^0{}_{sc}(k_1) - (2\pi/\hbar^2)\cdot$$

$$\cdot \sum_{\vec{k}_{p1}+\vec{k}_{p2}=\vec{k}_1+\vec{k}_2} * \ |T(\vec{k}_{p1},\vec{k}_{p2},\vec{k}_1,\vec{k}_2)|^2 n_p(\vec{k}_{p1}) n_p(\vec{k}_{p2}) \delta(\omega_p-\omega(\vec{k}_1)-\omega(\vec{k}_2)) . \tag{47a}$$

Using (41), we obtain the estimate:

$$\gamma_{sc}(k_1) - \gamma_{sc}{}^0(k_2) \cong \pi|TN_p|^2/k_p(\partial\omega/\partial k_p) \ . \tag{47b}$$

It is seen that at a reasonably high $N_p = N_{cr}$, secondary magnon damping can become negative. For $N_{cr}$, it follows from (47b) that:

$$|TN_{cr}|^2 \cong \gamma(k_1)k_p(\partial\omega/\partial k_p) \ . \tag{48}$$

Here into damping $\gamma$ all essential relaxation processes were included.

A more comprehensive analysis carried out in /24/ has shown that in ferromagnets, the instability (48) has a minimal threshold for magnons at the bottom of the spin-wave spectrum ($k_1 || M$, $k_p >> k_1$ $(10^2 \div 10^3)cm^{-1}$), for which damping $\gamma(k_1)$ is minimal. Experimentally the second-order kinetic instability was found in 1981 /24/. Its brightest effect is the electromagnetic radiation caused by secondary magnons, which onsets with increased pumping amplitude. This radiation is fairly monochromatic $\Delta\omega_{rad}/\omega_{rad}$ $(10^{-3} \div 10^{-2})$, its frequency

$$\omega_{rad} = \omega(\vec{k}_1) + \omega(-\vec{k}_1) \cong 2\omega_0 \tag{49}$$

does practically not differ from the doubled frequency of the spin-wave spectrum bottom $\omega_0$ and linearly depends on the external magnetic field H. For the spherical samples

$$\omega_0 = h(H-4\pi M_0/3) \ , \tag{50}$$

where $g = 2\pi \star 2.8$ Ghz/kOe is the magnetomechanical electron ratio. The emission mechanism is clear from (49): two secondary magnons with $\vec{k} = \vec{k}_1$ and $\vec{k} = -\vec{k}_1$ confluence to give a photon with $\vec{k} = 0$. The direct radiation caused by secondary magnons at a frequency $\omega_0$ is strongly suppressed due to the fact that $k_1 \cong 10^2 cm^{-1}$ by far exceeds the wavevector of the photon at a frequency $\omega_0 (k_{ph} \cong 10^{-1} cm^{-1})$. In /25/ and /26/ the nonlinear theory of kinetic instability in ferrites has been worked out, being in good agreement with the experiment on $Y_3Fe_5O_{12}$.

There are many other kinetic effects stimulated by the increasing divergence of magnons from the equilibrium. Their investigation has just begun. The author hopes that he will be able to deliver a lecture on them at one of the next Gorky schools on nonlinear waves.

## REFERENCES


1. V.S.L'vov. Nonlinear Spin Waves. Moscow, Nauka, 1987.
2. V.E.Zakharov, V.S.L'vov, S.S.Starobinets. Instability of monochromatic spin waves. FFT, 1969, 11, 2923-2332.
3. V.E.Zakharov, V.S.L'vov, S.S.Starobinets. New mechanism of spin wave amplitude limitation in parallel pumping, FFT, 1969, 11, 2047-2055.
4. H.Suhl. Theory of ferromagnetic resonance at high signal power. Phys.Chem.Sol., 1957, 1, 209-227.
5. F.R.Morgenthaler, Parallel-pumped magnon instabilities in a two-sublattic ferromagnetic crystal. J.Appl.Phys., 1960, 31, Suppl. 95S-97S.
6. V.E.Zakharov, V.S.L'vov, S.S.Starobinets. Stationary nonlinear theory of parametric wave excitation. Soviet Phys.-JETP, 1970, 59, 1200-1214.
7. V.E.Zakharov, V.S.L'vov, S.S.Starobinets. Turbulence of spin waves above the threshold of parametric excitation. Soviet Phys.-Uspekhi, 1974, 114, 4, 609-654.
8. V.S.L'vov. Solutions and nonlinear effects in parametrically excitation of spin waves. In: Solitons, ed. by S.E.Trullinger, V.E.Zakharov and V.L.Pokrovskii, Elsevier, B.V., 1986.

9. V.S.L'vov, L.A.Prozorova. Spin waves above the threshold of parametric excitations In: Spin Waves and Magnetic Excitations, p.1, ed. by A.S.Borovik-Romanov and S.K.Sinka, Elsevier B.V., 1988.

10. L.A.Prozorova, A.I.Smirnov. Nonlinear high-frequency properties of yttrium iron garnet at low temperatures. Soviet Phys.-JETP, 1974, 69, 2(8), 758-763.

11. G.A.Melkov, I.V.Krutsenko. Mechanisms of amplitude limitation of the parametrically excited spin waves. Soviet Phys.-JETP, 72, 2, 564-575.

12. B.B.Zautkin, V.S.L'vov, E.V.Podivilov. Distribution transformation in parametric resonance in ferrites. Soviet Phys.-JETP, (in press).

13. B.B.Zautkin, B.S.L'vov, S.S.Starobinets. On the resonance effects in the system of parametric spin waves. Soviet Phys.-JETP, 1972, 63, 182-190.

14. B.I.Orel, S.S.Starobinets. Radiofrequency magnetic susceptibility and the collective resonance of magnons in the parallel pumping. Soviet Phys.-JETP, 68, 1, 317-325.

15. V.V.Zautkin, V.S.L'vov, B.I.Orel, S.S.Starobinets. Large-amplitude collective oscillations of magnons and the double parametric resonance. Soviet Phys.-JETP, 1977, 72, 272-284.

16. T.S.Hartwick, E.R.Peressini, M.T.Weiss. Suppression of subsidiary absorption in ferrites by modulation techniques. Phys.Rev.Lett., 1961, 6, 176-177.

17. J.A.Monosov. Nonlinear Ferromagnetic Resonance. Moscow, Nauka, 1971.

18. V.S.L'vov, S.L.Moosher, S.S.Starobinets. The theory of magnetization self-oscillations in the parametric excitation of spin waves. Soviet Phys.-JETP, 1973, 64, 1084-1097.

19. V.L.Grankin, V.S.L'vov, V.I.Motorin, S.L.Moosher. Secondary turbulence of the parametrically excited spin waves. Soviet Phys.-JETP, 1981, 81, 2(8), 757-768.

20. V.V.Zautkin, S.S.Starobinets. Magnetization self-oscillations in the parallel pumping of spin waves. Soviet Phys.-JETP, 1974, 16, 3, 678-686.

21. V.S.L'vov. On damping of the monochromatic wave in nonlinear medium. Soviet Phys.-JETP, 1975, 68, 1, 308-316.

22. V.S.Lutovinov, G.E.Fal'kovich, V.B.Cherepanov. Nonequilibrium distribution of quasi-particles in the parametric excitation in antiferromagnets with a decay spectrum. Soviet Phys.-JETP, 1986, 90, 1781-1794.

23. B.J.Kostyuzhansky, L.A.Prosorova, L.E.Svistov. Studies on the electromagnetic emission of magnons parametrically excited in antiferromagnets. Soviet Phys.-JETP, 1984, 86, 1101-1116.

24. A.V.Lavrinenko, V.S.L'vov, G.A.Melkov, V.B.Cherepanov. Kinetic instability of a strongly nonequilibrium system of spin waves and transformation of ferrite emission. Soviet Phys.-JETP, 1981, 81, 1022-1036.

25. V.S.L'vov, V.B.Cherepanov. Nonlinear theory of kinetic excitation of waves. Soviet Phys.-JETP, 1981, 81, 1406-1422.

26. A.Yu.Taranenko, V.B.Cherepanov. Energy absorption in ferrites above the threshold of kinetic instability. Soviet Phys.-JETP (in press).

27. V.S.Lutvinov, G.A.Melkov, A.Yu.Taranenko, V.B.Cherepanov. Kinetic instability of first-order spin waves in ferrites. Soviet Phys.-JETP (in press).

# The Nonlinear Dynamics of Wave Systems with Spatio–Temporal Collapses

*A.G. Litvak, V.A. Mironov, and A.M. Sergeev*

Institute of Applied Physics, USSR Academy of Sciences, 46 Ulyanov Str., 603600 Gorky, USSR

The processes of wave energy concentration in nonlinear media are analyzed. A classification of nonlinear systems with wave collapses according to the character of the field energy localization is presented and the corresponding examples of the evolution are considered. A detailed analysis is given for the case of the bievolutional (spatio-temporal) behaviour of strong electromagnetic waves in media with nonlinearity in which an arbitrary energy flux can be trapped into contradicting filament or moving focus distributions.

Together with the structure interactions and the properties of spatio-temporal chaos, the processes of wave energy concentration have traditionally attracted attention in nonlinear dynamics. In particular, this attention has passed into the theory of stationary self-focusing and Langmuir wave collapses /1-3/, which have become classical examples of nonlinear physics. In any field of knowledge, the information gained by solving direct problems is accomplished by the formulation of inverse problems which require a certain level of generalization. For example, one of the most important problems in the physics of nonlinear processes is to determine the properties of a nonlinear medium that could ensure the maximum possible localization of the wave field energy. By energy we mean a conventional energy characteristic, the physical meaning of which depends on particular formulation of the problem. This can be the number of quanta involved into singularity, the energy flow localized in the cross-section of the beam, etc. Besides the obvious applications, the interest in this problem is due to the fact that in most systems investigated today, upper limits exist on the energy value involved, and concentrated by the wave field bunch, and the nonlinear systems demonstrate "flexibility" resisting the localization of a given arbitrary portion of wave energy. This resistance is manifested, as a rule, in the structural instability of wave collapses.

There are several universal types of the behaviour of nonlinear systems with collapses, which can conveniently be classified by the character of wave energy trapping into the singularity domain. Figure 1 shows schematically the scenarios of the evolution of collapsing field bunches, which we shall call strong, weak, fractal, disturbed and complete collapses.

Historically, the case of strong collapse, which occurs at stationary self-focusing of electromagnetic waves in media with local cubic nonlinearity, was first studied by BESPALOV and TALANOV /3/. In this case, the initial (boundary) field



Research Reports in Physics Nonlinear Waves 3
Editors: A.V. Gaponov-Grekhov · M.I. Rabinovich · J. Engelbrecht

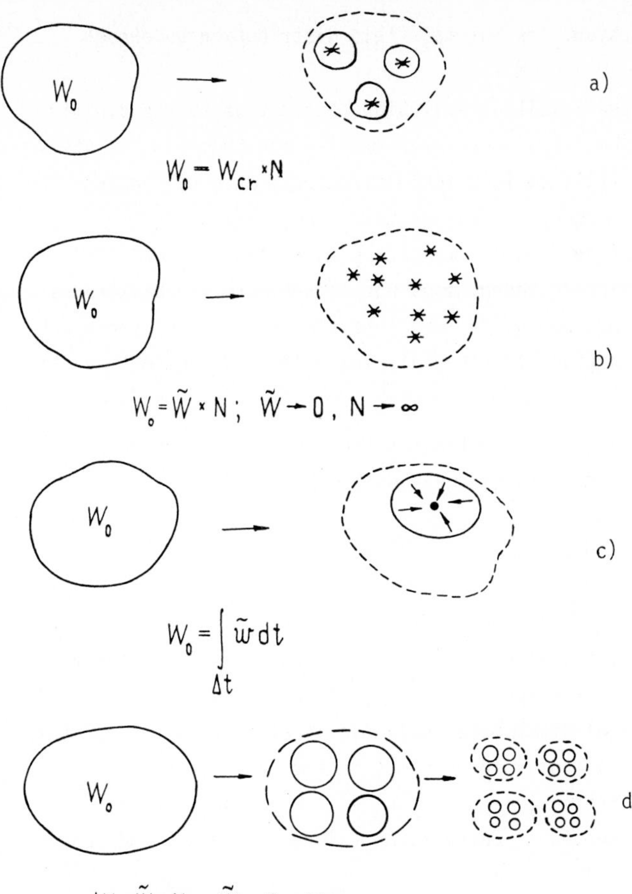


Figure 1. Schematical representation of a) strong, b) weak, c) distributed and d) fractal collapses.

distribution is divided into several secondary collapsing bunches (beams) with a definite, so called critical portion of energy $W_{cr}$ (critical power of self-focusing) involved into each bunch. Symbolically, this process can be represented as: $W_0 = W_{cr} \times N$, where $W_0$ is the intial value of the energy parameter and N is the finite number of secondary bunches (beams).

In the case of weak collapse /4,5/, the initial distribution is divided during its evolution into an infinite number of secondary bunches with a "hollow" field singularity arising in each bunch: a secondary bunch loses the whole of its acquired energy when approaching the singularity. With small-scale dissipation in bunches taken into account, a finite portion of energy $\tilde{W}$ is absorbed, but with decreasing spatial scale of damping (or increasing threshold field amplitude beginning from the absorption becomes essential) the locally dissipated energy decreases. Such a behaviour makes this case radically different from the case of strong collapse. As a result of its evolution, the initial field distribution undergoes multiple splitting: $W_0 = \tilde{W} \times N$, $\tilde{W} \to 0$, $N \to \infty$.

Similar to the above described result of evolution (but not as to the dynamic behaviour) is the fractal collapse /6/, where each small-scale bunch is, in its turn, unstable and splits into still smaller-scale structures when collapsing. The symbolic representation of this process is analogous to the previous one.

By a distributed collapse /7/ one usually means the case when a "weak" (hollow) singularity is formed, the collapsing energy does not escape into the background distribution of the wave field but begins to leak into the singularity (the black hole image), so that a finite portion of energy dissipates over a finite time t: $W_0 = \int_{\Delta t} \tilde{W} dt$.

A convenient mathematical model for illustrating the types of evolution described above, is the nonlinear Schrödinger equation:

$$-i \frac{\partial \psi}{\partial t} + \sum_{i=1}^{d} \alpha_i \frac{\partial^2 \psi}{\partial x_i^2} + |\psi|^2 \psi = 0 . \tag{1}$$

In two-dimensional geometry (d = 2), at the coefficient values $\alpha_i = 1$, Eq.(1) describes a strong collapse with critical energy $W_{cr} = \int |\psi|^2 dx_1 dx_2 \cong 11.9$ involved into each singularity. In a similar three-dimensional situation (d = 3) weak collapses take place. A possibility of distributed collapse is also predicted for this case. The processes of the fractal collapse develops in the case of opposite signs of the dispersion coefficients $\alpha_1 = \alpha_2 = 1$, $\alpha_3 = -1$ when the elementary act of evolution consists in two-dimensional compression (along $x_1$ and $x_2$) and longitudinal bunch splitting (along $x_3$).

Thus, in all the examples given above, it is not possible to localize arbitrary large energy into the singularity, i.e. to realize its maximum concentration. The question is in what systems a complete collapse is possible. This is exactly a short reformulation of the inverse problem mentioned above.

An answer (possibly not a single one) to this question is as follows: trapping of an arbitrary portion of energy into singularity occurs in media with nonlinear inertia. As an illustration, we shall consider a simple example with relaxation nonlinearity obtained from (1) through modification:

$$i \frac{\partial \psi}{\partial t} + \sum_{i=1}^{2} \frac{\partial^2 \psi}{\partial x_i^2} - n\psi = 0 , \tag{2.1}$$

$$\frac{\partial n}{\partial t} + n = -|\psi|^2 . \tag{2.2}$$

Unlike the case of local coupling ($\partial/\partial t = 0$ in (2.2)), the nonlinear parameter n of the medium (for definiteness let us call it the perturbation of matter density) during the collapse changes too slowly to reach the values corresponding to a similar field amplitude at inertialess nonlinearity. The higher the collapse rate, the stronger is the retardation of the density perturbations form the locally nonlinear ones, and therefore, the greater field amplitude in the bunch is needed to achieve the former level of the matter perturbation. It is easy to see that such distributions lead to the dependence of the collapse rate on the energy $\int|\psi|^2 d\vec{x} = W$ trapped into singularity. On the contrary, for each value of stored energy there is a collapse rate ensuring the trapping of the whole portion of energy. Within the framework of Eqs.(2.1)-(2.2), a complete collapse is described by the following approximate self-similar solutions /8/:

$$\psi = e^{pt}u(\vec{r}_{\perp}e^{pt})\exp(i\phi(t)r^2) \ , \quad n = e^{2pt}N(\vec{r}_{\perp}e^{pt}) \ .$$

At $p \to 0$, $W \to W_{cr}$; if $p \to \infty$ , W increases as p. The structure of a self-similar solution for the case of strongly nonlocal coupling (p = 100) is represented in Figure 2.

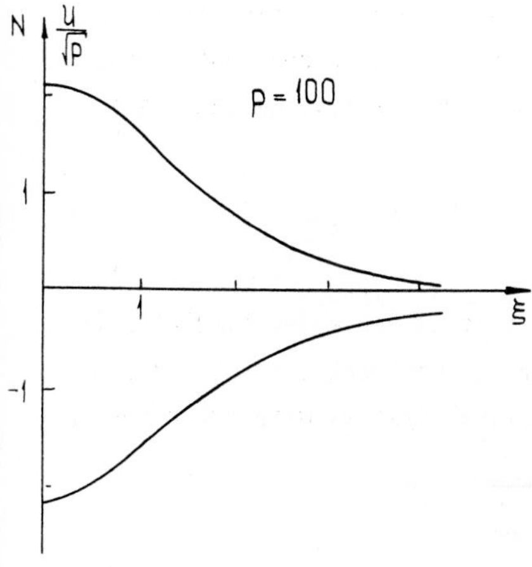


Figure 2

Like in the case of local nonlinearity, system (2.1)-(2.2), evolving in the course of time, has its own spatial analog. With a substitution $t \to z$, Eq.(2.1) describes the variation of the transverse structure in the wave field along the direction of the quasioptical beam propagation. Equation (2.2) takes into account the spatial nonlocality ("inertia") of the nonlinear response of the medium in the longitudinal coordinate. Such a situation can take place, for example, when intense waves propagate in a medium with a stationary flow of matter (beam self-focusing at longitudinal wind). Evidently, the spatio-temporal analogy is complete with exact transformation of the initial conditions of one problem into the boundary conditions of another problem.

This analogy suggests an idea that the conceptions of complete collapse can be generalized to a wide range of wave systems with bievolution behaviour. By the term bievolution we mean a unidirectional process along the temporal and one of the spatial coordinates, i.e. the simultaneous fulfilment of the temporal causality principle and the reflectionless spatial propagation condition for the wave field. The simplest generalization in the wave description is the use of the equation:

$$-i \frac{1}{V_{gr}} \frac{\partial \psi}{\partial t} - i \frac{\partial \psi}{\partial z} + \Delta_{\perp} \psi - n\psi = 0 \tag{3}$$

for the field complex amplitude, where $V_{gr}$ is the wave group velocity. The perturbations of density (n) can be found from relations evolving in only one of the independent variables. We shall consider the case of material coupling, inertial in time and local along the wave propagation direction z.

The following typical examples of material coupling can be used.

(i)

$$\partial n/\partial t = \nu(|\psi|)n. \tag{4}$$

This is the case of ionization nonlinearity. The effects of interest /9/ arises with decreasing dependence of the medium ionization frequency on the field amplitude. This occurs, for example, in the presence of superstrong electromagnetic fields in gases when the oscillatory energy of free electrons exceeds noticeably the molecule ionization potential /9/. Without loss of generality, we can assume $\nu \sim 1/|\psi|$.

(ii)

$$\partial n/\partial t = -n + F(|\psi|^2). \tag{5}$$

This example of local coupling with respect to all spatial variables simulates the simplest type of nonlinear relaxation of the medium. It was mentioned when discussing the purely temporal evolution of two-dimensional systems with cubic nonlinearity.

(iii)

$$\partial n/\partial t = \Delta_{\perp} n + \Delta_{\perp} F(|\psi|^2). \tag{6}$$

At diffusion relaxation the rate of the density perturbation onset depends on the beam width. Such a nonlinearity is typical of the medium heating in the field of a powerful electromagnetic wave and is realized, for example, in a weakly ionized collisionless plasma.

(iv)

$$\partial^2 n/\partial t^2 = \Delta_{\perp} n + \Delta_{\perp} F(|\psi|^2). \tag{7}$$

This type of coupling is due to the excitation of sound motions in the medium and is characteristic of a wide class of nonlinear processes, for example, in the case of laser radiation self-action in rarefield coronal plasma. In the case (5)-(7), the introduction of an arbitrary function F enables one to describe various situations, from the simplest dependence $F = |\psi|^2$ to the nonlinear saturation effects. Note that the transition to new time $\tau = t - z/V_{gr}$ will not change the material coupling structure but simplifies Eq.(3) by excluding the term with a temporal derivative:

$$-i \frac{\partial\psi}{\partial z} + \Delta_{\perp}\psi - n\psi = 0 \;. \tag{8}$$

The boundary condition at z = 0 retains, evidently, its form while the initial conditions must be set exactly when the pulse emitted at the boundary at t = 0 reaches a given point z. Since the nonlinearity is inertial in time, for an unperturbed initial state of the medium the initial (with respect to $\tau$) conditions correspond to a stationary diffractional field pattern in the linear problem with n = 0.

The main idea of what follows is that complete concentration of the energy flow in the cross-section of the beam is possible in the class of nonlinear systems (4)-(7) with inertial coupling of the medium density perturbations and the wave field amplitude. This can be demonstrated by a scheme based on the search for self-similar solutions with an arbitrary energy flow entrained into singularity, the analysis of stability of these solutions and the numerical illustration on spatio-temporal dynamics of the wave.

Note, first of all, that a common feature of all the types of inertial nonlinearity being discussed is the class of solutions in the form of homogeneous (in z), collapsing jets along which the trapped electromagnetic wave propagates. The Pointing vector is constant along the jet and depends, generally speaking, on the transverse structure of the mode and the collapse rate. The time of singularity formation on a finite-dimension set (either a straight line or a plane depending on the transverse form of the beam) is determined by the type of nonlinearity. Let us formulate some regularities of the jet scale decrease at the self-similar stage of collapse: $a \sim 1/t^2$ for a plane beam in a medium with the ionization defined by the effective frequency $\nu_i \sim 1/|\psi|$; $a \sim e^{-pt}$ $(p > 0)$ for the case of local relaxation, cubic nonlinearity and three-dimensional beam. In the analogous case for diffusion relaxation the blow-up behaviour takes place: $a \sim 1/\sqrt{t_0 - t}$, as well as for sound relaxation: $a \sim 1/(t_0 - t)$. For each of these laws it is possible to find a mode localized in the beam cross-section.

Nevertheless, the existence of appropriate self-similarities does not mean that the corresponding solutions will necessarily be realized at arbitrary initial and boundary conditions of the problem. For example, if a permanent source is given at the input to the nonlinear medium, then the question arises whether it is possible to match the stationary field distribution at z = 0 with the dynamic jet, (say, at $t \to \infty$). In a not so general by qualitatively analogous formulation the question is the following: is the solution in the form of collapsing jet an attracting set in the class of z-inhomogeneous structures?. The simplest analysis can be performed under the assumption of given self-similar) transverse form of the jet with the width being variable in z. It appears that homogeneous jet distributions with an infinite time of singularity formation ((4)-(5)) are stable manifold which attract beams (from the close vicinity at least) with the energy flow retained. In a similar analysis of jets with blow-up singularity formation (6), (7), it is concluded that they appear to be unstable and, therefore, cannot serve as solutions with the complete concentration of energy flow in the cross-section of the beam.

The existence of longitudinal instability in jets with blow-up behaviour makes one seek for z-homogeneous self-similar solutions which concentrate the energy flow at separate spatial points (moving, generally speaking, in the course of time). Let us discuss in more detail how such solutions in a medium with diffusion type relaxation of nonlinearity $F(|\psi|^2) = |\psi|^2$ are found.

Note, first of all, the general assertation that the structurally stable mode of collapse in a reference frame compressed together with the field distribution and removing its amplitude must correspond to a stable stationary localized structure. We now apply the inhomogeneous compression transformation to system (7)-(8):

$$\psi = \frac{1}{a(\eta)}\,\overline{\psi}\left(\frac{\vec{r}_\perp}{a},\eta,t\right)\exp\left(-i\,\frac{a_\eta}{4a}\,\vec{r}_\perp^{\,2}\right) ,$$
$$n = \frac{1}{a^2}\,N\left(\frac{\vec{r}_\perp}{a},\,\eta,t\right); \qquad \eta = z + \int v(t)dt, \qquad \vec{\xi} = \vec{r}_\perp/a , \tag{9}$$

where $v(t)$ is the arbitrary velocity at which the singularity propagates towards the incident radiation. For the self-similar functions $\overline{\psi}$ and N we have

$$ia^2\,\frac{\partial\overline{\psi}}{\partial\eta} = \Delta_{\vec{\xi}}\overline{\psi} - N\overline{\psi} - \frac{1}{4}\,a^3 a_{\eta\eta}\xi^2\overline{\psi} , \tag{10}$$

$$a^2(N_t+vN_\eta) - vaa_\eta(\vec{\xi}\nabla_{\vec{\xi}}N+2N) - \Delta_{\vec{\xi}}N = \Delta_{\vec{\xi}}|\psi|^2 . \tag{11}$$

Evidently, the presence of the last (lens) term in Eq.(10) excludes the existence of strictly localized modes (except for the case $a_{\eta\eta} = 0$ but then a complete self-similarity is not reached in (11)). It is possible, however, taking into account the known /10/ stabilizing properties of the "lens" term, to construct quasilocalized solutions with exponentially weak leakage of RF field quanta from the mode to the surrounding background. In these solutions $a = \sqrt{\eta_0-\eta}$, $\overline{\psi} = u(\vec{\xi})\exp\left[-i\gamma(t)\int a^{-2}d\eta\right]$, $N = N(\vec{\xi})$ and the transverse structure is completely analogous to the distribution in the homogeneous dynamic jet and can be found from the system:

$$\Delta_{\vec{\xi}}u - \gamma u - Nu = 0$$
$$\frac{1}{2}\,v(\vec{\xi}\nabla_{\vec{\xi}}N+2N) - \Delta_{\vec{\xi}}N = \Delta_{\vec{\xi}}u^2 . \tag{12}$$

To illustrate this, Fig.3 shows the main axisymmetric mode for strongly nonlocal nonlinearity (the high rate of collapse $v/\gamma$). Note that in such a situation the energy flow concentrated into singularity and the mode structure depend on parameter $v/\gamma$. The supercritical (as compared to stationary nonlinearity) localization

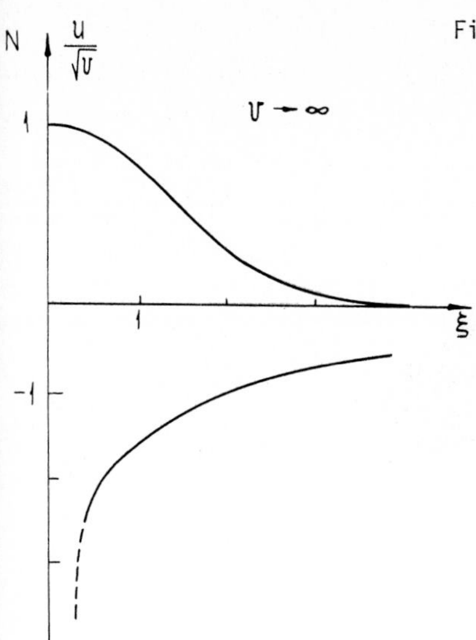

Figure 3.

requires the relations $v \gg \gamma \gg 1$ to be satisfied. If the singularity propagates at a velocity $v \sim 1$, then the energy flow into it is close to a critical one.

A similar consideration is possible in the case of sound relaxation of nonlinearity. Unlike the previous case, the field structure near the singularity is characterized by a conical form: $a = \eta_0 - \eta$, and the corresponding mode appears to be strictly localized because of the absence of the "lens" term from this type of self-similarity. As previously, an increase in collapse rate $p \sim v/\gamma$ leads to an increase in energy flow into the singularity propagating towards the radiation source.

Thus, the dynamic pattern of wave energy concentration can be radically different from the known self-focusing processes in media with stationary nonlinearity. In this context, it should be reasonable to ascertain whether the correspondence between the dynamic and the stationary regimes of self-action is retained or not during the long-term evolution of the electromagnetic wave emitted by a given source at the boundary with the medium. Evidently, the peculiarities of energy concentration are fully manifested in the case of sharp switching on the source for the time less than the linear relaxation time of the field in the medium. Exactly at this stage complete concentration of energy is possible with the appearance of singularities such as jets or moving foci. We should emphasize that under such conditions, supercritical localization of the energy flow in the wave beam is possible. At times much larger than the linear relaxation time, the system must inevitably evolve to the stationary limit. It should be interesting to consider a situation with a transition from complete dynamic concentration of the wave beam to the regime of its multiple splitting in the stationary case with supercritical

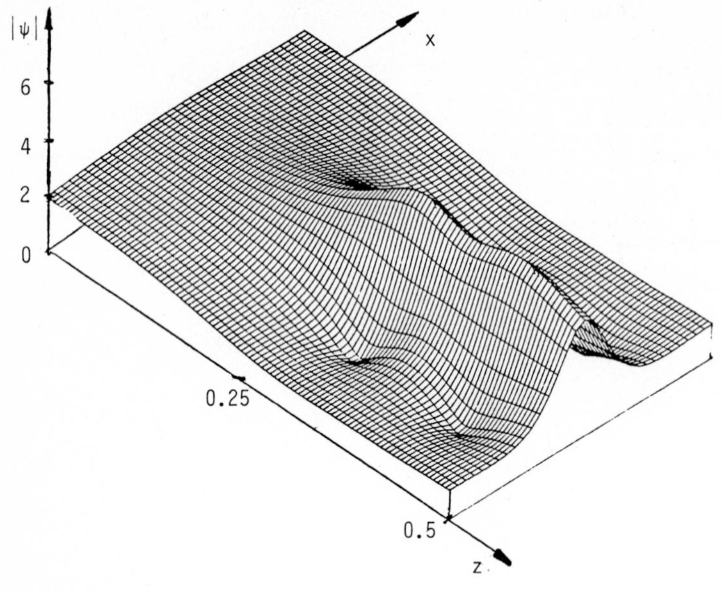


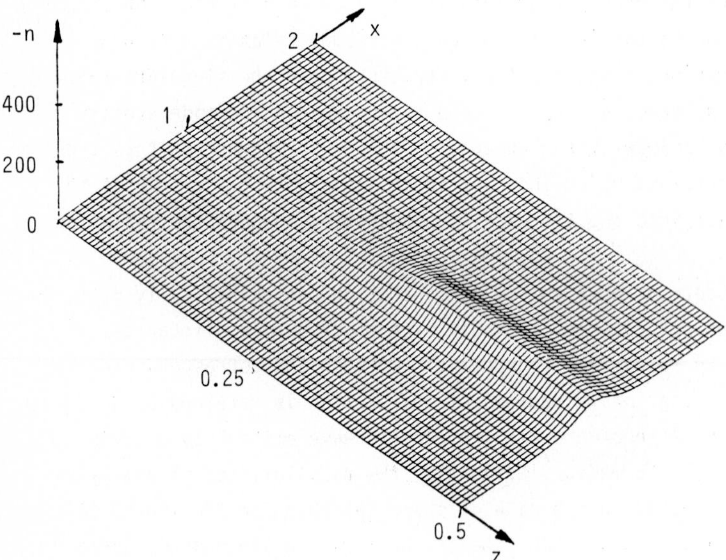


Figure 4a, t = 1.

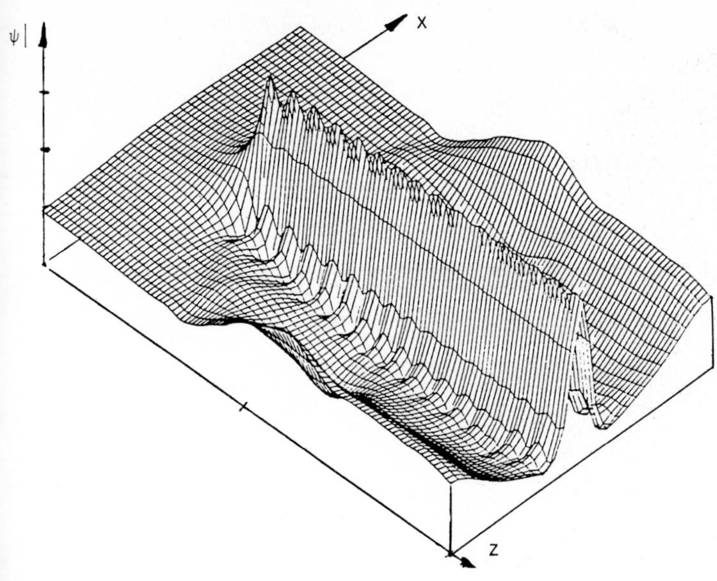


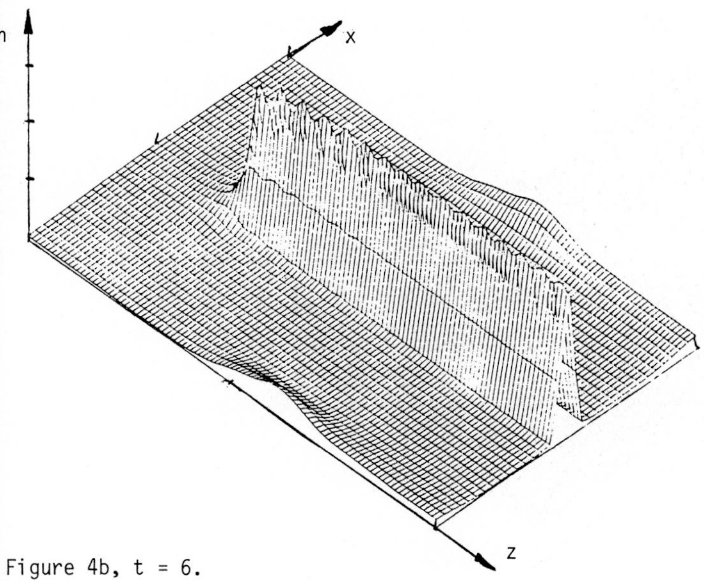


Figure 4b, t = 6.

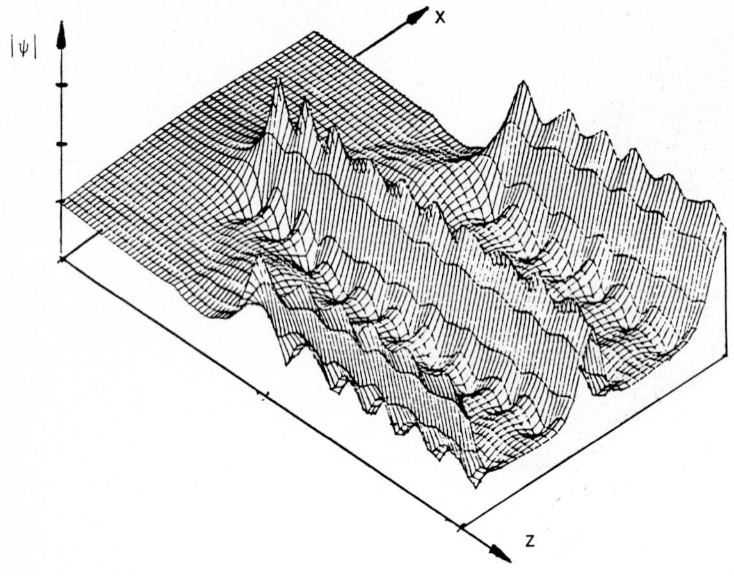


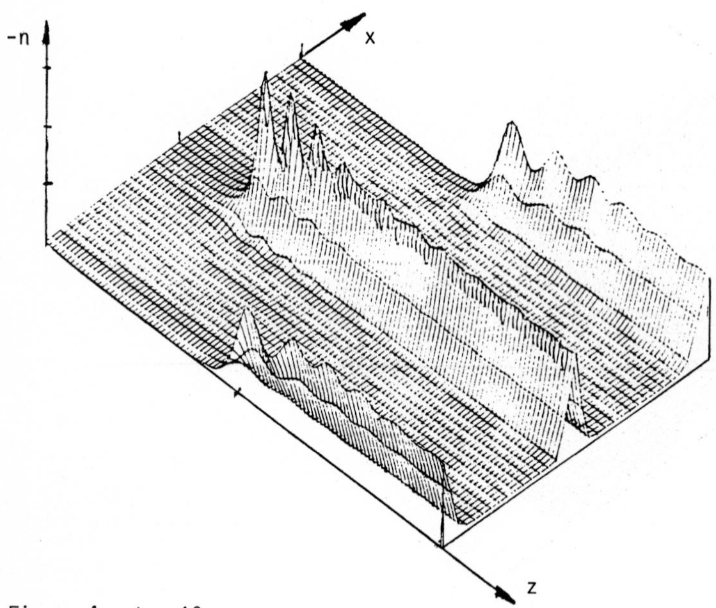


Figure 4c, t = 10.

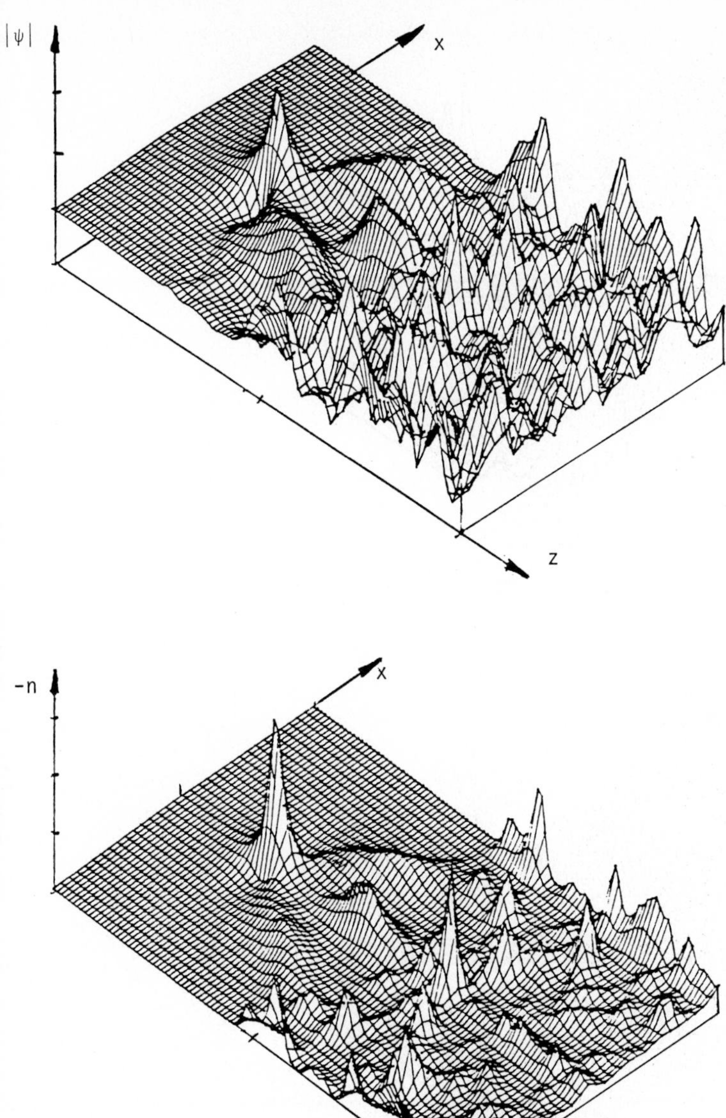


Figure 4d, t = 22.

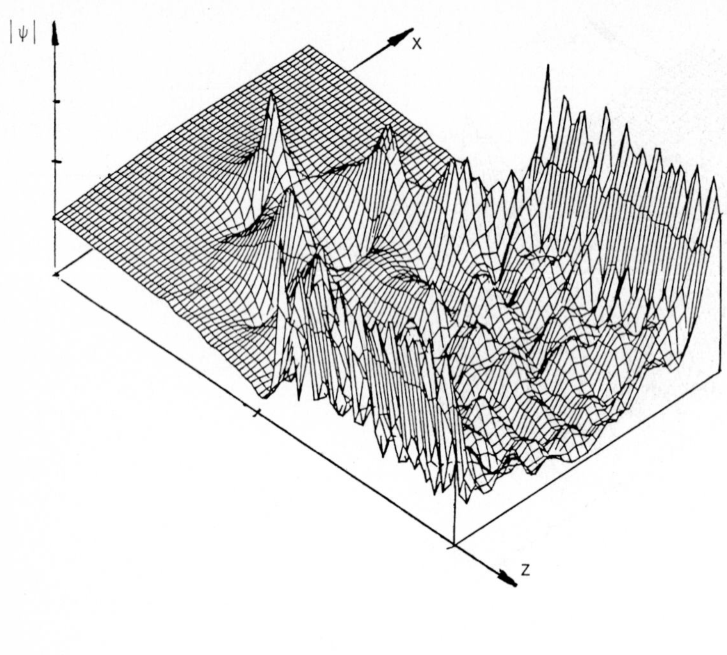


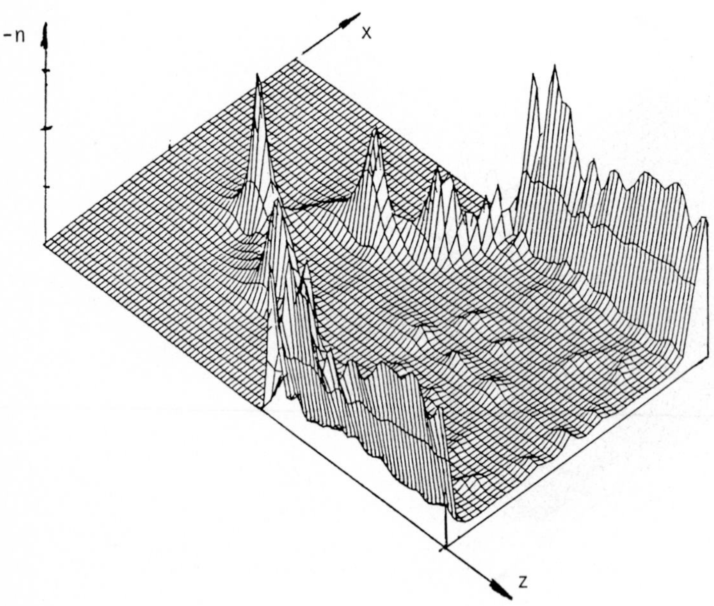


Figure 4e, t = 34.

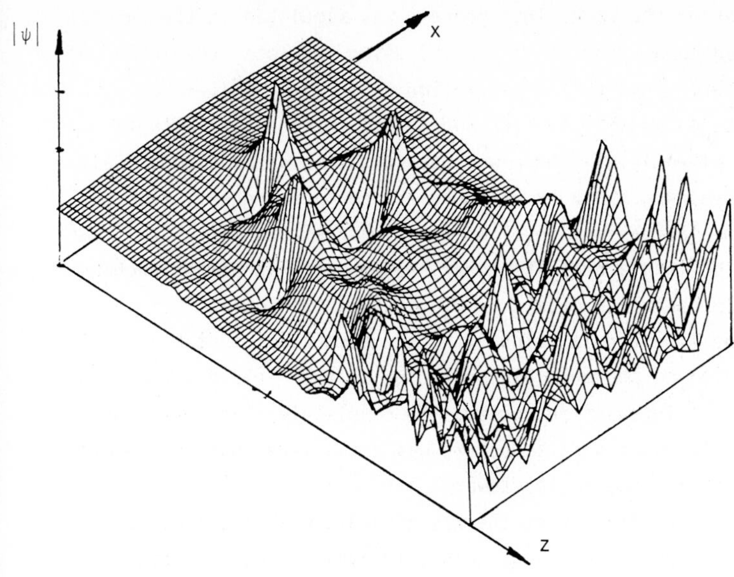


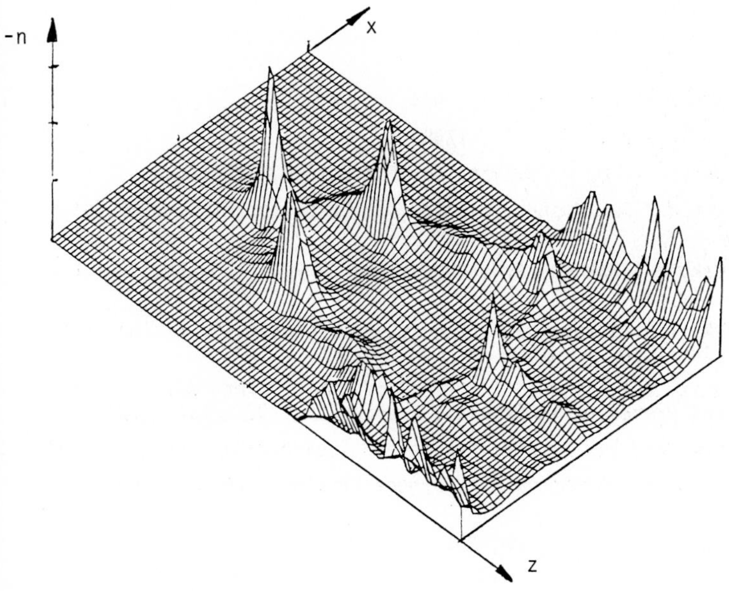


Figure 4f, t = 40.

power in the cross-section of the wave. This process was simulated in the case of self-action of a two-dimensional beam ($\Delta_{\perp} = \partial^2/\partial x^2$) using the local relaxation of nonlinearity to a level determined by its saturation $F(|\psi|^2) = |\psi|^4/(1+\alpha|\psi|^4)$ . At the initial dynamic stage (see Fig.4) (the source was switched on instantaneously and then maintained at a fixed level) there was the formation of an exponentially compressing plane jet entraining the bulk of the electromagnetic energy flow passed through the medium. The maximum level of the field reached $|\psi| \sim 1/\alpha^{1/4}$ and then the interaction passed into a quasistationary regime with an almost unchanging form of the jet and slow propagation of its origin towards the source. Only at times much larger than the linear relaxation time, the structural instability of the jet developed, which led to the multiple splitting of the field distribution and the turbulization of the interaction region. The completely stationary inetraction pattern with the filamentational structure was established due to the displacement of the dynamic turbulence region towards larger z.

To summarize, we stress once again that the use of media with inertial nonlinearity is attractive as a possible way to reach, in dynamic regimes, high levels of wave energy concentration, exceeding the corresponding values for stationary action of radiation on matter.

REFERENCES


1. V.I.Talanov. Izvestiya VUZov - Radiofizika, 1964, 7, 564.
2. V.E.Zakharov. Zh.Eksp.Teor.Fiz., 1972, 62, 1745.
3. V.I.Bespalov, V.I.Talanov. Pis'ma v Zh.Eksp.Teor.Fiz., 1966, 3, 471.
4. V.E.Zakharov, E.A.Kuznetsov, S.L.Musher. Pis'ma v Zh.Eksp.Teor.Fiz., 1985, 41, 125.
5. V.E.Zakharov, A.G.Litvak, E.I.Rakova, A.M.Sergeev, V.F.Shvets. Zh.Eksp.Teor.Fiz., 1988, 94, 107.
6. N.A.Zharova, A.G.Litvak, T.A.Petrova, A.M.Sergeev, A.D.Yunakovsky. Pis'ma v Zh. Eksp.Teor.Fiz., 1986, 44, 12.
7. S.N.Vlasov, L.I.Piskunova, V.I.Talanov. In: Proc. of the III Int. Workshop on Nonlinear and Turbulent Processes in Physics, Kiev, USSR, 1987, v.2, p.210.
8. V.E.Zakharov, A.F.Mastryukov, V.S.Synakh. Kvantovaya Elektronika, 1976, 3, 2557.
9. Ya.L.Bogomolov, S.F.Lirin, V.E.Semenov, A.M.Sergeev. Pis'ma v Zh.Eksp.Teor,Fiz., 1987, 45, 532.
10. G.M.Fraiman. Zh.Eksp.Teor.Fiz., 1985, 88, 390.

# Part III

# Dynamical Chaos and Self-Organization

# The Destiny of the Universality Hypothesis in the Theory of Developed Wave Turbulence

*G.E. Fal'kovich*

Institute of Automation, Siberian Branch of USSR Academy of Sciences,
630090 Novosibirsk, USSR

The structural instability of the isotropic Kolmogorov spectrum of a weak wave turbulence is discussed. The physical criterion for the existence of the drift Kolmogorov distribution is formulated. In the case of weak acoustic turbulence, universal two-flux spectra (bearing the energy and momentum fluxes) are obtained analytically. The universality hypothesis for the developed wave turbulence is formulated.

The assertion about the universality of the developed turbulence spectrum is usually interpreted in the following way: in the range of intermediate scales distant from both the pump and the damping region, the turbulence is isotropic and the energy distribution over scales depends on a single external parameter, viz. on the energy flux through k-space /1/.

On the basis of this hypothesis, isotropic stationary spectra (usually called after Kolmogorov) were found, both for vortex hydrodynamic turbulence /2,3/ and for wave turbulence in hydrodynamics, plasma physics, acoustics, etc. /4/. It should be pointed out, however, that whatever type of interaction, both waves and vortices conserve not only the total energy of the system but also its total momentum. Besides, any real source of turbulence is anisotropic and asymmetric, which results in the onset of some non-zero momentum in the system.

In the case of weak wave turbulence, stationary corrections $\delta n_k$ bearing the momentum flux R to the Kolmogorov solution $n_k$ which bears the energy flux P, were constructed in /5/. For waves with the power dispersion $w_k \sim k^s$ these, so called drift solutions, have a simple form:

$$\delta n_k/n_k \sim (Rk)w_k(Pk^2) \sim \cos \Theta_k k^{s-1} , \qquad (1)$$

which is a consequence of the fact that $y = (Rk)w_k/(Pk^2)$ is the only dimensionless parameter one could compose out of the values under consideration.

In the case of a decay-type dispersion law, value $s > 1$ and the contribution of anisotropic part of the spectrum increases with k, i.e. with going away from the source it goes deep into the inertial range. In the case of a nondecay-type dispersion law, a similar effect occurs with the Kolmogorov solutions transporting the flux Q (being also an integral of motion when the interaction is 4-wave) to a long-wave region. In this situation the contribution of the drift correction



Research Reports in Physics Nonlinear Waves 3
Editors: A.V. Gaponov-Grekhov · M.I. Rabinovich · J. Engelbrecht

$$\delta n_k/n_k \sim (Rk)/(Qk^2) \sim \cos\Theta_k k^{-1} \quad (2)$$

also increases with going deep into the inertial range, i.e. with decreasing k. It is proved in /5/ that drift corrections turn the linearized Stoss-term on the background of the isotropic Kolmogorov solution into zero. The question is: could solutions (1) and (2) be excited by an anisotropic pump?

It was shown in /7,8/ that the drift corrections are shaped, for example, for acoustic and capillary waves with the energy flux and they are not shaped for the spectrum of gravitational waves with the flux of the action. Using the general criterion of the stability of isotropic Kolmogorov spectra derived in /7/, one may prove the following theorem: the drift solutions can be shaped only if the momentum flux is transported by them in the same direction as the flux of the basic integral of motion. This condition has a plain physical sense: the momentum influx in the wave system is caused by the anisotropy of the pump. Mathematical proof is based on the proportionality between the momentum flux and the derivative of the Stoss-term with respect to scaling index of drift solution. This relation allows us to connect the sign of momentum flux with the sign of difference between the indices of drift and anisotropic solutions and the sign of the Stoss-term for anisotropic mode with the Kolmogorov index, which appear in the criterion of /7/.

Indeed, one may easily check up by straightforward computation that in each of three cases under discussion the momentum flux is directed towards large k, i.e. in the same direction as the energy flux for capillary and acoustic waves, and in the direction opposite to the action flux for gravitational waves.

Thus, if drift solutions (1) and (2) are shaped, the isotropic Kolmogorov spectrum is structurally unstable: even small anisotropy of the pump should lead to an essentially anisotropic spectrum deep within the inertial range.

The fact that the isotropic solution is unstable implies the absence of the simplest universality, i.e. one parameter is insufficient to determine the stationary distribution. How many parameters does it need?

In out opinion this question raises the interest mostly in the case of acoustic waves with small positive dispersion $w_k \sim k^{1+e}$, $e \ll 1$. As it was shown in /7,8/, in the presence of an anisotropic pump the corrections may be shaped not only in the form of the first harmonics (1) or (2), but also in the form of higher angular modes:

$$\delta n_k/n_k \sim P_m(\cos\Theta_k)k^{em(m+1)/2}, \quad em(m+1) < 1 \ , \quad (3)$$

where $P_m$ are the Legendre polynomials. If m = 1, then (3) coincides with (1). As is seen from Eq.(3), the higher the number of a harmonic, the more rapidly increases its contribution to the spectrum with k. Thus the acoustic turbulence represents the most unstable system since in this case the maximal number of angular harmonics ($m \cong e^{-1/2}$) is generated by anisotropic pump. For other systems which have been considered, harmonics with the order more than unity are not excited (see /8/). Anisotropic stationary solutions of kinetic equation for weakly dispersive waves were also obtained /6,9/. They have the following form:

$$n_m(k) = k^{-9/2+em(m+1)/4}\, P_m^{1/2}(\cos\Theta) \;. \tag{4}$$

These solutions were obtained in the framework of the differential approximation with respect to angular variables and they are correct in the domains where $P_m(\cos\Theta) > 0$ only. Formula (4) represents the set of independent jets with constant (along k) angul width $\pi/m$.

The existence of solutions (4) has lead earlier to the presumption that in the region of the angular distribution of the spectrum may have been chopped, for example, into a set of narrow jets, with the characteristic scale of the chopping being of the same order of magnitude as the interaction angle (see below (7)). Essentially, that could mean the absence of any universality of the spectrum in the inertial range. Later, however, a family of universal anisotropic solutions depending on two parameter (the fluxes of anergy and momentum) was constructed.

The kinetic equation has the following form for weakly dispersive waves:

$$\begin{aligned} \partial n_k/\partial t = \int k k_1 |k-k_1| \Big[ & \delta(w_k - w_k - w_{k-k})(n_k n_{k-k} - n_k n_{k-k} - n_k n_k) - \\ & - 2\delta(w_k - w_{k-k} - w_k)(n_k n_{k-k} - n_k n_{k-k} - n_k n_k)\Big] dk \;. \end{aligned} \tag{5}$$

Supposing $w = k^{1+e}$, $e \ll 1$, one may integrate (5) over the polar angle and, in the stationary case, the equation takes the following form in the variables $x = k_1/k$:

$$\begin{aligned} & \int_0^1 (1-x^{1+e})^{(2-e)/(1+e)} x^2 (n_1 n_2 - n_1 n_k - n_2 n_k)\, dx\, d\phi = \\ & = 2\int_1^\infty (x^{1+e}-1)^{(2-e)/(1+e)} x^2 (n_k n_2 - n_1 n_2 - n_1 n_k)\, dx\, d\phi \;. \end{aligned} \tag{6}$$

Here, in accordance with the delta-dunction in Eq.5, function $n_1 = n\left[k_1, \Theta_{kk_1}(k_1,\phi)\right]$ is considered on the surface determined by conditions

$$\begin{aligned} \cos\Theta_{kk_1} & = \left[1 + x^2 - (1-x^{1+e})^{2/(1+e)}\right]/(2x) \approx \\ & \approx 1 + e\left[x\ln x + (1-x)\ln(1-x)\right](1-x)/x \end{aligned} \tag{7a}$$

in the first integral and

$$\begin{aligned} \cos\Theta_{kk_1} & = \left[(x^{1+e}-1)^{2/(1+e)} - 1 - x^2\right]/(2x) \approx \\ & \approx 1 - e\left[x\ln x - (x-1)\ln(x-1)\right](x-1)/x \end{aligned} \tag{7b}$$

in the second one. As far as $n_2 = n(k-k_1)$ is concerned, one must put $\Theta_{kk_2} = x\Theta_{kk_1}/ /(x-1)$. It is obvious that the angles between interacting waves are small with the exclusion of narrow intervals about $x = 0,1$ and the infinitely remote domain $x \to \infty$. However, those intervals do not contribute to the interaction of waves because of the convergence of the integral (see below (9)).

Let us seek an axially symmetric solution of Eq.(6) which depends on k, P and $(\vec{R}\vec{k})$. In accordance with the dimensional relation $Pk \sim Rw_k$, this solution should take the form

$$n_k = P^{1/2}k^{-9/2}f\,((Rk)w_k/(Pk^2)) = P^{1/2}k^{-9/2}f(y)\ . \qquad (8)$$

Here f is an unknown function of the dimensionless variable y. Note that function $n_k = P^{1/2}k^{-9/2}$ is the isotropic Kolmogorov solution /4/. Because the dispersion is small, one can use the differential approximation with respect to y, since

$$y_1 = (Rk)w_1/(Pk_1{}^2) = \cos(kR)(k_1/k_a)^e \approx (k/k_a)^e \cos(kR) +$$

$$+ \sin(kR)\sin\phi\cdot\Theta_{kk_1} - \cos(kR)\Theta_{kk_1}{}^2/2 + e\cos(kR)\ln(k_1/k)\quad ,$$

i.e. $|y_1-y| \ll y$. Let us expand functions $f(y_1)$ and $f(y_2)$ in (6) up to the terms proportional to e. It is convenient to separate the second integral in (6) into two identical parts; then one of them is to be subjected to the substitution $x \to x^{-1}$, and the other to $x \to (1-x)^{-1}$ (the Zakharov transformations). After that, taking into account that $n_k \sim k^{-9/2}$ is an exact solution of (6), one can obtain the following equation in the first order in e:

$$\left[(df/dy)^2 + fd^2/dy^2\right]\sin^2(kR)(k/k_a)^e\int_0^1 x^2(1-x)^2\left[x\ln x+(1-x)\ln(1-x)\right]\cdot$$

$$\cdot\left[x^{-9/2}(1-x)^{-9/2} - x^{-9/2} - (1-x)^{-9/2}\right]dx = 0\ . \qquad (9)$$

The equation in f(y) is multiplied by a convergent integral. The solution is obtained trivially:

$$f(y) = \begin{cases} (ay+b)^{1/2}\ , & y > -b/a \\ 0\ , & y < -b/a \end{cases} \qquad (10)$$

According to (8) and (10), n should turn into zero on some surface in k-space. In reality, if $y \to -b/a$, then the derivatives of f(y) sharply increase and in the narrow neighbourhood of the surface (at $y+b/a < e^{1/2}$), the applicability conditions of the differential approximation are violated. Original equation (6) should have the solution f(y), which is smooth but sharply decreasing on the scale of a characteristic interaction angle (i.e. $e^{1/2}$). If $y \to -\infty$ , then $f(y) \to 0$. The constants of integration a and b may be included into the definition of the fluxes R and P. Constant a should be considered positive since the replacement $a \to -a$ is equivalent to the mere rotation of a coordinate system $\Theta \to \pi - \Theta$, whereas the two signs of b define two different families of solutions /9/:

$$n_k = k^{-9/2}(Rw_k \cos \Theta/k + P)^{1/2} , \tag{11a}$$

$$n_k = k^{-9/2}(Rw_k \cos \Theta/k - P)^{1/2} . \tag{11b}$$

Function (11a) corresponds to the spectrum which narrows when k increases. In particular, this solution describes a stationary distribution with a small momentum flux ($Rw_k << Pk$) generated by a weakly anisotropic pump. In this case, expanding (11a) into powers of $Rw/(Pk)$ at small k, one can obtain the isotropic Kolmogorov solution in the zeroth order, drift correction (1) in the first order, high modes (3) in the consequent orders. The contribution of high harmonics grows together with k. At large k most of the waves are located in the right hemisphere.

Solution (11b) describes an expanding spectrum. The angular width $\Delta\Theta(k)$ increases together with k according to the law $Rw_k \cos \Delta\Theta(k) = Pk$. If at the edge of the inertia range (at $k = k_0$) $Rw(k_0) \approx Pk_0$, then the boundary width $\Delta\Theta(k)$ can be very small. Value $\Delta\Theta(k)$ is bounded below by the interaction angle $e^{1/2}$, since the differential approximation is violated if given this width. Thus one can suppose that solution (11b) should be generated by a narrow source with the width $0 < \Delta\Theta < \pi/2$. In particular, a source finely chopped on the angle may generate a distribution with the intermediate asymptotics which looks like a set of narrow but expanding jets (11b) flowing together at large k into a single smooth distribution like (11). It is important that both solutions (11a) and (11b) coincide at $k \to \infty$ and $-\pi/2 < \Theta < \pi/2$. In this domain the spectrum is determined by the momentum flux only, with the angular shape of the spectrum independent of the boundary conditions (i.e. of the pump shape):

$$n_k \to k^{-9/2+e/2}(R \cos \Theta)^{1/2} .$$

In order to determine which kind of distribution takes place in the inertial range, the multi-jet solution (4) or the universal one (11), one should study the temporal evolution problem, which hitherto has not been tackled.

In spite of this we would like to suppose that the existence of solutions (11) is an evidence for a universal structure of the spectrum in the inertial range. These solutions admit boundary conditions in the source region to be quite arbitrary: from an isotropic to extremely narrow one with $\Delta\Theta \cong e^{1/2}$. Probably that means that any source generates a stationary distribution $n_k$ which has the universal form (11) at large k.

All above-stated leads to the following universality hypohesis for the spectrum of the developed wave turbulence. In the inertial range, the spectrum is two-parametric (certainly if it is stable) provided that the drift mode transports the momentum flux in the same direction as that of the flux of the energy (or the action). Otherwise, when moving in the k-space from source to damping, the spectrum becomes isotropic and tends to the conventional Kolmogorov solution (provided that there is also its global stability).

*Acknowledgement*

The author is grateful to A.M.Balk, V.E.Zakharov, E.A.Kuznetsov, A.M.Rubenichik, V.S.Lvov and A.V.Shafarenko for useful discussions.

REFERENCES

1. L.D.Landau, E.M.Lifshitz. Fluid Mechanics. M.:Nauka, 1987 (in Russian).
2. A.N.Kolmogorov. S.R.USSR Acad. Sci., 1941, 31, 538; A.M.Obukhov. Izvestiya USSR Acad.Sci., 1941, 4, 453.
3. V.I.Belincher, V.S.L'vov. Zh.Eksp.Teor.Fiz., 1987, 93, 533.
4. V.E.Zakharov. In: The Handbook of Plasma Physics, v.2, North-Holland, Amsterdam, 1984.
5. A.V.Katz, V.M.Kontorovich. Sov.Phys.-JETP, 1973, 37, 80.
6. V.S.L'vov, G.E.Fal'kovich. Sov.Phys.-JETP, 1981, 53, 2.
7. G.E.Fal'kovich, A.V.Shafarenko. Physica, 1987, 27D, 399.
8. A.M.Balk, V.E.Zakharov. In: Int.Workshop Plasma Theory, Nonlinear and Turbulent Processes in Physics, Kiev, 1987. Singapore, World Sci. Publ. Co. 1988.
9. G.E.Fal'kovich. Ph.D. thesis. Novosibirsk, 1984.
10. V.S.L'vov, G.E.Fal'kovich. Preprint IA&E, No.399, Novosibirsk, 1988 (in Russian).

# Renormalization Chaos in Period Doubling Systems

*S.P. Kuznetsov*

Institute of Radiotechnics and Electronics,
USSR Academy of Sciences, Saratov Office, USSR

The present paper gives an account of a recent conception of renormalization chaos in application to some scaling properties of period doubling systems (global structure of the Feigenbaum attractor, ratios of spectral amplitudes, response under periodical perturbation).

## 1. INTRODUCTION. RENORMALIZATION DYNAMICS AND RENORMALIZATION CHAOS

Renormalization group approach is one of the powerful tools of modern theoretical physics for analyzing systems with coexisting patterns with wide interval of space and time scales. This approach arised in the quantum theory of field and was developed and applied later in the theory of phase transitions and critical phenomena /1/. Recently, beginning from FEIGENBAUM's works /2-4/, the renormalization group approach is actively used for investigating the behaviour of nonlinear dynamical systems near the onset of chaos.

What does the renormalization group analysis of problems of nonlinear dynamics consist in? The first step is usually a discretization of time. Then the dynamics is described by some evolution operator $f_0$ connecting the consequent states of the system: $x_{n+1} = f[x_n]$. The most important is the next step. This is a transition to some greater interval of time discretization accompanied by suitable rescaling or, more general, by some change of dynamical variables. At this stage the new evolution operator is introduced which evidently can be represented through the old one: $f_1 = R[f_0]$, where R is the operator of renormalization.

One can repeat the above procedure many times and so a sequence of evolution operators $f_2$, $f_3$,... may be defined. The whole construction is similar to the known picture of Kadanoff's blocks in phase transition theory /1/. It leads to rapid results in such situations, when the system is characterized by similar or approximately similar behaviour on different time scales including very large time scales. This is just the situation which has taken place at the onset of chaos through period doubling bifurcations or through quasiperiodical regimes.

The equation which expresses the evolution operator through the previous one, is the renormalization group equation. One can say that this equation defines some kind of dynamics in space of operators called the renormalization dynamics. The number of steps of renormalization may be determined as renormalization time.



Research Reports in Physics Nonlinear Waves 3
Editors: A.V. Gaponov-Grekhov · M.I. Rabinovich · J. Engelbrecht

The traditional type of renormalization dynamics is a situation of a nonstable fixed point in operator space. This is just the case considered earlier in statistical mechanics and in the first investigations on renormalization group analysis of dynamical systems. However, the situations may exist in which the renormalization group equation leads to more complicated dynamics, such as periodic or chaotic dependence of evolution operators $f_n$ on the renormalization time n. As the author knows, such a possibility was discussed first by OSTLUND ed al /5/ for a problem of transition from quasiperiodicity to chaos and independently by CHIRIKOV and SHEPELYANSKY /6,7/ for the analogous problem in Hamiltonian systems. They also introduced the terminology used here.

What are the structural peculiarities of the phase space and the parameter space of dynamical systems which are connected with each type of renormalization dynamics? The transition to larger time scales during the described renormalization is associated with the consideration of more and more small objects in the phase space and the parameter space. So the nontrivial renormalization dynamics corresponds to structures similar to the Russian toy "matreshka". Such a toy consists of a sequence of hollow wooden puppets, the smaller one put inside the larger one. Three different types of renormalization dynamics are illustrated in Fig.1. A situation, when all figures are similar to each other (scaling), is associated with a fixed point (Fig.1a). The case of repeating through defined number of steps M corresponds to a renormalization cycle of period M (Fig.1b). At last, in renormalization chaos there is no full repeating excluding the repeating in approximate, mean or statistical sense (Fig.1c).

It is clear that the connection must exist between characteristics of renormalization dynamics and scaling properties of structures of the "matreshka"-type in the phase space and the parameter space. For the case of a fixed point this connection is well known. Each nonstable direction of the fixed point is associated

Figure 1. Russian toy "matreshka" as an illustration of three different types of renormalization dynamics.

with a relevant parameter of the system. The factor of scaling connected with this parameter is defined by the eigenvalue of linearized renormalization transformation. The case of renormalization cycle of some period M can be reduced to a case of fixed point by introducing a new renormalization group transformation consisting of M steps of the initial one. For renormalization chaos, however, a further investigation of a connection between its quantitative characteristics and scaling properties is needed.

For understanding the generalization of renormalization chaos it is desirable, of course, to have as many examples of it as possible. The situations of transition from quasiperiodicity to chaos /5-7/ are sufficiently complicated. Simpler examples can be presented by periodic-doubling systems and will be considered in this paper. Section 2 is devoted to the discussion of global scaling properties of the Feigenbaum attractor from the viewpoint of renormalization chaos. The approximate approach developed here casts a new light on some known results and is associated in some aspects with thermodynamical formalism /5,7/. In Sect.3 the conception of renormalization chaos is applied to problems of periodical perturbation and spectral properties of period doubling systems. This material is based particularly on our works with PIKOVSKY.

## 2. GLOBAL SCALING PROPERTIES OF THE FEIGENBAUM ATTRACTOR

### 2.1. The Generalized Renormalization Scheme

Let us consider the well-known one-dimensional map

$$x' = 1 - Ax^2 \tag{1}$$

and associated with it the geometrical image of the Feigenbaum tree or a graphical representation of x versus A (Fig.2). Each road up to the tree may be coded by a binary sequence. At each branching we use the following rule: the branch having the greatest deflections from the parent one is denoted by symbol 1 and the other by symbol 0. One can introduce also a code of moving down from the tree obtained by reading the previous code in backward order. For cycles of finite period $2^n$ each element has its own code consisting of n symbols. It is a remarkable obser-

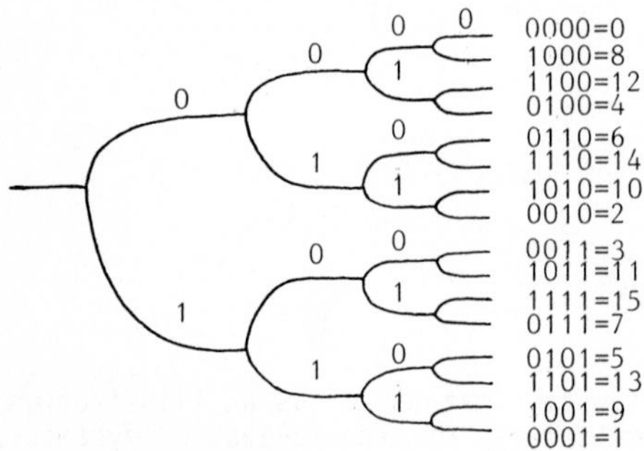


Figure 2. The Feigenbaum tree and a rule of coding of roads in it.

vation that the code of moving down from the tree correspondent to any element of the cycle is at the same time the binary representation of the number of this element in their natural time order. For the limit object formed by the infinite sequence of period doublings (the Feigenbaum attractor), the elements are coded by infinite binary sequences.

Let us remember the traditional Feigenbaum's renormalization group analysis. Performing twice an initial map $f(x)$, one obtains a map $f(f(x))$ having a characteristic form with two humps (Fig.3a). For further details the consideration, the central extremum of this map is selected. After rescaling of $x$ by factor $a < 0$, the new function is obtained: $f_1(x) = af(f(x/a))$. The described procedure is repeated many times. Exactly in the critical point of period doubling accumulation, the procedure leads to a fixed point of renormalization group equation, i.e. to the universal function denoted by $g(x)$. This function satisfies functional equation $g(x) = ag(g(x/a))$ solved by FEIGENBAUM.

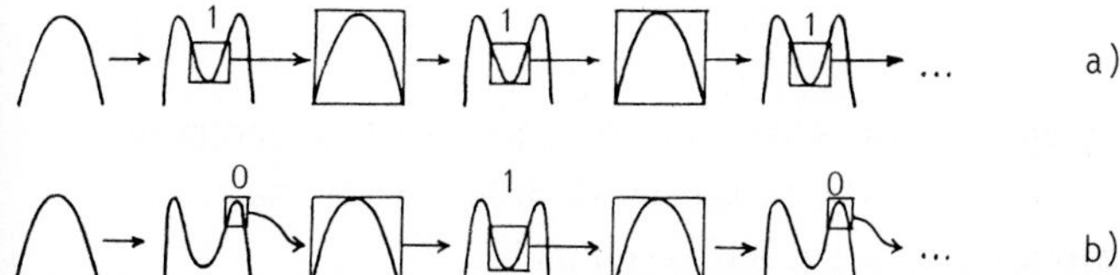

Figure 3. An illustration of the traditional Feigenbaum's renormalization transformation (a), and a generalized renormalization scheme (b).

It can be mentioned, however, that on each step of renormalization there are two possibilities: both the central and the right extremums of function $f(f(x))$ may be selected for further consideration. (The left extremum is not suitable because its vicinity is not visited by a system during its dynamics on attractor.) For the second variant the coordinate change must contain a shift of origin to a point of a new extremum, not only rescaling. We shall mark the first variant of transformation by symbol 1 and the second by symbol 0. Each binary sequence gives rise to its own renormalization scheme. Using such a construction we can achieve a hit of any element of $2^n$ cycle into a vicinity of the selected for detail consideration region of $x$ in n-th time renormalized function. For this purpose the code of renormalization scheme must coincide with the code of the road up to the tree to the the selected element.

Thus, with taking into account the whole set of possible binary sequences, one receives information about global scaling properties of the Feigenbaum tree while the traditional approach describes only a single branch coded by 11111... . This is an example of the most simple renormalization dynamics of a fixed point or a cycle of period 1. Other periodic codes give rise to more complicated renormalization cycles. For instance, a code 01010... corresponds to a cycle of period 2, code 011011... to a cycle of period 3 and so on. The stochastic binary sequences are associated with renormalization chaos.

## 2.2. The Approximate Description of Renormalization Dynamics

Let us concentrate our attention on the consideration of global scaling properties of the Feigenbaum attractor. For this purpose we shall consider here only the critical value of parameter in Eq.(1), A = 1,401155... . Then the shape of function $f^{2^n}(x)$ near its central extremum is described by the Feigenbaum function g(x) (for sufficiently large n). The key moment for further consideration is the fact that the shapes of this function near the other extremums are obtained from the central one by changes of the variable.

After the transition of origin to a point of extremum and rescaling it, provided the function at this point equals to 1, the change of variable must be close to identity. One can put

$$x \leftarrow (1-u)x + ux^2 \; , \tag{2}$$

where u is a parameter dependent on selected extremum. This parameter is proposed to be small and to be taken into account only in the first order. From the point of view of renormalization dynamics, u is a dynamical variable.

Performing the change (2) in the map x' = g(x), one obtains

$$x' = g_u(x) = g(x) + u\, g'(x)x^2 - (g(x))^2 - u\, g'(x)x - g(x) \; . \tag{3}$$

Let us undertake a step of renormalization transformation for the map (3). We perform it twice and transform the resulting map by an appropriate change of variables to the primary form with a new value of u. This is accomplished by two different ways: we can select the vicinity of the central (code 1) or side (code 0) extremum for consideration. The result is the recurrent equation

$$u' = \begin{cases} u/a & \text{(code 1)} \\ p + su & \text{(code 0)} \end{cases} \tag{4}$$

and the rule of rescaling of the variable:

$$\Delta x' = \begin{cases} a(1-u+u/a)\Delta x & \text{(code 1)} \\ (1+uc)(1-c)^{-1}\Delta x & \text{(code 0)} \end{cases} \tag{5}$$

where a = -2.5029, p = 0.659, s = 0.155, c = 0.8323. Constants in Eqs.(4), (5) are defined through the function g(x) and its derivatives. The known polynomial approximation of g found in /2/ was used for calculations.

Taking an arbitrary code of the road up to the tree consisting of n symbols, we can find the representation of function $f^{2^n}(x)$ in the normalized form in the vicinity of the correspondent extremum. Starting from $u_0$ = 0 (for the initial map

(1)), we must iterate Eq.(4) selecting one or another formula depending on symbols of code sequence. The shape of the function is given by (3) where parameter u is a result of iterations. The ratio of scales of initial and renormalized variables is obtained by the product of coefficients in (5).

Thus, in our approximation the renormalization dynamics is described by one-dimensional map plotted in Fig.4. The points showing results of computations are in good agreement with (4). Let us briefly explain the idea of the computation. All interesting extrema of function $f^{2^n}(x)$ may be numbered by m such as the variable change $x' = f^m(x)$ transforms this extremum into a central one. Correspondent codes are determined by binary representation of m reading from the right to the left. Performing the computations for n and n+1, we find two values u' for each u. One can see from Fig.4 that u is always smaller than 0.1; so the above proposition is justified.

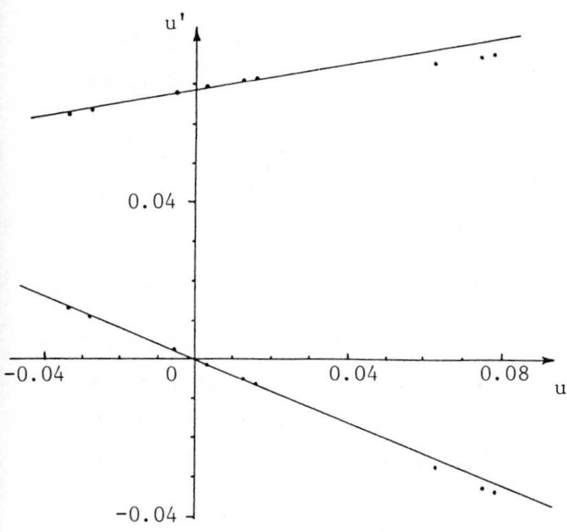


Figure 4. The one-dimensional map describing approximately the renormalization dynamics of parameter u. The exact numerical data are shown by points.

It is clear from (4) and Fig.4 that initial perturbations of u decrease under successive iterations. In other words, a value of u and so a form of $g_u(x)$ near some extremum are determined by the last symbols of the code. This is just a property of universality in a renormalization chaos. The quantitative characteristic of it is a negative Lyapunov exponent of the map (4). For the random code it is equal to $(1/2)\ln|(s/a)| = -1.39$.

### 2.3. The Feigenbaum Scaling Function $\sigma$

FEIGENBAUM /3/ has proposed a function $\sigma(t)$ for the description of global scaling properties of the attractor in the critical point of onset of chaos. It is determined through the elements of the large period cycles existing in a critical point $A_0$:

$$\sigma(t) = \lim_{n,N\to\infty} \sigma(\tfrac{m}{2n}) = \lim \frac{x_m^{(2N)}-x_{m+N}^{(2N)}}{x_m^{(N)}-x_{m+N/2}^{(N)}} \ , \quad N = 2^n \ , \tag{6}$$

where the superscripts denote the period of the cycle and subscripts the number of an element in time sequence. Let us show how to calculate this function from our renormalization dynamics.

Having some code moving down from the Feigenbaum tree a,b,c,... and reading it from the right to the left, we find from Eq.(4) the corresponding value of u. It makes sense for infinite codes as well as for finite ones because the main contribution is brought by several first symbols of the code a,b,c,... . Further we can consider the map (3) for this u and determine the distance between two elements of its cycle of period 2:

$$\Delta x(u) = 1.146(1+0.293u) \ . \tag{7}$$

Further, one can add symbol 1 or 0 to the code and find a next value of u' from Eq.(4). Then, substituting (7) into (6), with taking into account (5), we obtain the values of (t) for t = 0.01abc... and t = 0.00abc... :

$$\sigma(0.01abc...) = (1+u-u/a)\frac{\Delta x(u')}{a\Delta x(u)} = \frac{1+0.989u}{a} \ , \tag{8a}$$

$$\sigma(0.00abc...) = (1-c)(1-uc)\frac{\Delta x(u')}{\Delta x(u)} = 0.171(1-1.080u) \ . \tag{8b}$$

The results of calculation of function $\sigma$ are compared with Feigenbaum's numerical data /3/ in Fig.5. The agreement is very good. Most important is the correct description of the fine (fractal) structure of the function. So the renormalization dynamics given by Eq.(4) is just the mechanism that provides this fine structure.

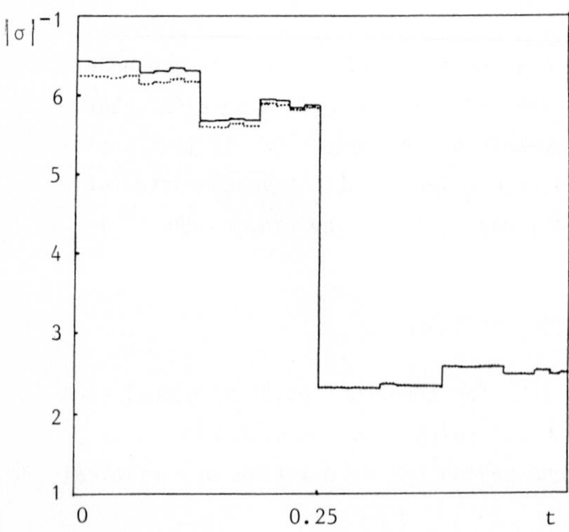


Figure 5. Feigenbaum's function $\sigma$. The dotted line corresponds to exact numerical data and the solid one to our approximate solution.

The most essential divergence takes place in the left part of the picture because this is just the region where u is maximal while the theory is constructed in the first order of accuracy in u.

## 2.4. Generalized Dimensions and Spectrum of Scaling Indices

The computations listed in the title of this section have recently been introduced in /10,11/ and are used for the description of multifractals or complicated sets which are imagined as interwoven variety of fractal (scaling invariant) sets. The Feigenbaum attractor is one of popular examples of multifractals. Spectra of generalized dimensions and scaling indices for this object are calculated in /11/. These characteristics as well as function $\sigma$ give some form of a description of global scaling properties of the Feigenbaum attractor but they are presented by smooth functions.

Let us remember the algorithm of calculations proposed earlier /11/. Firstly, the sequence is determined $x_0$, $x_1$, $x_2$, ... beginning from the point of extremum $x_0 = 0$ and consisting of iterations of this point by the map (1). One can consider the Feigenbaum attractor as a limit object of construction shown in Fig.6. It is similar to the construction of the Cantor set but the rule of decreasing of interval length is more complicated. At the n-th level of resolution we have $2^n$ intervals of different length $l_i$ with equal probability of visiting $p_i = 2^{-n}$. The sum of values $p_i^q/l_i^\tau$ is considered depending on two parameters q and $\tau$. Then the condition is adopted that this sum is equal to 1 and so an equation connecting q and $\tau$ is obtained in the limit of large n:

$$q = \lim \frac{1}{n} \log_2 \sum_{i=1}^{2^n} |x_i - x_{i+2^n}|^{-\tau} \quad . \tag{9}$$

Values $D_q$, f and $\alpha$ are determined through q and $\tau$:

$$D_q = \frac{\tau}{q-1} , \quad \alpha = \frac{d\tau}{dq} , \quad f = \alpha q - \tau \quad . \tag{10}$$

The dependence D vs. q is determined as a spectrum of generalized dimensions, and

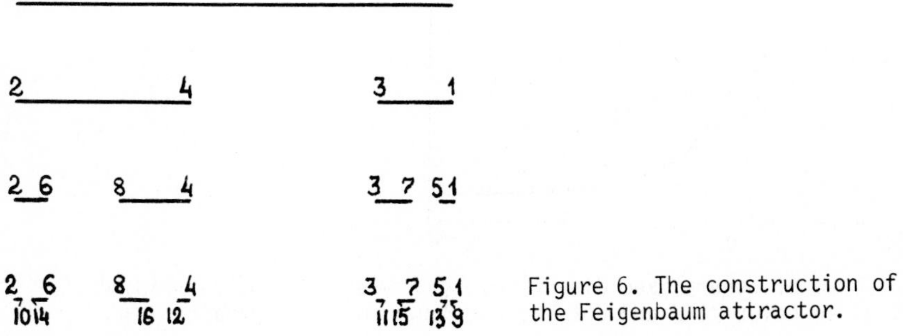


Figure 6. The construction of the Feigenbaum attractor.

f vs. $\alpha$ is determined as a spectrum of scaling indices of $f(\alpha)$-spectrum. Below we show how to obtain these spectra through our renormalization dynamics technique.

Let us consider the n-th level of resolution of attractor structure. It can be proved that the lengths of intervals $l_i$ are

$$l_i = |f^{2^n}(\bar{x}_{i-1}) - f^{2^n}(f^{2^n}(\bar{x}_{i-1}))| , \tag{11}$$

where $\bar{x}_{i-1}$ is the point of extremum of a function $f^{2^n}$, for which the code of moving down from the Feigenbaum tree is just the binary representation of i-1. By constructing two sums

$$Z_n = \sum_{i=1}^{2^n} l_i^{-\tau} \text{ and } S_n = \sum_{i=1}^{2^n} u_i l_i^{-\tau} , \tag{12}$$

we can put a question: how do these sums change at the next level of resolution, i.e. for n+1? Instead of each interval $l_i$ we have now two intervals with lengths

$$l_i\beta(1+\mu u_i) , \quad \beta = |a|^{-1} , \quad \mu = 1-a^{-2} \tag{13a}$$

for an added symbol 1 in the code and

$$l_i\gamma(1+\nu u_i) , \quad \gamma = (1-p/a)(1-c) , \quad \nu = -c + (1-s)/a \tag{13b}$$

for an added symbol 0. These formulae are obtained using the representation of a function $f^{2^n}(x)$ near the correspondent extremum by Eq.(3) and scaling rules (5) in the first order in u. Introducing (13) into (12) and neglecting the terms of the second power of u, we obtain the following matrix equation:

$$\begin{pmatrix} Z \\ S \end{pmatrix}_{n+1} = \begin{pmatrix} \beta^{-\tau}+\gamma^{-\tau}-\tau(\mu\beta^{-\tau}+\nu\gamma^{-\tau}) \\ p\gamma^{-\tau}(s-p\nu\tau)\gamma^{-\tau}+\beta^{-\tau}/a \end{pmatrix} \begin{pmatrix} Z \\ S \end{pmatrix}_n \tag{14}$$

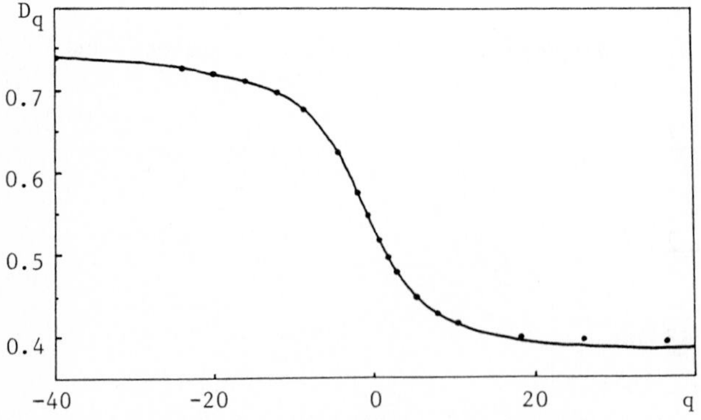


Figure 7. The spectrum of generalized dimensions $D_q$. The solid line corresponds to the exact numerical solution /11/ and dots to our approximate approach.

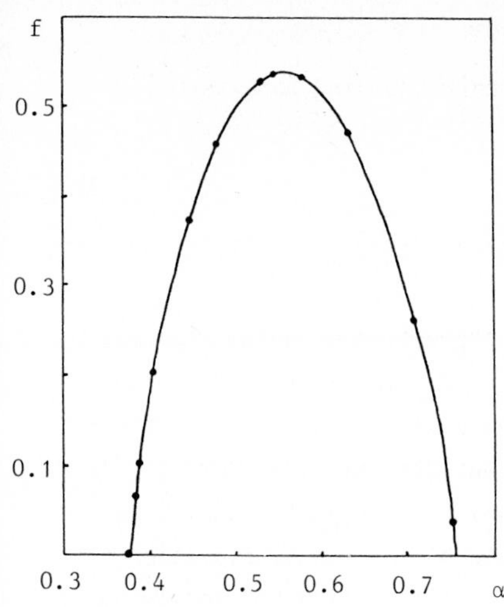


Figure 8. The spectrum of scaling indices. The solid line corresponds to the exact numerical solution /11/ and dots to our approximate approach.

For large n the sums $Z_n$ and $S_n$ will change proportionally to $\lambda(\tau)^n$, where $\lambda$ is the larger eigenvalue of the matrix. So it follows from (9) that $q(\tau) = \log_2\lambda(\tau)$. Values $D_q$, $\alpha$ and f are then obtained from Eqs.(10).

Figures 7 and 8 show the spectra of generalized dimensions and scaling indices of the Feigenbaum attractor. The solid lines correspond to calculations by method described in /11/ and points - to our approximate solution. Some special values of the generalized dimensions are compared in Table 1. The good agreement of data in Figs.7 and 8 as well as in Table 1 means that the considered approximation for renormalization dynamics correctly reflects the principle of constructing the Feigenbaum attractor as a multifractal set.

Table 1

| Dimension | Exact numerical /11/ | Our approximation |
|---|---|---|
| Fractal $D_0$ | 0.53804 | 0.5375 |
| Information $D_1$ | 0.51710 | 0.5161 |
| Correlation $D_2$ | 0.49780 | 0.4964 |
| Minimal $D_\infty$ | 0.37776 | 0.3727 |
| Maximal $D_{-\infty}$ | 0.75551 | 0.7555 |

## 3. FURTHER EXAMPLES OF RENORMALIZATION CHAOS

For a wider understanding of the conception of renormalization chaos we shall briefly consider its manifestations in two other problems also connected with period doubling systems.

### 3.1. Periodical External Perturbations of the System

Let us introduce an additional term into Eq.(1) corresponding to external periodical perturbation

$$x' = 1 - Ax^2 + b\cos 2\pi nw \ , \tag{15}$$

where b is an amplitude, w is a frequency of the perturbation and n is the discrete time.

The empirical computer investigation of scaling properties of model (15) was undertaken in /13/. In /14/ the authors developed the renormalization group approach to the problem. An attempt of renormalization analysis was also undertaken by ARNEODO /15/. However, he did not take into account the principal moment of a change of the frequency parameter under renormalization which just leads to renormalization chaos. According to /14/, the map describing a system evolution through $2^m$ units of time has for small amplitudes of perturbation the following form

$$x' = f_m(x) + \frac{1}{2}b\ \phi_m(x)\exp(2\pi i n w_m) + \text{c.c.} \ , \tag{16}$$

where $f_m(x)$ characterizes an evolution without external perturbation while the renormalization dynamics of function $\phi(x)$ and parameter w is determined by equations

$$\phi_{m+1}(x) = a\ g'(g(\tfrac{x}{a}))\phi_m(\tfrac{x}{a}) + \phi_m(g(\tfrac{x}{a}))\exp(2\pi i w_m) \ , \tag{17a}$$

$$w_{m+1} = 2w_m \ , \quad \text{mod } 1 \ . \tag{17b}$$

The dynamics described by Eq.(17b) may be periodical (for frequencies w represented by periodical binary fractions) and chaotical (for typical irrational w). Consequently, the scaling properties of dynamical regimes on (A,b)-plane are also determined by binary representation of w.

The first example of a picture of dynamical regimes on (A,b) -plane is shown in Fig.9a. It corresponds to a case of rational $w = 1/3 = 0.010101...$ and represents a renormalization cycle of period 2. White regions denote periodical regimes while shaded ones are domains of chaos. A form of each second region repeats periodically the previous form. Constants of scaling for A and b are correspondently 21.8 (the Feigenbaum's constant $\delta$ squared) and 58.96 which is found in /14/ by numerical solution of (17) as an eigenvalue of renormalization transformation (17b) through the period of renormalization cycle for $w = 1/3$. For other rational w coded by period periodical binary fractions the repeating of forms takes place through each p steps where p is a period of binary fraction.

For typical irrational w the renormalization chaos is realized and there is no exact repeating of forms (see Fig.9b taken from /16/). One can, however, understand a special form of scaling in a statistical sense. It is determined by the mean scaling index $\langle \ln|\phi_{m+1}/\phi_m| \rangle \approx 1.83$. The approximate repetition of forms must be observed under condition

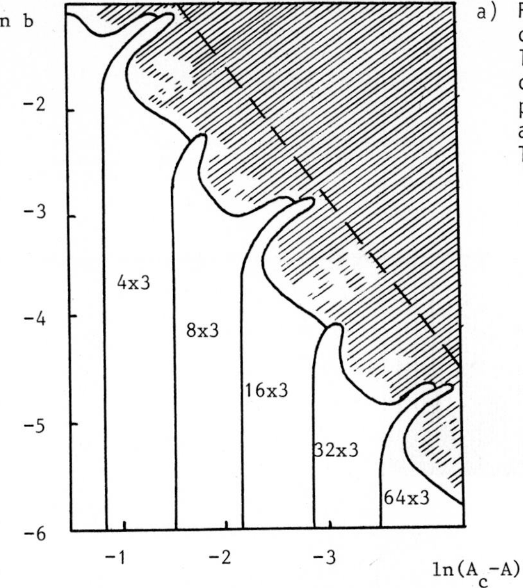


a) Figure 9. The domains of different dynamics in (A,b)-plane in the logarithmical scale for period-doubling system under external periodical perturbation for: a) w = 1/3 and b) w = $(5^{1/2}-1)/2$. The second one is taken from /16/.

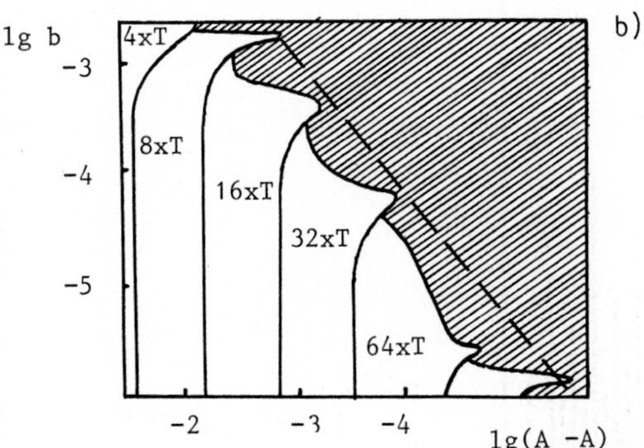


b)

$$b \sim (\Delta A)^{-\chi} \;, \quad \chi \approx 1.83/\ln\delta \approx 1.19 \;, \tag{18}$$

giving in the logarithmic scale a straight dotted line in Fig.9b. One can see that the border between quasiperiodical and chaotical regimes actually lies along this line. Using Eq.(18) one can evaluate a number of "torus doubling bifurcations" /16, 17/ observed in a system when A is increasing with constant b ($n \sim -\frac{1}{\chi}\log_\delta b$).

### 3.2. Spectrum of the Feigenbaum System and Renormalization Chaos

It was noted by PIKOVSKY that a similar to (18) renormalization group equation may may be used for the analysis of scaling laws in spectrum generated by a period doubling system. For this aim the next system of two maps may be considered

$$x' = f(x) \;, \quad C' = C + \phi(x)\exp(2\pi i n w) \;, \tag{19}$$

where the initial function $\phi(x) = 1$. It is clear that for sufficiently large n the

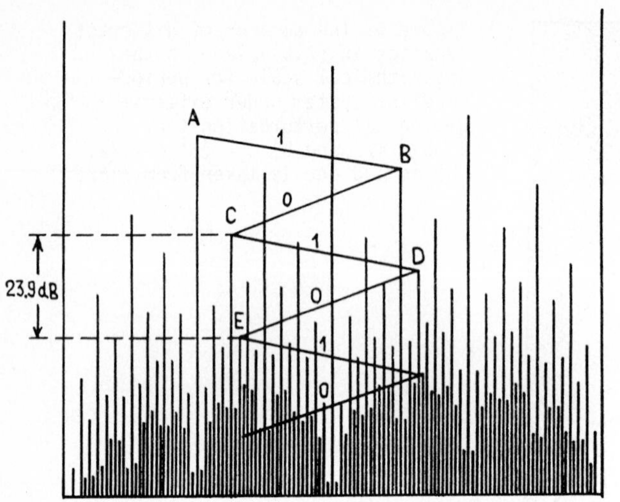


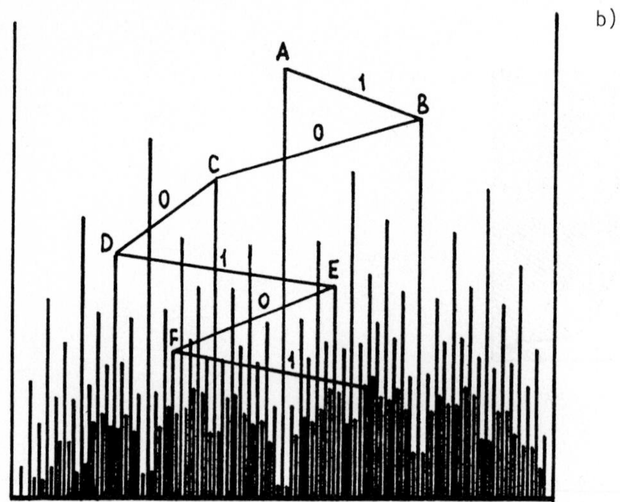


Figure 10. The renormalization cycle (a) and renormalization chaos (b) in spectral amplitudes of period-doubling system obtained for the sequences of binary shifts of the frequencies w = 0.0101010 and w = 0.1010011, correspondingly.

value $C_n$ will be proportional to the amplitude of the spectral component with frequency w. Performing the Feigenbaum's doubling procedure to Eqs.(19) many times, we obtain the following recurrent renormalization group equations

$$\phi_{m+1}(x) = \frac{1}{2}\,\phi_m(x/a) + \phi_m(g(x/a))\exp(2\pi i w_m) \quad , \tag{20a}$$

$$w_{m+1} = 2w_m \;, \quad \bmod 1 \;. \tag{20b}$$

Analyzing these equations allows us to make conclusions about ratios of amplitudes of spectral components. These ratios are found to be determined by the structure of w as a binary fraction.

Let us take any binary fraction w = 0.abcd...e, and consider a sequence of w's obtained by binary shifts: $w_1$ = 0.bcd...e, $w_2$ = 0.cd...e etc. Then we proclaim that the ratios of spectral amplitudes over frequencies of this sequence change periodically in a case of periodical structure of code abcd...e and chaotically for random structure of it. Figure 10 illustrates both situations. For the renormalization cycle of period 2 (Fig.10a) the amplitudes of spectral components decrease by 0.0638 times or -23.9 dB per period of the cycle. The same value was obtained in /24/ by numerical solution of (20) for w = 1/3 as an eigenvalue of renormalization transformation through the whole cycle. For random codes (as in Fig.10b), one can determine a mean scaling factor. According to calculations of /14/ it is equal to

$$\langle 20 \lg |\phi_{m+1}/\phi_m| \rangle \approx -13.9 \text{ dB} .$$

group. Illustrations of chaotical renormalization dynamics in this paper are simpler than those considered earlier in /6-7/ and so may be useful for a better understanding of this subject as well as of the approximate approach to the description of renormalization dynamics developed in the present paper.

## REFERENCES


1. R.Balescu. Equilibrium and Nonequilibrium Statistical Mechanics, Vol.1. New York, London, 1975.
2. M.Feigenbaum. J.Stat.Phys., 1979, 21, 6, 669.
3. M.Feigenbaum. Commun.Math.Phys., 1980, 77, 1, 65.
4. M.Feigenbaum, L.Kadanoff, S.Shenker. Physica, 1982, 5D, 2, 370.
5. S.Ostlund et al. Physica, 1983, 8D, 3, 303.
6. B.V.Chirikov, D.L.Shepelyansky. Physica, 1984, 13D, 395.
7. B.V.Chirikov, D.L.Shepelyansky. Proc.Conf. "Renormalization Group", Dubna, 1986. World Sci., Singapure, 1988, 221.
8. E.B.Vul, Ya.B.Sinai, K.M.Khanin. Sov.Math.Usp., 1984, 39, 3, 3 (in Russian).
9. D.Bensimon, M.Jensen, L.Kadanoff. Phys.Rev., 1986, 33A, 5, 3622.
10. H.Hentchel, I.Procaccia. Physica, 1983, 8D, 3, 435.
11. T.Halsey et al. Phys.Rev., 1986, 33A, 2, 1141.
12. B.G.Levi. Physics Today, 1986, April, 17.
13. S.P.Kuznetsov. Sov.Phys.-JETP Lett., 1984, 39, 3, 133.
14. S.P.Kuznetsov, A.S.Pikovsky. Preprint No.168, Inst.Appl.Phys., Gorky, USSR, 1987 (in Russian).
15. A.Arneodo. Phys.Rev.Lett., 1984, 53, 1240.
16. K.Kaneko. Progr.Theor.Phys., 1984, 72, 2, 202.
17. V.S.Anishchenko, T.E.Letchford, M.A.Safonova. Izvestiya VUZ'ov - Radiofizika, 1985, 28, 9, 1112 (in Russian).

# On Resonances, Pendulum Equations, Limit Cycles and Chaos

*A.D. Morozov*

Gorky State University, 23 Gagarin Ave., 603600 Gorky, USSR

The contribution of pendulum equations is well-known in the oscillation and wave theory /1,2/. However, so far only the results for special cases have been obtained for non-conservative pendulum equations. This paper formulates general statements on limit cycles. On the basis of them, more general class of "self-oscillatory" pendulum equations can be investigated. It is shown that equations describing the topology of resonance zones in non-conservative systems similar to the two-dimensional Hamiltonian ones belong to this class. The scenario of the appearance of a quasi-attractor with a developed chaos with moderate values of external force amplitude is described, which is based on global cycle bifurcations, i.e.invariant curves of Poincaré maps.

## 1. RESONANCES AND PENDULUM EQUATIONS

Let us consider the system

$$\dot{x} = H_y(x,y) + \varepsilon g(x,y,\nu t)$$
$$\dot{y} = -H_x(x,y) + \varepsilon f(x,y,\nu t) \qquad (1)$$

where H, g, f are rather smooth functions of x and y in some region D ($D \subset R^2$ or $D \subset R^1 \times S^1$) and are continuous and periodic over $\phi = \nu t$ with the period $2\pi$, $\nu$ is a parameter, $\varepsilon$ is a small parameter. Under certain hypotheses /3/, system (1) can be expressed in variables I (action) and $\Theta$ (angle):

$$\dot{I} = \varepsilon F(I,\Theta,\phi), \quad \dot{\Theta} = \omega(I) + \varepsilon R(I,\Theta,\phi), \quad \dot{\phi} = \nu \quad , \qquad (2)$$

where $F = fx_\Theta' - gy_\Theta'$, $R = -fx_I' + gy_I'$ are periodic functions of $\Theta$ and $\phi$ with period $2\pi$.

In addition to (2) let us consider the autonomous system

$$\dot{I} = \varepsilon B_0(I) \, , \quad \dot{\Theta} = \omega(I) + \varepsilon Q(I) \, ,$$

$$B_0 = \langle F \rangle_{\Theta,\phi} \, , \quad Q = \langle R \rangle_{\Theta,\phi}$$

defined in the ring $\Delta \times S$ , which can also be derived from the system

$$\dot{x} = H_y + \varepsilon\bar{g}(x,y) \, , \quad \dot{y} = -H_x + \varepsilon\bar{f}(x,y) \qquad (3)$$

after the transition to variables I, $\Theta$ and averaging over $\Theta$. Here $\bar{g} = \langle g \rangle_\phi$, $\bar{f} = \langle f \rangle_\phi$. It is well known (see e.g. /4/) that there exist structurally stable limit cycles in system (3) if the Poincaré-Pontryagin generating equation $B_0(I) = 0$ has simple roots in the interval $\Delta$. Let us denote $B_1(I) = B_0'(I)$. While dealing with three-dimensional



Research Reports in Physics Nonlinear Waves 3
Editors: A.V. Gaponov-Grekhov · M.I. Rabinovich · J. Engelbrecht

system (2), resonances are of the greatest importance. Level $I = I_{pq}$ is called the resonance one, if $\omega(I_{pq}) = q\nu/p$, $p,q \in N$. The neighbourhood $\{(I,\Theta): I_{pq}-c\sqrt{\varepsilon}<I<I_{pq}+c\sqrt{\varepsilon}, 0 \leq \Theta < 2\pi\}$ is called the resonance zone. The topology of non-degenerate resonance zone ($b = \omega'(I_{pq}) \neq 0$) in non-conservative systems of type (2) is determined in the case of the "general position" using a pendulum equation /3-5/

$$\frac{d^2v}{d\tau^2} - bA_0(v,I_{pq}) = \mu^2\sigma(v,I_{pq})\frac{dv}{d\tau}, \tag{4}$$

where $\mu = \sqrt{\varepsilon}$, $\tau = \mu t$,

$$A_0 = \frac{1}{2\pi p}\int_0^{2\pi p} F(I_{pq}, v+q\phi/p, \phi)d\phi$$

$$\sigma = \frac{1}{2\pi p}\int_0^{2\pi p} (f_y+g_x)\Big|_{\substack{x=x(I_{pq},v+q\phi/p)\\ y=y(I_{pq},v+q\phi/p)}} d\phi$$

Functions $A_0$, $\sigma$ which are periodic over v with the smallest period $2\pi/p$ are given by $A_0 = A_*(v,I_{pq}) + B_0(I_{pq})$, $\sigma = \sigma_*(v,I_{pq}) + B_1(I_{pq})$. Note that the number of harmonics in $A_0$, $\sigma$ is defined by the number of harmonics (of $\phi$) in functions f, g. As, according to the definition, functions $A_0$, $\sigma$ can be dependent on various parameters, the harmonics with different numbers may be predominant in them.

The general classifications of resonance zones of individual levels I = const, is given in /3/. Here we express more exactly the case of impassable resonance level, when $B_0 = 0$. In this case we get the following standard equation:

$$\ddot{x} + \sin x = \varepsilon\dot{x}(\cos nx + a), \tag{5}$$

where a is a parameter, n = 0,1,2,... and where for the sake of convenience the original notation of variables and small parameter $\varepsilon$ is used. Along with (5), let us consider a standard equation:

$$\ddot{x} + \sin x = \varepsilon\dot{x}\cos nx. \tag{6}$$

Equations of type (5), (6) appear in some applied problems /6,7/, they are also of great interest for the theory of non-linear oscillations, having much more possibilities than the well-known Van der Pol and Duffing equations.

## 2. LIMIT CYCLES IN PENDULUM EQUATIONS

First let us consider equation (6). When $\varepsilon = 0$, this equation has the integral $\dot{x}^2/2 - \cos x = h$. When $h \in (-1,1)$, there exists the zone of oscillatory motion of the pendulum, and when $h > 1$, there is a zone of rotary motion. Equation (6) is expressed as (3) and a degenerating function $B_0(h(I))$ can be defined. Denote $B_0 = F_n^{(s)}(\rho)$, where s = 1 corresponds to the zone of oscillatory motion, $\rho = (1+h)/2$ and s = 2 to the zone of rotary motion, $\rho = 2/(1+h)$. Then, by definition of $B_0$, we have

$$F_n^{(s)}(\rho) = \frac{1}{2\pi}\int_1^{2\pi} \cos(nx^{(s)})x^{(s)}(x^{(s)})_\Theta{}' d\Theta , \tag{7}$$

where $x^{(s)}(\Theta,\rho)$ is an undisturbed solution. According to /8/, the elliptical integral in (7) is reduced to the standard form:

$$F_n^{(s)}(\rho) = C_n^{(s)}\ P_n^{(s)}(\rho)K(\rho) + Q_n^{(s)}(\rho)E(\rho) \quad , \tag{8}$$

where $C_n^{(1)} = 16/(2n+1)!!$, $C_n^{(2)} = 8/\ (2n+1)!!\rho^{n+1/2}$ , $P_n^{(s)}$, $Q_n^{(s)}$ are the polynomials of the power n, $(n \geqq 1)$, K, E are the full elliptical integrals, $k = \rho^{1/2}$ is their modulus.

For the sake of convenience, we introduce functions

$$F_n^{(1)}/16 \qquad F_n^{(1)} , \quad F_n^{(2)}\rho^{1/2}/8 \qquad F_n^{(2)} .$$

Thus, the problem of limit cycles in Eq.(6) leads to the solution of two classes of special functions $F_n^{(s)}(\rho)$ , $s = 1,2$. In addition to (7), (8) we also give the properties of functions $F_n^{(s)}(\rho)$.

*The Properties of Functions* $F_n^{(1)}(\rho)$ */9/*

1° Functions $F_n^{(1)}$ satisfy the Gauss hypergeometrical equation

$$4\rho(1-\rho)(F_n^{(1)})'' + \lambda_n F_n = 0 , \quad \lambda_n = 4n^2 - 1 , \quad n = 0,1,\ldots .$$

2° $F_n^{(1)}(\rho) = 4\pi\rho F(\frac{1}{2} + n, 2; \rho)$,

where F is the hypergeometrical function.

3° $F_n^{(1)}$ allows for the analytical contribution on the complex plane with a cut along the real axis from $z = 1$ to $z = \infty$, which is denoted as $C^*$.

4° There exist the following relations at the boundary points of the interval (0,1):

$$F_n^{(1)}(0) = 0 , \quad F_n^{(1)}(1) = (-1)^{n+1}/\ _n ,$$

$$(F_0^{(1)}(0))' = 4\pi , \quad \lim_{\rho\to 1}(F_n^{(1)}(\rho))' = (-1)^n\infty$$

5° The recurrent formula is valid

$$(3+2n)F_{n+1}^{(1)}(z) + 4n(2z-1)F_n^{(1)}(z) + (2n-3)F_{n-1}^{(1)}(z) = 0 , \quad z \in C^*$$

$$F_0^{(1)}(z) = (z-1)K(z) + E(z) , \quad F_1^{(1)}(z) = \left[(1-z)K(z) + (2z-1)E(z)\right]/3$$

6° Functions $F_n^{(1)}(z)$, $n \geqq 1$ have exactly n-1 simple zeros in the interval (0,1) of the real axis. Zeros $F_{n+1}^{(1)}$, $F_n^{(1)}$ alternate with each other.

7° Functions $F_n^{(1)}(\rho)$ constitute the Chebyshev system on $\left[0,1\right]$.

*The Properties of Functions* $F_n^{(2)}(\rho)$ */10/*

1° $F_n^{(2)}(\rho)$ satisfies the equation

$$4\rho(1-\rho)(\rho(F_n^{(2)})')' + (\rho-\mu_n)F_n^{(2)} = 0\ , \quad \mu_n = 4n^2, \quad n = 0,1,\dots\ .$$

2° $F_n^{(2)}(\rho) = C_n\rho^n F(-\frac{1}{2} + n, \frac{1}{2} + n, 1+2n; \rho)\ , \quad C_n = \text{const}\ .$

3° The recurrent formula is valid

$$(2n+3)zF_{n+1}^{(2)}(z) + 4n(2-z)F_n^{(2)}(z) + (2n-3)zF_{n-1}^{(2)}(z) = 0\ , \quad z \in C^*$$

$$F_0^{(2)}(z) = E(z),\ F_1^{(2)}(z) = 2(z-1)K(z) + (2-z)E(z)\ /3z$$

4° Functions have no zeros in $C^*$, different from $z = 0$.

5° $F_n^{(2)}(0) = 0\ , \quad F_n^{(2)}(1) = F_n^{(1)}(1)\ .$

From properties 6° of functions $F_n^{(1)}$ and 4° of functions $F_n^{(2)}$ we have:

*Theorem 1.* There exists such $\varepsilon_*(n) > 0$, which is small enough that under any $|\varepsilon| \in (0,\varepsilon_*)$ Eq.(6) with $n \geq 1$ in the zone of oscillatory motion has exactly n-1 structurally stable limit cycles (of the 1st kind). In the zone of rotary motion there are no limit cycles (of the 2nd kind).

From the above mentioned characteristics of functions $F_n^{(s)}(\rho)$, $s = 1,2$ we have:

*Theorem 2.* There exist such values of parameter a and small enough values of $\varepsilon$, when Eq.(5) has exactly n limit cycles in the zone of oscillatory motion and has no limit cycles in the zone of rotary motion. If $a = a_* = (-1)^n/\lambda_n$, then one of the n cycles is of a saddle type.

*Remark 1:* In /7/ Eq.(6) was treated in an oscillatory zone. However, the theoretical investigation was carried out not for (6), but for a quasi-linear equation ($\sin \to x$).

*Remark 2:* The fitness of the limit cycle number in (3) when H, f, g are polynomials, follows from /11/.

## 3. THE EFFECT OF LIMIT CYCLES ON THE GENERATION OF QUASI-ATTRACTORS WITH A DEVELOPED CHAOS

Let us consider equation

$$\ddot{x} + \sin x = \varepsilon x(\cos nx + a)(1 + c \sin \nu t) \tag{9}$$

which, on the one hand, takes into account non-autonomous terms in the system, describing the structure of resonance zones, and on the other hand, is of great importance itself, being a special case of system (1).

When $c = 0$ and $|\varepsilon|$ is small, theorem 2 describes the qualitative behaviour of Eq.(7). Numerical calculations show the validity of theorems 1, 2 with $|\varepsilon|$ being not small (see Figs.1,2, where $\varepsilon = 1$, $a = a_*(n)$; $n = 1$ in Fig.1 and $n = 3$ in Fig.2). Using

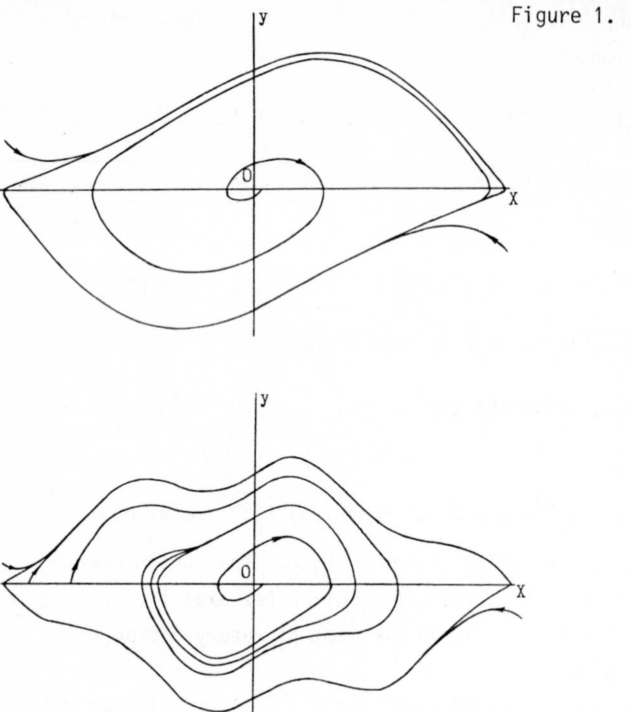


Figure 1.

Figure 2.

this fact, let us consider the scenario the appearance of attracting a non-trivial set (quasi-attractor-chaos) in Eq.(7) when $c \neq 0$.

Let $\varepsilon = 1$, $a = a_{\star} = (-1)^n/(4n^2-1)$, $\nu = 5$ and let parameter c change. In this case we shall see the trajectories of the Poincaré mapping on the display of the personal computer (see Figs.3-5, where the region $|x| \leq \pi$, $|y| \leq 3$ is shown).

When $n = 1$ and c increases from 0, a quasi-attractor appears instead of a stable saddle limit cycle (Fig.3). We call this quasi-attractor a saddle one. Then let us consider the case $n = 3$. If c changes from 0 to $c_{\star} \approx 0.025$, a saddle quasi-attractor exists in (7). When $c \in (c_{\star}, c_{\star\star})$, where $c_{\star\star} \approx 0.72$, there is no saddle quasi-attractor though we have here the Poincaré homoclinic structure, and Eq.(5) has a stable saddle limit cycle (see Fig.4, where $c = 0.5$ and a stable resonance regime of the period $12\pi/5$ is an attracting set of points 1-6). This fact is explained by the existence of two more limit cycles in Eq.(5). With the increase of c, the width of the neighbourhood of a homoclinic contour becomes also larger. When $c = c_{\star}$ it absorbs an unstable cycle and becomes unstable inside. Further, when $c = c_{\star\star}$, this neighbourhood absorbs the internal stable cycle and becomes stable again. In this case Eq.(7) has a saddle quasi-attractor with a developed chaos (a fractional part of the Lyapunov dimension ~0.6 when $c = 1$; see Fig.5). The values $c_{\star}$, $c_{\star\star}$ denote the crisis of bifurcation /12/. When $n > 3$ is odd, the bifurcations of a quasi-attractor obey the scenario described

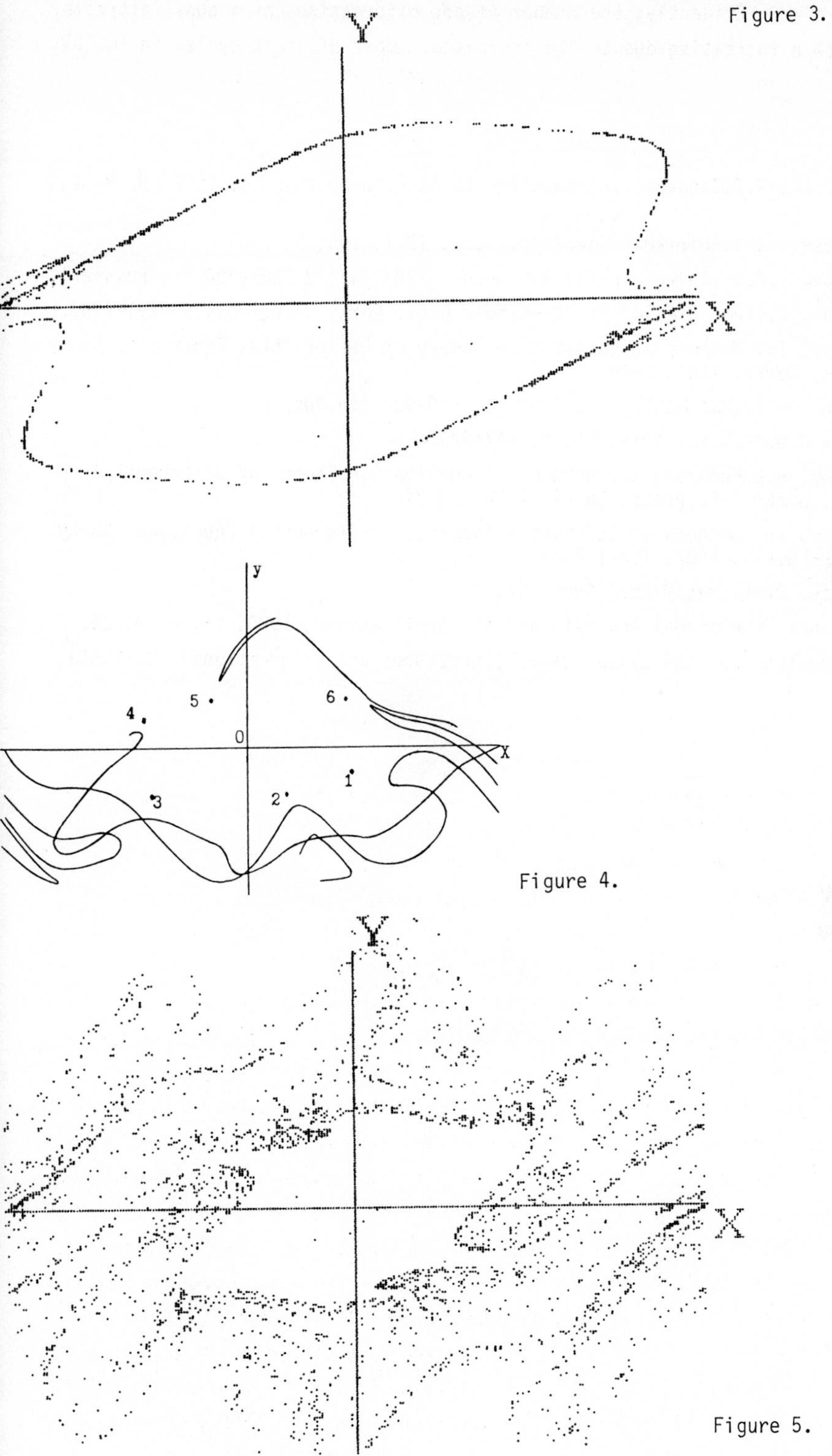


Figure 3.

Figure 4.

Figure 5.

for the case n = 3. Evidently, the number of the bifurcations of a quasi-attractor increases with n increasing due to the increased number of limit cycles in Eq.(5).

REFERENCES


1. G.M.Zaslavski, R.Z.Sagdeev. Introduction to Non-Linear Physics. USSR, M.:Nauka, 1988.
2. G.M.Zaslavski, B.V.Chirikov. Sov.Phys.-Usp., 1971, 105, 1, 3-39.
3. A.D.Morozov, L.P.Shilnikov. Prikl.Mat.Mekh., 1983, 47, 3, 385-394 (in Russian).
4. A.D.Morozov. Systems Similar to Non-Linear Ones. Gorky Univ.Press,Gorky, 1983.
5. A.D.Morozov. In: Methods of Qualitative Theory of Differential Equations. Gorky Univ.Press, Gorky, 1980, 3-16.
6. N.N.Bautin. Prikl.Mat.Mekh., 1969, 33, 6, 969-988 (in Russian).
7. C.M.Leech. J.Mech.Sci., 1979, 21, 9, 517-525.
8. A.D.Morozov, E.L.Fedorov. In: Methods of Qualitative Theory of Differential Equations. Gorky Univ.Press, Gorky, 1982, 20-34.
9. A.D.Morozov. In: Methods of Qulitative Theory of Differential Equations. Gorky, Univ.Press, Gorky, 1987, 113-127.
10. A.D.Morozov. Prikl.Mat.Mekh., 1989, 53, 4.
11. A.N.Varchenko. Functional Analysis and its Applications, 1984, 18, 2, 14-25.
12. V.S.Afraimovich. In: Non-Linear Waves (Structures and Bifurcations). M.:Nauka, 1987, 189-213.

# Synergetics. New Trends

*S.V. Ershov, S.P. Kurdyumov, G.G. Malinetskii, and A.B. Potapov*

M.V. Keldysh Institute of Applied Mathematics, USSR Academy of Sciences, Miusskaya Sq. 4, 125047 Moscow, USSR

In the present paper several trends in synergetics are discussed: (i) the inner properties of nonlinear media and the laws of organization of dissipative structures; (ii) comparison of theoretical predictions with experimental data for systems with chaotic behaviour; (iii) new basic models of nonlinear phenomena.

## 1. INTRODUCTION

In the early period of synergetics it was discovered that the behaviour of many open nonlinear physical, chemical, biological and hydrodynamic systems have common features. Among them are formation of dissipative structures /1/, effective reduction of the number of degrees of freedom that describe processes in nonlinear media (the existence of order parameters) /2/, chaotic regimes near the instability threshold (few-mode chaos) /3/ and others.

The development of these concepts showed that they are not only of heuristic value but can help in creating mathematical formalism for the analysis of different nonlinear equations. Besides, they promote the creation of new computational methods for investigating nonlinear systems and processing experimental data. New mathematical approaches in their turn led to a paradoxical view on some classical probelms and to the formulation of some principal questions that probably will be under active investigation in the nearest future.

In this paper we pay attention to three trends which attract more and more investigators. The first trend deals with the inner properties of nonlinear media and the laws of organization that determine how the simplest dissipative structures can be united into complex ones. The second trend is connected with the problem of comparing theoretical predictions and experimental data for systems with chaotic behaviour. Its solution can provide efficient application of synergetics to specific systems. The third trend is a search for new basic models of nonlinear phenomena. We shall discuss some results connected with each trend.

## 2. SPATIAL ORDERING

Most investigations on self-organization beginning with the classical paper by TURING /4/ on the modelling of morphogenesis traditionally lead to the formulation of some questions. What is a complete set of structures which can exist in a medium

Research Reports in Physics **Nonlinear Waves 3**
Editors: A.V. Gaponov-Grekhov · M.I. Rabinovich · J. Engelbrecht

under investigation? Do they depend on boundary conditions or are they determined only by the inner properties of the system? How are the simplest structures in this medium related to the more complex ones? These questions are especially interesting for the spatially multidimensional case. Yet for most reaction-diffusion models these problems remain unsolved. However, for two nonlinear systems these questions were investigated thoroughly and some important results were obtained.

The first system is often called the model of heat structures /5,6,7/. In the simplest case this describes the combustion of a nonlinear medium, the heat conductivity coefficient k(T) and nonlinear bulk source Q(T) of which being powers of temperature

$$T_t = \mathrm{div}(k(T)\nabla T) + Q(T)\ , \quad T(\vec{r},0) = T_0(\vec{r})\ ,$$
$$k(T) = k_0 T^{\sigma}\ , \quad Q(T) = q_0 T^{\beta}\ , \quad k_0,\ q_0,\ \beta,\ \sigma > 0\ , \quad \beta > \sigma + 1\ . \tag{1}$$

In such a medium the nonstationary dissipative structures can arise which develop in a peaking regime. Their amplitude infinitely increases during finite time $t_f$ which is the peaking time. In the two-dimensional case they are described by self-similar solution

$$T(\vec{r},t) = g(t)\cdot y^{\alpha}(\xi)\ , \quad \xi = \vec{r}\cdot\psi(t)\ , \quad \alpha = (\sigma+1)^{-1}\ .$$

Function y defines the form of structure and satisfies the nonlinear elliptic equation

$$\alpha\Delta y - \frac{\beta-\sigma-1}{2t_f(\beta-1)}\cdot(\xi,\nabla(y^{\alpha})) + y^{\alpha\beta} - \frac{1}{t_f(\beta-1)}\cdot y^{\alpha} = 0$$
$$y(0) < \infty, \quad y \to 0 \text{ and } |\nabla y| \to 0 \text{ as } |\xi| \to \infty \tag{2}$$

To construct configurations describing complex structures (i.e. with several maxima), it is natural to use numerical methods. This approach, however, requires good initial approximation which can be found by means of the method proposed in /8,9/ and based on the conception of order parameters.

The characteristic form of the solution of equation (2) is shown in Fig.1. Within the region where $y(\xi)$ is non-monotonous, it slightly oscillates about the plane "plateau" and its behaviour is characterized by the linearized equation. There is also the asymptotic region where $y \to C(\phi)\cdot\xi^{-}$, $p = 2(\sigma+1)/(\beta-\sigma-1)$. To determine function $C(\phi)$ by traditional methods, one must solve a free boundary problem. The hypothesis about the symmetry of the self-similar solution and the existence of the finite number of order parameters enables not to consider continuous function $C(\phi)$ but to match the solution of linearized equation with asymptotics on several rays. This method provides the hierarchy of approximate solutions

$$\tilde{y}(\xi,\phi_j) = \begin{cases} 1 + \sum_i \gamma_i\cdot R_m(\xi)\cdot\cos(im\phi_j)\ , & \xi < \xi_{0j} \\ C_j\cdot\xi^{-p} & , \ \xi > \xi_{0j} \end{cases} \qquad j = 0,1,\dots,k-1\ , \tag{3}$$

Figure 1.

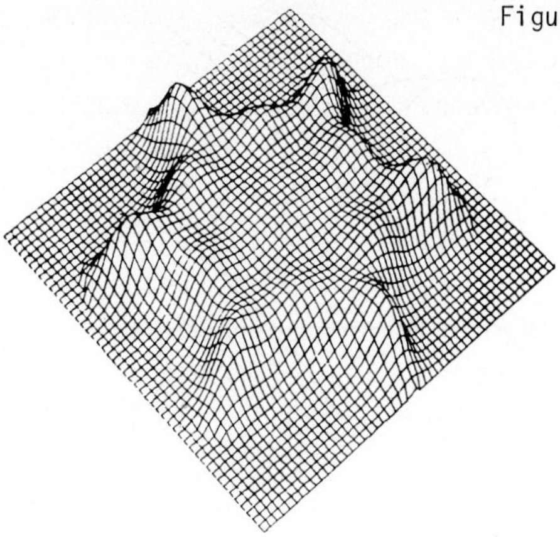

where $\phi_j$ define the directions of k rays and $\tilde{y}(\xi,\phi_j)$ is the value of the approximate solution on them. Outside of the rays the solution is determined by interpolation. This approximation provides useful information about the class of two-dimensional solutions of (2) and enables to construct the significant number of solutions numerically /9/. It is interesting that the increase of the number of harmonics k for $k \geqq 2$, in contrast with the case of the Fourier series of the Galerkin approximations, does no improve but breaks the accuracy of approximation /10/.

Thus, in the simple nonlinear medium the complex organization can exist. There is the finite number of configurations preserving their form during the evolution process. They can be interpreted as several simple single-maximum structures with different heights united into one complex structure. The principles of such a unification define the solutions of the problem (2). One cannot construct more complex ordering in the medium with given $\beta$ and $\sigma$ values. In some problems this effect is of fundamental importance because the setting of initial data can be considered as the way of control of processes in nonlinear media.

As a rule, in physical research, self-similar solutions can be determined from a system of ordinary differential equations. In spatially multidimensional case a more complex situation occurs - one must solve partial differential equations (here problem (2)). Similar problems arise in the course of investigating autowave processes in nonlinear media. Apparently, such solutions will be under active investigation in near future. Let us also note that the conception of order parameters proved to be useful not only for evolutionary problems but for the elliptic equations.

The laws of organization of dissipative structures in spatially multidimensional case were found for the class of trigger media which are described by the reaction-diffusion systems

$$\vec{u}_t = D\Delta\vec{u} + \vec{F}(\vec{u}) , \quad \vec{u}(\vec{r},0) = \vec{u}_0(\vec{r}) , \tag{4}$$

where $\vec{}$ is the vector with components $u_1,\ldots,u_N$, D is a diagonal matrix. Let us sup-

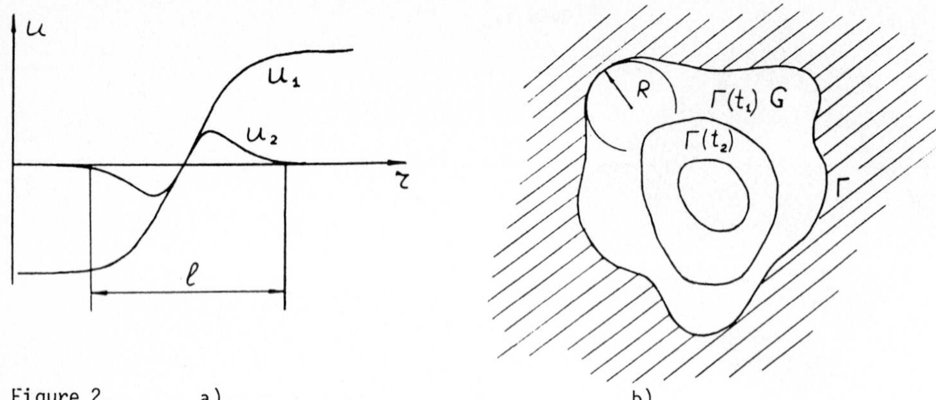


Figure 2. a) b)

pose that $\vec{F}(\vec{u}) = -\vec{F}(-\vec{u})$, and if $u_i = \text{const}$, $i = 1,\dots,N$, the system has two stable solutions (backgrounds) $\vec{u}_s$ and $-\vec{u}_s$ and one unstable $u = 0$.

Earlier the one-dimensional problem has been investigated and the set of stationary structures has been constructed /11/. The simplest structure called elementary is the region of transition of $\vec{u}_{\sim}(x)$ from one stable background to another ($\vec{u}_{\sim} \to \pm\vec{u}_s$, $x \to \pm\infty$, see Fig.2a). It arises as a result of evolution of initial data $\vec{u}_0(x) = \vec{u}_s \cdot \sin(x)$. Qualitatively more complex structures are the set of elementary ones separated by regions where $u(x)$ is close to one of the stable backgrounds.

In the multidimensional case the situation is substantially different. Let $\vec{u}_0 = \vec{u}_s$ within region G and $-\vec{u}_s$ outside of it (Fig.2b). For the sake of simplicity, let us call the background $\vec{u}_s$ black and that of $-\vec{u}_s$ white. Numerical calculations show that when the characteristic size of G is significantly greater than $\ell$ (Fig.2a) along the contour $\Gamma$, the transition area arises. Along the normal to the contour $\Gamma$ it is close to the elementary structure $\vec{u}_{\sim}(r)$. The contour $\Gamma(t)$ (where $u_1(\vec{r},t) = 0$) decreases and $\vec{u} = -\vec{u}_s$ in the whole space when $t > t'$. Using the closeness of the transitional area to the elementary structure, one can estimate $t'$. The shortening of contour $\Gamma$ occurs because of different conditions for the black and white backgrounds and the term $R^{-1}\cdot\partial\vec{u}/\partial R$ in diffusion operator (written in local polar coordinates, R is the contour curvature radius) proves to be noncompensated. It enables one to get an approximate equation for the motion of contour $\Gamma(t)$ points

$$\dot{\vec{r}} = -\vec{n}\cdot R^{-1}\cdot C\,, \quad \vec{r}(t) \in \Gamma(t)\,, \tag{5}$$

where C is the parameter depending on the medium properties, $\vec{n}$ is the external normal to the contour.

In particular, expression (5) implies that the existence time of a black circle with the initial radius $R_0$ on white background depends on $R_0$ as $R_0^2$ if $R_0 \gg \ell$. In the three-dimensional space a black ball would exist during twice less time; that is in good correspondence with numerical results.

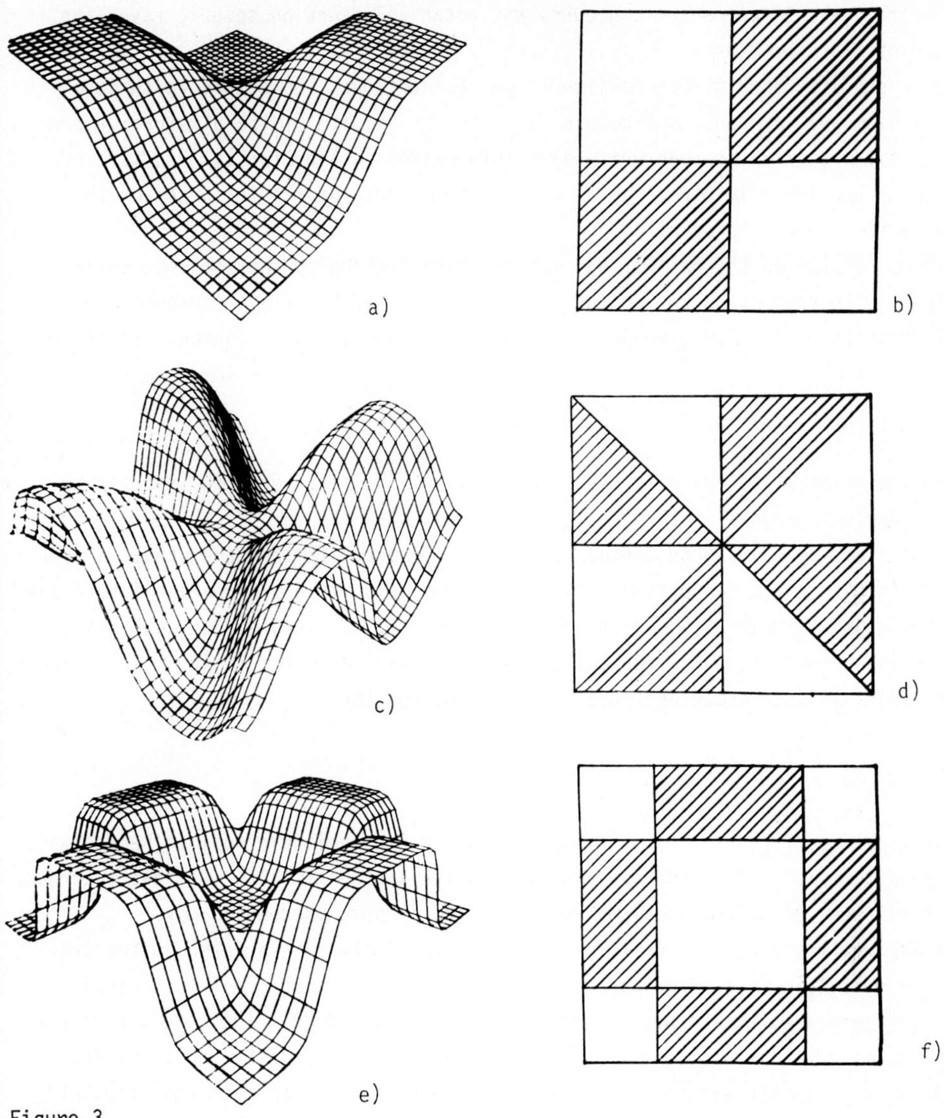

Figure 3.

In this medium stationary structures can also exist. System (4) does not change its form under transformation $\vec{u}_s \rightarrow -\vec{u}_s$ (black on white). Let us consider for the sake of simplicity a square region and let us set black and white into the same conditions, e.g. like in Fig.3b. The convergence to black or white in this case would mean breaking the symmetry of equation (4). Numerical calculations show that evolution of such initial data results in the formation of the stationary structure "the Cross" (Fig. 3a). Other initial data shown in Fig.3c,e result in the formation of the structures "the Mill" and "the Parquet" (Fig.3c and 3e). The system "forgets" details of initial data if they have the definite symmetry. To describe the form of such structures one can linearize equations near backgrounds and match solutions on several rays like in the case of heat structures.

Let us pay attention to the configuration "Parquet". It shows that in this nonlinear medium one can construct complex structures. Despite of the dissipative properties of the medium, one can write down information setting initial profiles of different type. The only condition is the correspondence of the influence on the medium to its internal properties.

We have considered the medium that can be characterized by the simplest colour symmetry - with respect to the change of black on white. Media with more colours are of great interest. For example, if in a two-component medium sources are of the following form

$$F_1 = (u_1-u_2)(3-2u_1^2-u_2^2) , \quad F_2 = (u_1+u_2)(3-2u_2^2-u_1^2) ,$$

then for describing it, one must have four colours. The structures in it prove to be more complex.

The possibility to use the methods of colour symmetry /12/ for the study of dissipative structures is unexpected. It can be supposed that in the nearest future the nonlinear media with more complex types of symmetry will be found. The possibility to use this approach also for the analysis of autowave processes is of great interest. Some examples of such ordering types can be found in /13/.

## 3. DYNAMICS

Above we have discussed complex spatial ordering in nonlinear media. The analysis of complex temporal order and chaos in dissipative systems is also of great interest. As we shall see, these two problems are connected in some unexpected way.

Chaotic regimes are very sensitive to initial conditions. Two neighbouring trajectories quickly diverge from each other. Let us suppose that the investigated phenomenon possesses this property and we have its ideal model. Then the difference between theoretical predictions and experiment will also increase with time. The reason of it is in the nature of investigated phenomenon rather than the drawbacks of the model. The question arises, how to compare theory and experiment in this case.

For the description of such phenomena a number of qualitative characteristics can be used. In the simplest cases power spectrum provides enough information about the system. In more complex cases one must calculate fractal dimensions of the attractor of the system. For this purpose the standard procedure of attractor reconstruction can usually be used: one expands the time series $x(t)$ into a vector time series

$$\xi(t) = \{x(t),x(t+\Delta t),...,x(t+(n-1)\Delta t)\}$$

and then determines the fractal dimension of the set $\{\xi\}$ for sufficiently large n /14/.

In such a way the great number of different experiments have been processed. As a rule, the correlation exponent v was calculated using the formulae

$$C(\varepsilon) = N^{-2}\cdot\sum_i \sum_j \nu(\varepsilon-|\xi_i-\xi_j|) \ , \quad C(\varepsilon) \cong \varepsilon^{v} \text{ as } \varepsilon \to 0 \ , \tag{6}$$

where $\nu$ is the Heaviside step function /15/.

To compute $C(\varepsilon)$ it is possible to apply several efficient numerical methods /16/. But then a number of serious problems arise. For example, in paper /17/ the following estimate of the length of time series N is given. To calculate v with the 5% accuracy one must have $N \cong 42^M$ points where M is an integer part of attractor dimension d. That is, if $d \cong 3$, it is necessary to have about 70 000 data points, but if $d \cong 10$ (which is characteristic of many hydrodynamic systems), then $N \cong 10^{16}$. The size of available time series is sufficiently less. Such a situation is in methodology (weather and climate attractor /18,19/) and in geophysics (determination of geodynamic attractor from the data on inhomogeneity of Earth rotation /20/). The number of points does not exceed several hundreds. The equation of what information can be extracted from small time series is still unsolved. Besides, the presence of even small noise makes this problem significantly more complex /20/.

The fractal dimension characterizes geometric properties of the attractor but does not give clear idea of the dynamics in the system. The fundamental problem is how to construct a mathematical model of a phenomenon from experimental data and attractor dimension. New ideas and qualitative conceptions about chaotic regimes in nonlinear systems are necessary.

The calculations show that in some reaction-diffusion models and differential-delay equations one can use a simplified description of the system. For example, it can be divided into a number of weakly interacting subsystems with chaotic behaviour.

The same problem as in the analysis of spatial structures arises here - to understand how several subsystems with known dynamics can be "matched" together to form a complex chaotic attractor.

## 4. STATISTICAL PROPERTIES

Most of low-dimensional systems having been studied recently are obtained as a truncation of an original infinite dimensional systems. Besides, there are models admitting no satisfactory approximations by few ordinary differential equations or mappings in a low dimensional phase space. So it is important to find out qualitative differences between low- and infinite-dimensional systems. The following questions arise. We have said that for some systems there is no satisfactory low-dimensional approximation. Nevertheless, perhaps there is an approximation describing statistical properties of the original system rather than its dynamics? Then how the behaviour becomes complicated if the dimension of an attractor increases infinitely as one varies the model parameter? Is there any simple order in such a behaviour?

For the study of such problems we have chosen the differential-delay equation

$$\varepsilon\dot{u}(t) + u(t) = f(u(t-1)) , \quad 0 \le t < \infty . \tag{7a}$$

It is equivalent to the infinite-dimensional mapping

$$\varepsilon\dot{u}_{n+1}(t) + u_{n+1}(t) = f(u_n(t)) , \quad u_{n+1}(0) = u_n(1) , \quad 0 \le t < 1 , \tag{7b}$$

where $u_n$ denotes the solution u(t) on the interval $\left[n,n+1\right]$. We have chosen $f(x) = 1 - a|x|$.

Such equations describe systems with delayed feedback that respond to the input stimulus with some delay. They arise in a control theory, economics, biology and nonlinear optics /21,22/.

Integration of equation (7) is relatively easy while its behaviour is rather complicated, i.e. a lot of Lyapunov exponents are positive and the dimension of the attractor increases as $O(1/\varepsilon)$ /22/.

Slightly altering the basic equation (7), we obtain a model with external excitation

$$\varepsilon\dot{u}_{n+1}(t) + u_{n+1}(t) = f(u_n(t)) , \quad u_n(0) = z_n , \quad 0 \le t < 1 . \tag{8}$$

It was called so because of the fact that if $z_n$ = const, the solution tends to stable "stationary function": $u_n(t) \xrightarrow{t\to\infty} u_*(t)$ and chaos can exist only if the "external force" is applied.

Now let us discuss the basic model (7) when $\varepsilon \to 0$. It is obvious from Fig.4 that the smaller $\varepsilon$, the more ragged the solution $u_n(t)$. Moreover, it is possible to prove that $||u_n||_{\mathbb{C}} = O(1/\varepsilon)$. However if we consider the solution in a "stretched" time

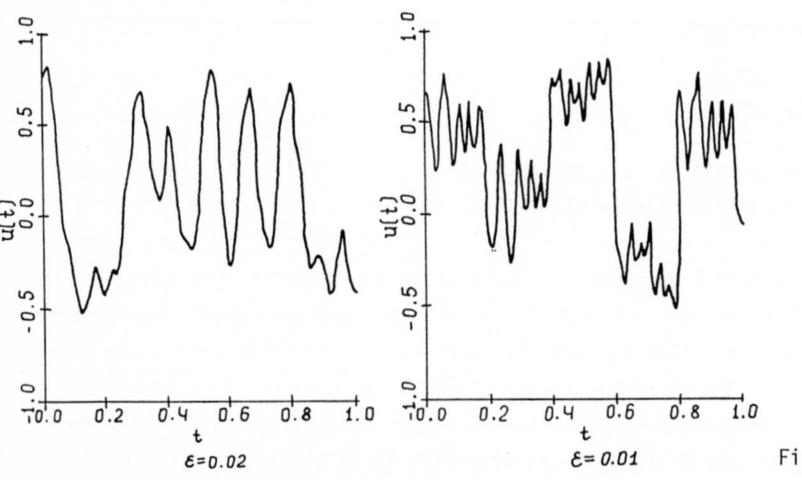


Figure 4.

$t' = t/\varepsilon$ , a number of its statistical properties would be independent from $\varepsilon$ for $\varepsilon \to 0$. For example, let $p_{k.\varepsilon}(u(t_j),\dots,u(t_{j+k}))$ be the density of the invariant distribution of k variables $\{u(t_j),\dots,u(t_{j+k})\}$. Then, if the time intervals $\Delta t_j = t_j - t_{j+1}$ are chosen in the $\varepsilon$-scale (i.e. $\Delta t_i = \nu_i\varepsilon$), some theoretical investigations lead to the estimate

$$||p_{k,\varepsilon} - p_{k,\varepsilon'}|| = O(\exp\{-C/\max(\varepsilon,\varepsilon')\}) \; . \tag{9}$$

It is in a good agreement with numerical results: the distribution p(u(t)) (Fig.5a) as well as p(u(t),u(t+2ε)) (its level lines are plotted in Fig.5b) seems to be almost idependent from ε.

Such a feature results from the fact that with a good accuracy $u_{n+1}(t)$ can be evaluated from $u_n(t')$ for, say, $t-10\varepsilon \leq t' \leq t$ :

$$u_{n+1}(t) = \int_0^{10} e^{-\xi} f(u_n(t-\varepsilon\xi))d\xi + O(e^{-10}) \; . \tag{10}$$

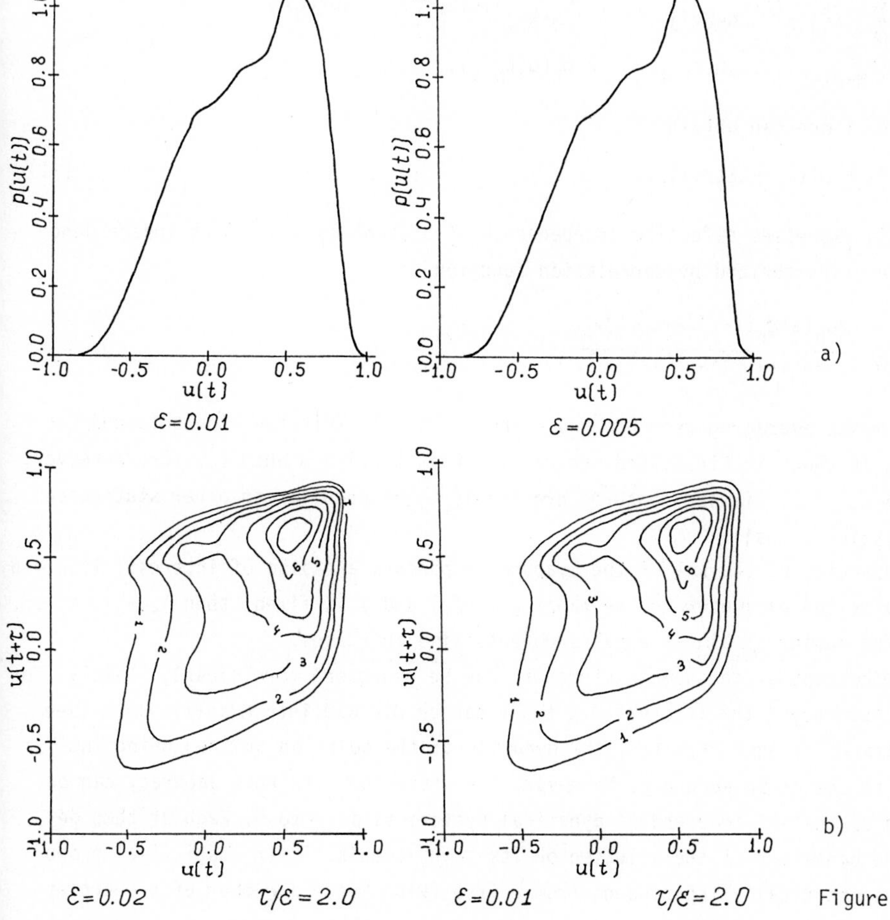


Figure 5.

If we replace t by $\tau \equiv t/\varepsilon$ and $u_n(\tau)$ by $v_n(\tau) \equiv u_n(\tau\varepsilon)$, we will be able to exclude $\varepsilon$ from equation (10). Hence, for $\varepsilon$ small enough the behaviour of the solution on a distant time intervals is almost independent and the "global" interval is virtually partitioned into $O(1/\varepsilon)$ domains.

To describe the domain structure quantitatively, let us divide the interval $[0,1]$ into N ones $[t_i,t_{i+1}] = ih$, $i = 1,\ldots,N$, of the length h and consider two domains $[t_{N-2k-j},t_{N-k-j}]$ and $[t_{N-k},t_N]$ separated bv a time interval $\Delta t = jh$. Now let

$$p_k(u(t_{N-2k-j}),\ldots,u(t_{N-k-j})) \quad \text{and} \quad p_k(u(t_{N-k}),\ldots,u(t_N))$$

be the distribution over each other of the domains (function $p_k$ is the same for all domains). Then let

$$p_{j,k}(u(t_{N-2k-j}),\ldots,u(t_{N-k-j}); \quad u(t_{N-k}),\ldots,u(t_N))$$

be the joint distribution over both domains. Then the error of its approximation by independent probability distribution is

$$s_{j,k}(') \equiv (p_{j,k}(u(t_{N-2k-j}),\ldots,u(t_{N-k-j});u(t_{N-k}),\ldots,u(t_N)) -$$
$$- p_k(u(t_{N-2k-j}),\ldots,u(t_{N-k-j})) \times p_k(u(t_{N-k}),\ldots,u(t_N))$$

and for $\varepsilon \ll 1$ one can obtain:

$$||s_{j,k}|| = O(\exp\{-C\Delta t/\varepsilon\}) \ . \tag{11}$$

This formula expresses effective independence of distant domains. Such independence can also be characterized by correlation function

$$C_\varepsilon(\tau) = \frac{\langle u_n(t)u_n(t-\tau)\rangle - \langle u_n(t)\rangle^2}{\langle u_n(t)^2\rangle - \langle u_n(t)\rangle^2}$$

where $\langle'\rangle$ means averaging over $0 \leq n \leq \infty$ and $0 \leq t \leq 1$. This function obtained numerically, is shown in Fig.6, and one can see that $C_\varepsilon(\tau) \cong 0$ when $\tau \geq 10\varepsilon$. Moreover, $C_\varepsilon(\tau) \xrightarrow{\varepsilon\to 0} C_*(\tau/\varepsilon)$. Similar features are likely to be peculiar to other statistical characteristics as well.

The properties of the set of the Lyapunov exponents are also of interest. If $\varepsilon \to 0$, $a = 1.9$ (a is the parameter in the mapping f (7)) and i is fixed, then $\lambda_i \to \lambda_* \cong 0.3$. Besides, the number of almost equal exponents also increases.

Now the concept of the domain structure can be expressed more clearly. Namely, for any given accuracy $\delta$ the interval $0 \leq t \leq 1$ can be divided into $M \cong 1/|\varepsilon\cdot\ln\delta|$ (see (11)) domains of length $\Delta t = 1/M$, the dynamics of the solution on them being independent with the given accuracy. Moreover, the attractor with this accuracy can be considered as that of independent dynamical systems similar to M. Each of them describes the behaviour of the solution on its "own" domain. Their statistical properties are asymptotically independent from $\varepsilon$ (see (9)). Hence a number of properties

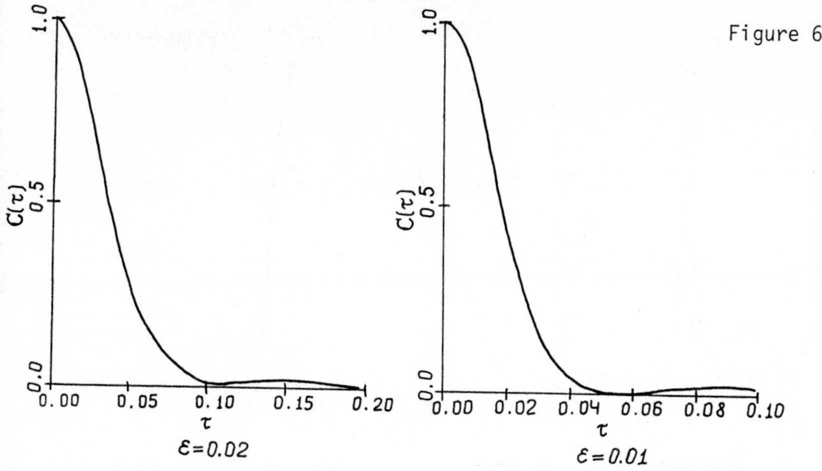


Figure 6.

of the model (7) for $\varepsilon$ small enough (when the attractor dimension $D = O(1/\varepsilon)$ is sufficiently large) can be estimated from its investigation for greater $\varepsilon$, when the dimension is smaller and the numerical analysis is possible. It simplifies the study of such systems.

Now let us pay attention to the fact that the difference between models (7) and (8) is only in the "initial conditions" which according to (10) will be "forgotten" when $t \gg \varepsilon$. So one can expect that if $t \gg \varepsilon$, the statistical properties of $\{u_n(t)\}$ in both systems will be similar. To be convinced of this, it is enough to show that the specific character of excitation $\{z_n\}$ is not important when $\varepsilon \to 0$. Equation (7b) is formally model (8) with the excitation $z_n = u_{n-1}(1)$. If for given $\{z_n\}$, one varies $\varepsilon$, then

$$||p_{k,\varepsilon} - p_{k,\varepsilon'}|| = O(\exp\{-C/\max(\varepsilon,\varepsilon')\}) , \tag{12}$$

where $p_{k,\varepsilon}(u(t_{N-k}),\dots,u(t_N))$ is the density of $\{u(t_{N-k}),\dots,u(t_N)\}$ distribution. The typical dependence of $p(u(1))$ on $\varepsilon$ is shown in Fig.7a, where $z_n$ is the uniformly distributed random noise. One can see that for the fixed excitation amplitude $\delta z = \max z_n - \min z_n$ this distribution is paractically independent from . On the contrary, if $\varepsilon$ is fixed while $\delta z \to 0$, the distribution can tend to another limit (see Fig.7b).

It should be stressed that model (8) is not only the transformer of chaos that converts the "input" (i.e. at $t = 0$) noise signal $\{z_n\}$ into its "own" chaotic $\{u_n(t)\}$ on "output" (i.e. at $t \cong 1$). It is also an "amplifier of chaos" because the "input amplitude" satisfies $\delta z \ll 1$ while the "output amplitude" is about 1, the oscillations being amplified gradually. The average amplitude $(Du(t))^{1/2} = (\langle\{u_n(t)-\langle u_n(t)\rangle\}^2\rangle)^{1/2}$ increases as $Du(t) \propto \delta z\cdot\exp(t/\varepsilon)$ when $t/\varepsilon \gg 1$ and $Du(t) \gg 1$. This exponential law holds till $Du(t)$ becomes large enough and then "saturation" occurs. It means that the properties of the "output" oscillations are determined by the internal structure of the system rather than the properties of "input signal".

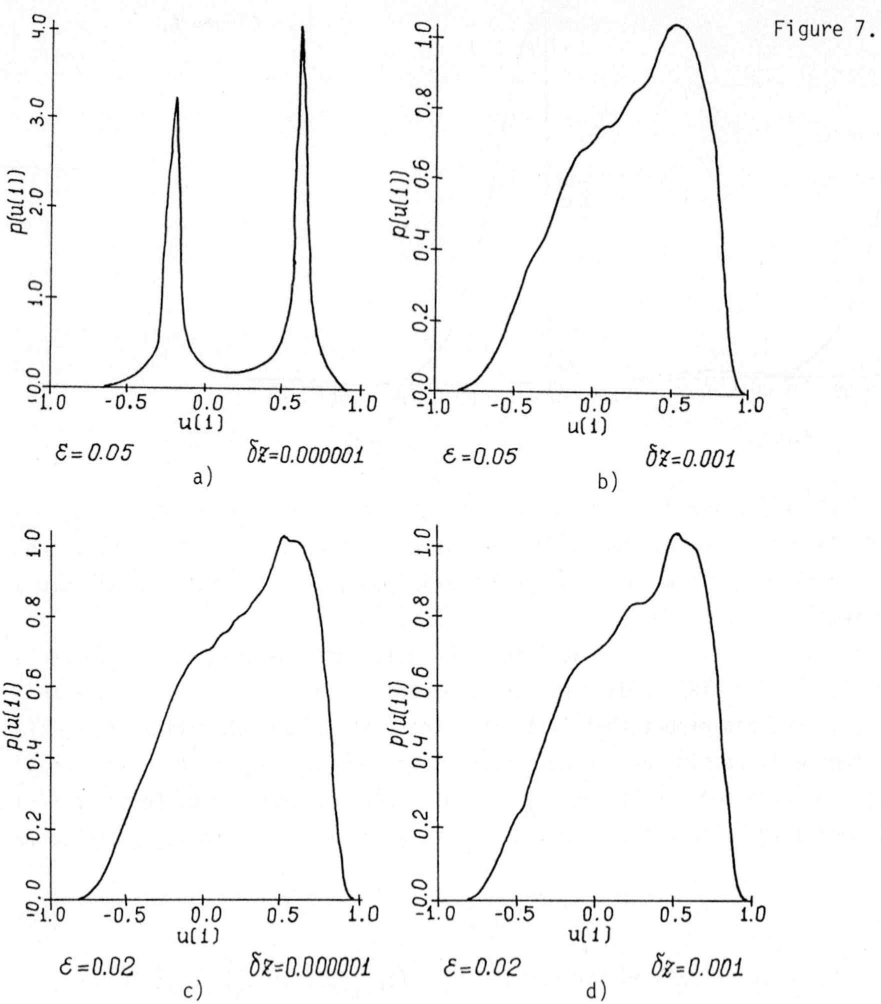


It can be expected that similar features would be found in a number of spatio-temporal systems with a chaotic behaviour if the spatial correlations decrease quickly.

## REFERENCES

1. G.Nicolis, I.Prigogine. Self-Organization in Non-Equilibrium Systems. New York, J.Wiley, 1977, 512 p.

2. H.Haken. Synergetics. In Introduction. Berlin etc., Springer, 1978, 404 p.

3. Strange Attractors. Moscow, Mir, 1981, 256 p. (in Russian).

4. A.Turing. The Chemical Basis of Morphogenesis. Phil.Trans.Roy.Soc. London, 1952, B237, 37-72.

5. A.A.Samarskii, V.A.Galaktionov, S.P.Kurdjumov, A.P.Mikhailov. Regimes with Peaking in the Problems for Quasilinear Parabolic Equations. Moscow, Nauka, 1987, 480 p. (in Russian).

6. S.P.Kurdjumov. The eigenfunctions of nonlinear medium combustion and constructive laws of its organization. In: Contemporary Problems of Mathematical Physics and Computational Mathematics. Moscow, Nauka. 1982, 217-243 (in Russian).

7. Reviews on Science and Technology. Contemporary Problems of Mathematics. Newest Achievements. V.28, Moscow, VINITI, 1987, 316 p. (in Russian).

8. S.P.Kurdjumov, E.S.Kurkina, A.B.Potapov, A.A.Samarskii. The architecture of multidimensional heat structures. Dokl.Akad.Nauk SSSR, 1984, 274, 5, 1071-1075 (in Russian).

9. S.P.Kurdjumov, E.S.Kurkina, A.B.Potapov, A.A.Samarskii. Complex multidimensional structures of nonlinear medium combustion. Zh. Vychisl.Matem. i Matem.Fiz., 1986, 26, 8, 1189-1205 (in Russian).

10. A.B.Potapov. The Combustion of Two-Dimensional Nonlinear Medium Eigenfunctions. Preprint No.8, M.V.Keldysh Inst. of Appl. Math., USSR Acad. Sci., Moscow, 1986, 26 p. (in Russian).

11. S.P.Kurdjumov, G.G.Malinetskii, Yu.A.Poveschenko et al. Dissipative structures in trigger media. Differentsialnye Uravneniya, 1981, 17, 10, 1875-1886 (in Russian).

12. L.M.Zamorzaev, E.I.Galyarskii, A.F.Palistrant. Colour Symmetry, Its Generalizations and Applications. Kishinev, Shtiintsa, 1978, 280 p. (in Russian).

13. T.S.Akhromeyeva, G.G.Malinetskii. On the symmetric solutions of Kuramoto-Tsuzuki equation. Differentsialnye Uravneniya, 1984, 20, 7, 1281-1283 (in Russian).

14. F.Takens. Detecting strange attractors in turbulence. In: Dynamical Systems and Turbulence. Warwick 1980. Berlin etc., Springer, 1981, 366-381.

15. P.Grassberger, I.Procaccia. Measuring the strangeness of strange attractors. Physica D, 1983, 9, 1-2, 189-208.

16. G.G.Malinetskii, A.B.Potapov. On the computation of strange attractors dimensions. Zh.Vychisl.Matem. i Matem.Fiz., 1988, 28, 7, 1021-1037 (in Russian).

17. L.A.Smith. Intrinsic limits on dimension calculations. Phys.Lett. A, 133, 6, 283-288.

18. C.Nicolis, G.Nicolis. Is there a climatic attractor? Nature, 1984, 311, 529-532.

19. K.Fraedrich. Estimating the dimension of weather and climatic attractors. J.Atmospher.Sci., 1986, 43, 5, 419-432.

20. S.M.Gizzatulina, G.G.Malinetskii, A.B.Potapov et al. The Dimension of Geomagnetic Attractor from the Data on the Length of Day Variations. Preprint No.95, M.V.Keldysh Inst. of Appl.Math., USSR Acad.Sci., Moscow, 1988, 25 p. (in Russian).

21. K.Ikeda, K.Matsumuto. High-dimensional chaotic behaviour in systems with time-delayed feedback. Physica D., 1987, 29, 223-235.

22. J.D.Farmer. Chaotic attractors of an infinite-dimensional dynamical systems. Physica D., 1982, 4, 366-393.

# Cellular Automata as Dynamical Systems

*V.S. Afraimovich and M.A. Shereshevskii*

Gorky State University, 23 Gagarin Ave., 603600 Gorky, USSR

Cellular automata (CA) are considered as discrete dynamical systems on a compact metric space. The separating trajectories of one-dimensional CA are proved to have positive entropy. A class of one-dimensional CA which are isomorphic to topological Markov chains is described.

## 1. INTRODUCTION

Recently cellular automata (CA) have been extensively investigated as the simplest mathematical models of nonequilibrium media /1-3/. The questions of growth, interactions and stability of structures have been explored (as a rule, numerically). But the attention paid to the fact that CA can be considered as a discrete dynamical system (see below) seems to be insufficient. Whereas, taking this possibility into account, we should be able to use such powerful tools as the theory of dynamical systems. The results obtained in the paper are based on this approach. Here we deal mainly with the "one-dimensional" CA, but some of our results admit a generalization to the higher dimensions.

## 2. DEFINITION OF THE CA

Let us connect with every integer point j on a line a variable $x_j$ (often called the site) which runs the finite set $\{0,1, \ldots, p-1\}$. Suppose that the value $x_j(n+1)$ of the site $x_j$ at time n+1 is determined only by the values $x_{j-r}(n), \ldots, x_j(n), \ldots, x_{j+r}(n)$ of 2r+1 nearest neighbours ($r \geq 1$) at the previous time n, moreover, this dependence is the same for all j:

$$x_j(n+1) = F(x_{j-r}(n), \ldots, x_j(n), \ldots, x_{j+r}(n)) . \tag{1}$$

Function F is called a rule. Formula (1) defines a map T of the space X of bilateral infinite sequences $\underline{x} = (\ldots, x_{-1}, x_0, x_1, \ldots)$. We provide X with the distance

$$\mathrm{dist}(\underline{x},\underline{y}) = \sum_{j=-\infty}^{+\infty} |x_j - y_j| 2^{-|j|} .$$

Note that map T is continuous, but , in general, irreversible. We call the dynamical system $\{T^n\}_{0 \leq n < +\infty}$ (or $\{T^n\}_{-\infty<n<+\infty}$ if T is invertible) on the compact X the one-dimensional cellular automaton, corresponding to the rule F.



Research Reports in Physics **Nonlinear Waves 3**
Editors: A.V. Gaponov-Grekhov · M.I. Rabinovich · J. Engelbrecht

Analogously, we connect with every point (i,j) of the two-dimensional integer lattice the variable $x_{ij}$ which runs over the set {0,1,...,p-1}. We introduce the distance

$$\mathrm{dist}(\underline{x},\underline{y}) = \sum_{i,j=-\infty}^{+\infty} |x_{ij}-y_{ij}|2^{-|i|-|j|}$$

on space X of infinite matrices $\underline{x} = (x_{ij})_{-\infty<i,j<+\infty}$ . Any function (rule)

$$x_{ij}(n+1) = F(x_{k\ell}(n);|k-i| \leq r_1,|\ell-j| \leq r_2) \tag{2}$$

($r_1,r_2 \geq 1$) defines a dynamical system on the compact X. Such a dynamical system is called the two-dimensional cellular automaton, corresponding to the rule F. Similarly one can define the three-dimensional CA etc..

## 3. CA AS A WIDE CLASS OF DYNAMICAL SYSTEMS

One could think that by limiting the number of "neighbours", upon which depends the next value of site $x_j$ (see the formulae (1), (2)), we fairly restrict the set of maps generating CA. However, the authors proved the following assertion:

Let X be the space of any dimension constructed in the previous section, T be a continuous map of X into itself. If T is "autonomous" in regard to the space coordinates (a mathematician would say that T commutes with the shifts along space coordinates) then T generates a cellular automaton of a corresponding dimension.

To illustrate the autonomity assumption we note that in the one-dimensional case when only one space coordinate is present and any map T may be given in the form

$$x_j(n+1) = F_j(\dots,x_j(n),x_{j+1}(n),\dots),$$

the autonomity of T means

$$F_{j+1}(\dots,x_j(n),x_{j+1}(n),\dots) = F_j(\dots,y_j(n),y_{j+1}(n),\dots) ,$$

where $y_j(n) = x_{j+1}(n)$.

## 4. DYNAMICAL CHAOS IN A CA

A dynamical system $\{T^n\}$ on a compact metric space X is said to be separating trajectories /4/, if there exists a positive constant $\varepsilon > 0$ such that the distance between any two trajectories which start from distinct points $x,y \in X$ becomes greater than $\varepsilon$ in some time n · dist $(T^n x,T^n y) > \varepsilon$ (when T is invertible, n may be negative) Value $\varepsilon$ is called the separating constant.

In accordance with the widespread approach, a dynamical system may be regarded as chaotic when it has positive Komogorov-Sinai entropy /5/ with respect to some invariant measure. There is another popular notion of entropy in the mathematical theory of dynamical systems. It is the so-called topological entropy (see /5/). Its value does not depend on the particular distribution in the phase space. It is a well-known fact proved by Goodwyn (see /5/ and references therein) that the topological entropy is the exact upper bound of the Kolmogorov-Sinai entropy taken over all invariant measures on the phase space.

There exist examples of smooth dynamical systems separating trajectories, which have zero topological entropy. But for the one-dimensional cellular automata the following statement is valid.

*Theorem 1.* Let $\{T^n\}$ be an one-dimensional CA separating trajectories. Then it has positive topological entropy: $h_{top}(T) > 0$.

We do not know how to recognize by means of rule F whether the CA generated by F, separates trajectories. However, below we formulate a certain sufficient condition for this property. The condition can be efficiently tested.

Let $F(x_{-r},\dots,x_0,\dots,x_r)$ be a rule (cf. (1)) and suppose that

(a) for any two strings $(x_1,\dots,x_{3r})$, $(x'_1,\dots,x'_{3r})$ of length 3r such that $x_i = x'_i$ for $1 \leq i \leq 2r$ and $x_{2r+1} \neq x'_{2r+1}$ there exists j, $r+1 \leq j \leq 2r$ such that

$$F(x_{j-r},\dots,x_j,\dots,x_{j+r}) \neq F(x'_{j-r},\dots,x'_j,\dots,x'_{j+r}) \ ;$$

(b) for any two strings $(x_{-3r},\dots,x_{-1})$, $(x'_{-3r},\dots,x'_{-1})$ of length 3r such that $x_i = x'_i$ for $-2r \leq i \leq -1$ and $x_{-2r-1} \neq x'_{-2r-1}$ there exists j, $-2r \leq j \leq -r-1$ such that

$$F(x_{j-r},\dots,x_j,\dots,x_{j+r}) \neq F(x'_{j-r},\dots,x'_j,\dots,x'_{j+r}) \ .$$

Then the CA corresponding to rule F separates trajectories.

Note that every rule belonging to class M (see the following section) satisfies the conditions (a) and (b) above.

## 5. CA AND THE TOPOLOGICAL MARKOV CHAINS

It is easy to see that the phase space X of a CA is homeomorphic to the Cantor set. There is the well studied calss of dynamical systems with the phase space of this kind. It is the so called topological Markov chains (TMC) /6/. It would be fruitful to pich out TMC among CA, and to make use of results of the TMC theory. We shall describe a class of cellular automata which are isomorphic to TMC with maximal (in some sense) chaotic properties.

We say that a rule F (and the corresponding CA) belongs to the class M, if for any $(x_{-r},\dots,x_{r-1})$ and $(x'_{-r+1},\dots,x'_r)$ functions $x \to F(x_{-r},\dots,x_{r-1},x)$ and $y \to F(y,x'_{-r+1},\dots,x'_r)$ are one-to-one correspondences.

It is not difficult to verify that the CA defined by the formula

$$x_j(n+1) = f(x_j(n)) + x_{j-1}(n) + x_{j+1}(n) \tag{3}$$

where f is any function of set $\{0,1,\dots,p-1\}$ into itself and the addition (mod p) is implied, belongs to class M.

*Theorem 2.* If rule F belongs to class M then the corresponding map T is topologically conjugated (see /4/) with the shift of a TMC, which has $p^{2r+1}$ states and the matrix permissible transitions A such that every component of the matrix $A^2$ is equal to $p^{2r-1}$.

The proof of Theorem 2 is based on the construction of Markov partition (cf. /7/) of the phase space X. Such a partition appears to be the one consisting of all possible cylinders

$$C\, x^*_{-r},\dots,x^*_r = \{\underline{x} \in X : x_j = x^*_j,\ |j| \leq r\} \ .$$

A number of helpful results follow from Theorem 2.

Let $T^n$ be a CA belonging to class M.

*Corollary 1.* Map T is topologically mixing.

*Corollary 2.* The topological entropy $h_{top}(T) = 2r\log p$.

*Corollary 3.* There exists the unique measure $\mu$ of maximal entropy (i.e. the measure, for which the Kolmogorov-Sinai entropy coincides with the topological one). Moreover,

$$\mu\{\underline{x} : x_i = x^*_i,\ j \leq i \leq j+\ell\} = p^{-\ell}$$

for any $(x^*_j,\dots,x^*_{j+\ell-1})$.

One can show that topological entropy of any (one-dimensional) CA is not greater than 2rlog p. Hence, the CA from class M, having the maximal entropy amongst CA, are (in some sense) the most stochastic ones.

## 6. CONCLUDING REMARKS

(i) In paper /8/ the so-called symplectic CA, obtained by the space and time discretization from the Hamiltonian system with $\sigma$-function force at the integer times, have been studied numerically. The one-dimensional CA corresponding to the rule

$$F(x_{j-1},x_j,x_{j+1}) = K(x_j) \pm (x_{j-1}-2x_j+x_{j+1}) \pmod 3,$$

where $x_j \in \{-1,0,1\}$ and $K : \{-1,0,1\} \to \{-1,0,1\}$ is any map, is considered. It is easy to see that such a CA belongs to class M (cf. (3)) and, therefore, the results of the last section are applicable to it. It must be mentioned, however, that index j runs in /8/ only over finite set $\{1,\dots,N\}$.

(ii) It is not difficult to pick out a class of two-(three- etc.) dimensional CA similar to class M and to generalize Theorem 2 to the multidimensional case. The only difference would be that in this case the partition consisting of the suitable "rectangles" would be not of Markov type but pre-Markov type. It means that the image of each "rectangle" may cover another "rectangle" more than once (see /6/).

*Acknowledgement*

We wish to thank M.I.Rabinovich for drawing our attention to this topic and for useful remarks he made during the preparation of the manuscript.

REFERENCES

1. J. von Neumann. Theory of Self-Reproducing Automata. A.W.Burks, ed. Univ. of Illinois, Urban, 1966.
2. S.Wolfram, ed. Theory and Applications of Cellular Automata. World Scientific, Singapore, 1986.
3. S.Wolfram. Statistical mechanics of cellular automata. Rev.Mod.Phys., 1983, 55, 601-644.
4. Z.Nitecki. Differentiable Dynamics. MIT Press, Cambridge-Massachusetts-London, 1971.
5. N.F.G.Martin, J.W.England. Mathematical Theory of Entropy. Addison-Wesley, Massachusetts, 1981.
6. V.M.Alekseev. Symbolic Dynamics. 11th Summer Math. School, Naukova Dumka, Kiev, 1976.
7. Ya.G.Sinai. Construction of Markov partitions. Funct.Anal. and its Appl., 1968,2, 3, 70-80.
8. K.Kaneko. Symplectic cellular automata. Phys.Lett., 1988, 129A, 9-16.

# An Approach to the Computation of the Topological Entropy

*A.L. Zheleznyak*

Institute of Applied Physics, USSR Academy of Sciences,
46 Ulyanov Str., 603600 Gorky, USSR

An approach for deriving the topological entropy of dynamical systems generated by one-dimensional piecewise-continous and piecewise-monotonous maps of an interval is proposed. The technique is based on the application of kneading theory and allows to reduce the calculations for finding a root of some polynomial which is the kneading determinant. Examples of computing the topological entropy including those from the experimental data are presented; the comparison with the Lyapunov exponent is given.

## 1. INTRODUCTION

As it is known, the topological entropy $h_{top}$ is one of the quantitative characteristics of chaos of dynamical systems. It was introduced in /1-2/ and characterizes the system's complexity in a sense of diversity of possible trajectories in contrast to the metric entropy $h_\rho$ (invariant of Kolmogorov-Sinai, see /3/). The topological entropy is the topological invariant of a dynamical system and possesses a number of properties, which may be useful while studying concrete applied problems. Thus, for a broad class of maps, the topological entropy depends continuously on governing parameters (/4,5/), whereas the singularity of the invariant measure can lead to discontinuous irregular behaviour of metric entropy under the action of small perturbations of the system (when the system is perturbed). According to /1,3/, $h_{top}(f) = \sup_\rho h_\rho(f)$ , where  is the invariant relative to f, normalized Borel measure, therefore the comparison of topological and metrical entropies may provide important information on the measure generated by the given dynamical system and its proximity to the measure with maximal entropy.

However, the topological entropy is rarely used in applications (it is possible to find $h_{top}$ directly from the definition only for model systems); it is associated with the lack of effective algorithms even for systems of small dimensions.

This paper proposes an approach to the computation of the topological entropy for dynamical systems, generated by piecewise-monotonous and piecewise-continuous maps of an interval. The type of a map can be given directly or can be reconstructed from one-dimensional time data. This approach is based on the results of kneading theory which can be found in /4-7/ and allows to reduce the calculating process to finding the minimum positive root of some polynomial which is the kneading determinant.

Research Reports in Physics **Nonlinear Waves 3**
Editors: A.V. Gaponov-Grekhov · M.I. Rabinovich · J. Engelbrecht

## 2. THE ALGORITHM OF COMPUTING THE TOPOLOGICAL ENTROPY

Consider a piecewise-monotonous and piecewise-continuous map $f : I \to I$ of the interval $I = [0,1]$. Let us divide I in m subintervals by points of the set $C = \{C_0=0, C_1,\dots,C_{m-1},C_m=1\}$ so that map f is strictly monotonous and continuous at every $I_j = (C_{j-1},C_j)$, $1 \leq j \leq m$. We shall put some fixed numbers $\gamma_j$ in accordance with subintervals $I_j$, then the sequence $\psi(x) = \{\psi_n(x) | \psi_n(x) = \gamma_j$, if $f^n(x) \in I_j$, $n \geq 0$ will correspond to any point $x \in I/D$, where $D = \bigcup_{n\geq 0} f^{-n}(C)$. Assume for any $x \in C$

$$\varepsilon(x) = \begin{cases} +1, & \text{if } f(x) \text{ increases in point } x, \\ -1, & \text{if } f(x) \text{ decreases in point } x; \end{cases}$$

For $x \in I/D$ we define, following /6/, the kneading sequence $K(x) = \{K_n(x) | K_n(x) = \varepsilon_n(x)\psi_n(x)$, where $\varepsilon_n(x) = \prod_{i=0}^{n-1} \varepsilon(f^i(x))$, $n \geq 1$, $\varepsilon_0(x) \equiv 1\}$ . This kneading sequence K(x) sets a map from I/D to the Bernoulli one-sided scheme $\Omega^+_{2m}$ of 2m symbols $\pm\gamma_1,\dots, \pm\gamma_m$.

Let

$$K^+(x) = \lim_{\substack{y \downarrow x \\ y \in I/D}} K(y), \quad x \neq 1;$$

$$K^-(x) = \lim_{\substack{y \uparrow x \\ y \in I/D}} K(y), \quad x \neq 0$$

(these limits exist for any point $x \in I$). We assume 2m different kneading sequences $K^+(0), K^-(C_1),\dots,K^+(C_{m-1}),K^-(1)$ and on their basis we define the formal power series of t: $\tilde{K}^+(0,t)$, $\tilde{K}^-(C_1,t),\dots,\tilde{K}^+(C_{m-1},t),\tilde{K}^-(1,t)$ according to formula

$$\tilde{K}(x,t) = \sum_{n=0}^{\infty} K_n(x)t^n, \text{ where } K(x) = \{K_n(x),\ n \geq 0\} \in \Omega^+_{2m} .$$

If we make further m-1 formal series $A_i(t) = \tilde{K}(C_i,t) - \tilde{K}^-(C_i,t)$, $1 \leq i \leq m-1$ and represent each of such series as $A_i(t) = N_{i1}(t) \cdot \gamma_1 + N_{i2}(t) \cdot \gamma_2 + \dots + N_{im}(t) \cdot \gamma_m$, we shall get kneading matrix $(N_{ij}(t))$ of the order of $(m-1) \times m$ with elements of the ring of power series of t with integer coefficients.

Let $M_i(t)$ be the determinant of $(m-1) \times (m-1)$ matrix obtained from kneading matrix $(N_{ij}(t))$ by eliminating the i-th column, then the expression

$$M_f(t) = \frac{(-1)^{i+1} M_i(t)}{1-\varepsilon(I_i) \cdot t}$$

does not depend on the choice of i and is a power series with bounded coefficients. The power series is called the kneading determinant of map f.

The possibility of computing the topological entropy of map f is based on the following two statements (see /6,7/), which relate the kneading determinant $M_f(t)$ to the Artin-Mazur dzeta-function

$$\zeta_f(t) = \exp\left(\sum_{n=1}^{\infty} \frac{p_n t^n}{n}\right)$$

where $p_n$ is the number of fixed points of map $f^n$:

(i) The power series $M_f(t)\cdot\zeta_f(t)$ is a holomorphic function of the complex variable t in the open unit circle;

(ii) The radius of convergence r(f) of dzeta-function $\zeta_f(t)$ is related to the topological entropy $h_{top}(f)$ by the ratio $h_{top}(f) = -\ln r(f)$. Thus, for computing the topological entropy, it is sufficient to find a minimum positive root t* of kneading determinant $M_f(t)$, its coefficients are defined by the algorithm mentioned above. Therewith $h_{top}(f) = -\ln t^*$.

## 3. NUMERICAL ANALYSIS

In this section we shall give the results of the numerical analysis of two dynamical systems, carried out with the application of the proposed approach. Alongside with the topological entropy, we shall calculate the Lyapunov characteristic exponent $\Lambda$, defined from formula $\Lambda = \lim_{\to\infty} \frac{1}{n}\ln|T^n x|$ where $T^n x = \prod_{i=0}^{n-1} T(f^i x)$, and the operator $T \equiv d/dx$. Since, according to /8/, $h_\rho \leqq \Lambda$, where is the ergodic measure with the compact support (for physical measures the strict equality - the Pesin identity often takes place), then by comparing topological entropy and the Lyapunov exponent, we can judge in what degree the invariant measure realizing itself in the given dynamical system differs from the measure with maximal entropy.

1. Let us consider one-dimensional map $f : x \to x$ of interval I into itself:

$$(x^2+r^2)^{1/2} - \left(\left(\frac{1-\bar{x}}{K}\right)^2 + r^2\right)^{1/2} = 4rN, \tag{1}$$

where

$N = \left[\frac{(x^2+r^2)^{1/2} - r}{4r}\right] > 0$ is the integer, $I = a,1$ ,$a = 1-2\sqrt{6}rK$; $r > 0$, $K > 0$ are the governing parameters. The map (1) arises from the study of an electromechanical Gipp clockwork /9/. It is easy to show that a quantity of subintervals of the map (1) is exactly equal to N (see Fig.1). Notice that this fact is convenient for controlling the efficiency of the technique for different N.

The system (1) can demonstrate both regular and chaotic modes, the structure of one-dimensional limit sets can be different. Figure 2 shows the dependence of the topological entropy $h_{top}$ and the Lyapunov exponent $\Lambda$ upon parameter r when parameter k is fixed. One can see that $h_{top}$ and $\Lambda$ are qualitatively identical but there is a region of values of parameter r for every k, when the distinction is sufficient enough and it is possible to affirm that the invariant measures of the dynamical system generated by the map (1), differs from the measure with maximal entropy.

A more detailed study of the system (1) and the discussion of the problems dealing with the numerical realization of the proposed approach can be found in /10/.

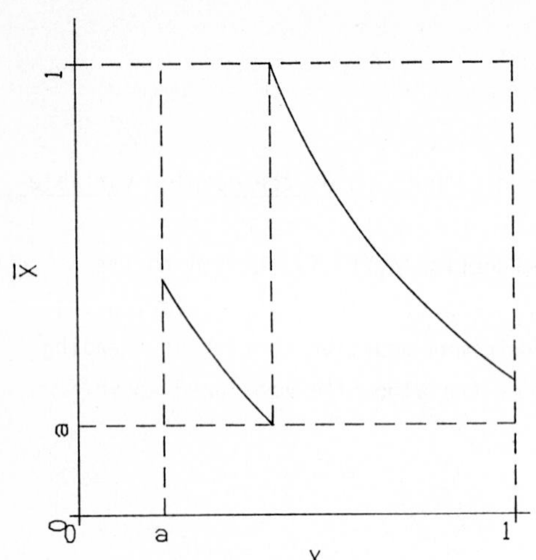


Figure 1. The form of mapping (1) at N = 2.

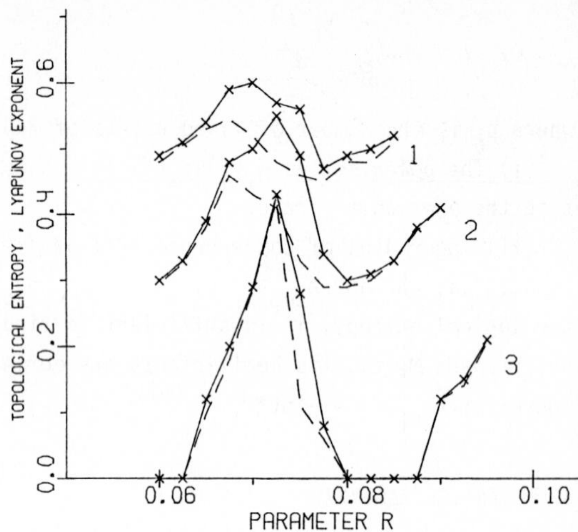


Figure 2. The dependence of the topological entropy $h_{top}$ (solid line ) and the Lyapunov exponent $\Lambda$ (broken line) on parameter r: 1 - k = 1.4 ; 2 - k = 1.2; 3 - k = 1.0.

2. The essential practical interest lies in the possibility of determining the topological entropy of dynamical systems directly from the experimental data. The proposed approach allows to calculate the topological entropy of systems for which one can choose such a step of discretization of one-dimensional data that the corresponding return map f: $x(t_i) \to x(t_{i+1})$ appears to be one-dimensional. The class of such physical systems is broad enough (see, e.g. /11/ and the references there).

For a model example we consider the dynamical system, given by differential equation

$$\alpha\dot{V} = j_e(t) - j(T,V)$$
$$\beta\dot{T} = j(T,V) \cdot V - (T-T_0). \tag{2}$$

Here $j = 2T^{1/2}e^{-1/T}\text{sh } V/T$; $\alpha$, $\beta$, $T_0$ are the positive parameters. System (2) describes the dynamics of nonlinear response of multilayer selectivity doped heterostructures to external periodical microwave signal $j_e(t) = \delta \cos \omega t$ (/12/).

As the numerical simulation has shown, the return map, constructed along V(t), $f_{\Delta t}$: $V(t_i) \to V(t_{i+1})$, $1 \leq i \leq M$, is practically one-dimensional (see Fig.3), if the step of discretization $\Delta t$ of one-dimensional data (e.g. V(t)) coincides with the period of external signal, i.e. $\Delta t = 2\pi/\omega$. If the length of data is large enough ($M \sim 10^3 \div 10^4$), then one can define the coordinates of critical points of the constructed return map $f_{\Delta t}$ with high precision and one can estimate the topological entropy $h_{top}(f_{\Delta t})$ on the basis of the proposed technique. Therewith the topological entropy of the initial dynamical system $h_{top}$ is related to the computed characteristics by the relation $h_{top} = 1/\Delta t \cdot h_{top}(f_{\Delta t})$. It is necessary to emphasize that the topological

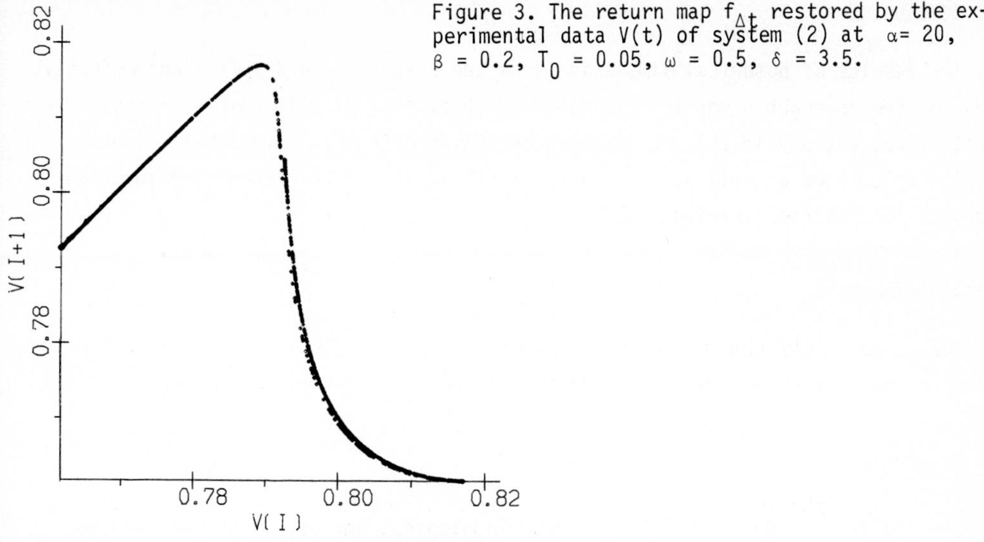

Figure 3. The return map $f_{\Delta t}$ restored by the experimental data V(t) of system (2) at $\alpha = 20$, $\beta = 0.2$, $T_0 = 0.05$, $\omega = 0.5$, $\delta = 3.5$.

entropy of piecewise-continuous and piecewise-monotonous map f depends continuously of the perturbations of f in C -topology (/5/), therefore the small errors in the determination of f lead to the small errors of computing the topological entropy.

Figure 4 shows the results of computing $h_{top}$ and the maximal Lyapunov exponent $\Lambda_1$ (see /13/) for different $\delta$. The essential distinction of $h_{top}$ from $\Lambda_1$ testifies that invariant measure of system (2) realizing at numerical simulation is far from the measure with maximal entropy.

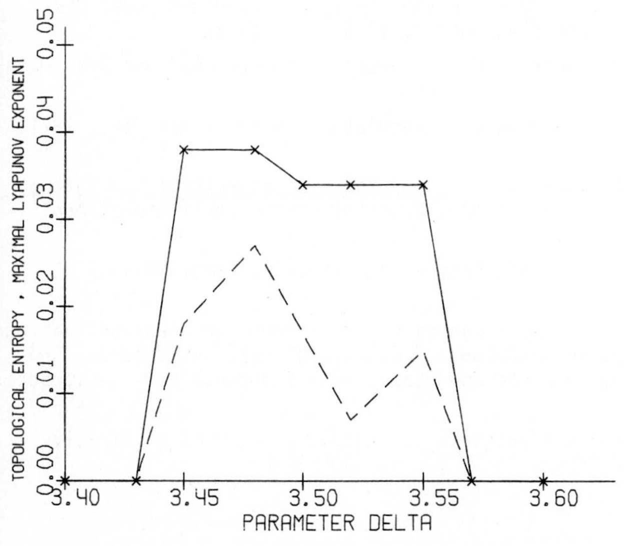

Figure 4. The dependences of the topological entropy $h_{top}$ (solid line) and the maximal Lyapunov exponent $\Lambda_1$ (broken line) of system (2) on parameter $\delta$.

## 4. CONCLUSION

The results of numerical simulation show the simplicity and sufficient effectiveness of the approach given in this paper. We hope that it will enable to include the topological entropy in the row of characteristics of chaotic dynamical systems which can be calculated as well as to understand better the properties of the invariant measure of concrete dynamical systems.

*Acknowledgements*

The author would like to thank V.S.Afraimovich, M.I.Malkin, I.L.Zheleznyak for fruitful collaboration and M.I.Rabinovich for the interest in this work.

## REFERENCES

1. R.L.Adler, A.G.Konheim, M.H.McAndrew. Topological entropy. Trans.Am.Math.Soc., 1965, 114, 2, 309-319.
2. R.Bowen. Methods of Symbolic Dynamics. M.: Mir, 1979 (in Russian).
3. U.N.Kornfeld, Ya.G.Sinai, S.V.Fomin. The Ergodic Theory. M.: Nauka, 1080 207 (in Russian).
4. J.Guckenheimer. The growth of topological entropy for one-dimensional maps. Lect. Notes in Math., 1980, 819, 216-223.
5. M.I.Malkin. On the continuity of the entropy of discontinuous maps of an interval In: Methods of Qualitative Theory of Differential Equations. Gorky Univ., 1982, 35-38 (in Russian).
6. J.Milnor, W.Thurston. On Iterated Maps of Interval. Preprint, Princeton Univ., 1977.
7. M.I.Malkin. Methods of Symbolic Dynamics in the Theory of One-Dimensional Discontinuous Maps. Cand.Sci. Dissertation, Gorky, 1985 (in Russian).
8. D.Ruelle. An inequality for the entropy of differentiable maps. Bol.Soc.Bras.Math. 1978, 9, 83-87.
9. L.A.Komraz. Dynamic models of Gipp's balance regulator. Zh.Prikl.Mat.Fiz., 1971, 35, 1, 148-162 (in Russian).
10. M.I.Malkin, A.L.Zheleznyak, I.L.Zheleznyak. Computational Aspects of the Entropy Theory of One-Dimensional Dynamical Systems. Preprint of the Institute of Appl. Physics N222, Gorky, 1988.
11. H.L.Swinney. Observations of order and chaos in nonlinear systems. Physica D, 1983, 70, 1-3, 3-15.
12. A.M.Belyantsev, A.L.Zheleznyak et al. Dynamics of avalanche-like electron heating oscillations and chaotic response behaviour of multilayer heterostructures. 3rd Conf. on Physics and Technology of GaAs and Other III-V Semiconductors. Abstracts. CSSR, 1988, p.51.
13. I.Shimada, T.Nagashima. A numerical approach to ergodic problem of dissipative dynamical systems. Prog.Theor.Phys., 1979, 61, 6, 1605-1616.

# Stochastic Dynamics of Pattern Formation in Discrete Systems

*V.A. Antonets, M.A. Antonets, and I.A. Shereshevsky*

Institute of Applied Physics, USSR Academy of Sciences,
46 Ulyanov Str., 603600 Gorky, USSR

A dynamic model of ordered systems growing from discrete elements is considered. The model is used for the analysis of the blood flow regimes in vascular systems. The order relation is shown to affect significantly the properties of the Gibbs distribution of an ordered system of interacting elements.

## 1. Introduction

The investigation of pattern formation in nonequilibrium media is a current topic of different branches of science: physics, chemistry and biology. A continuous description of such media using a nonlinear diffusion equation revealed a profound analogy between the generation of spatio-temporal structures and the dynamics of a nonlinear oscillator /1/. In particular, this approach allows to describe the structures which are formed from a great number of elements in discrete systems of the lattice type using a limit transition.

However, this approach leaves out a very wide class of dendritic patterns that are often formed as a result of particle aggregation, i.e. in animate and inanimate growth /2/. Such structures were most frequently analyzed in the study of polymerization processes (see /3/), i.e. the aggregation of monomers resulting in the formation of branching structures. Polymers, however, are not the most wide-spread objects of this kind. A dendritic structure is typical of a great number of objects such as, for example, trees of communication systems of animate organisms providing their life and integrity. The circulatory system /4/, a peripheral nerve network /5/ and other systems have such a structure. A dendritic pattern is characteristic of self-developing transport systems in a continuous medium. A duct network which is formed when a liquid is filtered through a porous medium may be, for example, such a system. In this paper we shall consider the dynamic models for the growth of dendritic systems and relate the parameters of these systems in the steady state to the parameters characterizing their dynamics. First, we shall analyze as an example the blood flow in the vascular system and then /6/ show how the dynamic equations that at first seem to be specific of dendric systems, may be used to assign the dynamics of various discrete systems, the evolving configurations of which remain continuous to a definite (although not always apparent) sense. The principle of retaining the configuration continuity is demonstrated on an example of the critical phenomena which owe their existence to the forbidden discontinuous configurations.

## 2. A MODEL OF CIRCULATION IN A VASCULAR SYSTEM

Let us assume that a vascular system has a dendritic structure (Fig.1) which is formed by the branching of each vessel into two parts /4/ and the system can be plotted as a dichotomic branching tree. Assume also that each vessel may be either open or closed for blood circulation. We shall describe the state of the vessels by random logic variables depending on a discrete time t. Let the state of the vessel a at a given moment of time be described by the variable x(a,t). If vessel a is open, x(a,t) is true and if it is closed, then x(a,t) is false.

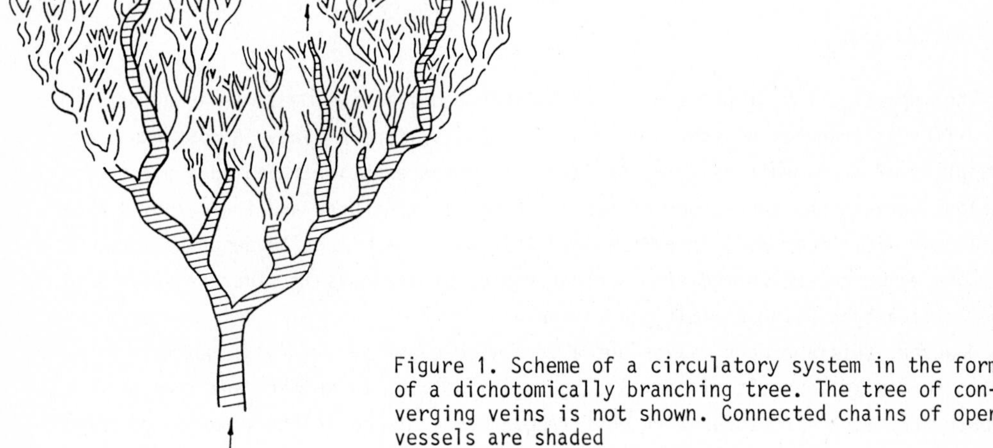

Figure 1. Scheme of a circulatory system in the form of a dichotomically branching tree. The tree of converging veins is not shown. Connected chains of open vessels are shaded

The equations of vessel dynamics will be formulated under the assumption that each vessel in the vascular system has active properties: it may open and close due to the presence of a muscle's wall and the variation of the flow parameters. These active properties may be described by random logic variables $\eta(a,t)$, $\xi(a,t)$. Variable $\xi(a,t)$ denotes the opening of vessel a, hence, x(a,t+1) is true unless it contradicts other possible conditions. Variable $\eta(a,t)$ denotes the closing of vessel a, hence, x(a,t+1) is false if this does not contradict other conditions either. The interaction of neighbouring vessels is understood, as in earlier paper /7,8,9/, as the dependence of the activity and the state of vessel a on the state of the neighbouring vessels and on the flow characteristics in them. It is supposed that the intersection occurs only with close neighbours: the ancestor, $\hat{a}$, two descendants, $a_0$, $a_1$ and $a'$, and the vessel of the same generation (Fig.2).

We assume also that the events which take place in the vascular system do not contradict the condition of flow continuity. This condition is taken in its natural meaning: if variable x(a,t) is true, then $x(\hat{a},t)$ is also true.

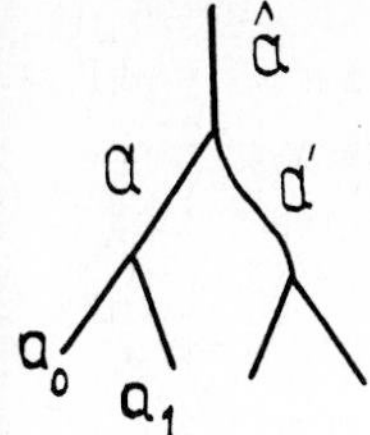


Figure 2. Scheme of vessel interaction.

In accord with the conditions formulated above, the dynamics of the variables x(a,t) is described by recurrent relations by means of ordinary logic operations:

$$x(a,t+1) = x(a_0,t) \vee x(a_1,t) \vee (x(a,t) \wedge \bar{\eta}(a,t)) \vee (\bar{x}(a,t) \wedge$$

$$\wedge\ x(\hat{a},t) \wedge \xi(a,t)) \tag{1}$$

where "V" means or, "Λ" - and, subscript "—" - not.

It can be shown that (1) will not contradict the condition of flow continuity in the system if

$$\xi(a,t) \Rightarrow \bar{\eta}(a,t)\ . \tag{2}$$

It follows from the condition of continuity that at each moment of time the open vessels form a spot, i.e. all vessels on the only tract connecting it with the root vessel, aorta, which is considered to be always open, are open simulataneously. Therefore, all probability characteristics of the vascular system at the moment of time are determined completely by the probability G(J,t) of the validity of $\bigwedge_{a \in J} x(a,t)$ where J is a random finite spot in the tree.

The dynamic equation (1) can yield the evolution equations for G(J,t). The probability distributions for the spots in the tree that correspond to the time-independent solutions of these equations are steady states of the vascular system.

The steady-state solution of Eq. (1) can be shown to have a form

$$G_{st}(J) = \nu_1^{|J|-1} \tag{3}$$

where $\nu_1$ is a minimal positive root of the equation

$$(1-\nu)(\nu^2-\nu+\beta) = 0, \tag{4}$$

where $\beta = r/p$, ris the probability of the event $\xi(a,t)$ and p is the probability of the event $\eta(a,t)$.

It follows from the condition of continuity that

$$p \leq (1-r)^2\ . \tag{5}$$

When $\beta > 1/4$, system (4) has the only solution $\nu = 1$ which corresponds to the state of the system with completely open vessels. When $\beta < 1/4$, two positive roots appear in (4), the smaller of which, $\nu = 1/2 - \sqrt{1/4 - \beta}$, gives a solution to (3). In this case $G_{st} \to 0$ as $|J| \to \infty$ . Thus, the steady-state consequence of (1), which assumes that the state of the vessel "a" changes only due to its own activity while the interaction with neighbouring vessels reduces to meeting the continuity conditions, is either a fully filled system or a system containing only finite filled clusters.

In the case of a more complex interaction which is not described by (1), the steady states can be considered as follows. Let us introduce the probability distribution of a triplet of vessels a, $a_0$ and $a_1$, provided that vessel a is open. This distribution may be specified by the following conditional probabilities: $\Pi_{11}$ is the probability that both the descendants, $a_0$ and $a_1$, of the open vessel a are open; $\Pi_{10}$ is the probability that only descendant $a_0$ is open; $\Pi_{01}$ is the probability that only descendant $a_1$ is open; and $\Pi_{00}$ is the probability that both the descendants are closed. The normalization relation

$$\Pi_{11} + \Pi_{10} + \Pi_{01} + \Pi_{00} = 1 \tag{6}$$

is, apparently, fulfilled.

Consider now the sequence of random values $n_1$, $n_2$, ..., $n_k$ and ... where $n_k$ is the number of open vessels in the k-th generation of the descendants of the vessel a connected with them by a chain of open vessels. This sequence of random values forms a branching process /14/ assuming that the state of a pair of descendants depends on the state of their ancestors. The generating function F(s) of this process i.e. the generating distribution function of the number of open descendants of the nearest generation has a form

$$F(s) = \Pi_{00} + (\Pi_{01} + \Pi_{10})s + \Pi_{11}s^2 \tag{7}$$

where s is the argument.

According to the known theorem /15/, the probability of the degeneration $\theta$ of such a process (in our case, the probability that any chain of open vessels that begins with vessel a consists of a dinite number of elements) is equal to the smalles positive root of the characteristic equation F(s) = s:

$$\Pi_{11}\theta^2 + (\Pi_{01} + \Pi_{10})\theta + \Pi_{00} = \theta \quad . \tag{8}$$

Hence $\theta = \min\{1, \Pi_{00}/\Pi_{11}\}$ . Then, for $\Pi_{00}/\Pi_{11} > 1$ the process degenerates with the probability 1. If $\Pi_{00}/\Pi_{11} < 1$, the probability of degeneration, i.e. of the formation of a finite cluster, is equal to $\Pi_{00}/\Pi_{11}$, while the probability of nondegeneration, i.e. the formation of an infinite cluster, is equal to

$$P = 1 - \Pi_{00}/\Pi_{11} \ . \tag{9}$$

Since the average number $A_1$ of the filled close descendants of the open vessel a is

$$A_1 = 1 \cdot \Pi_{01} + 1 \cdot \Pi_{10} + 2\Pi_{11} = 1 + (\Pi_{11} - \Pi_{00}) \tag{10}$$

then

$$P = \frac{A_1 - 1}{\Pi_{11}} \tag{11}$$

i.e. $P$ has a threshold with respect to $A_1$. For a nonzero probability of an infinite cluster, value $A_1$ must be larger than unity.

One can find an average number of open vessels in a finite cluster:

$$N = \sum_{k=0}^{\infty} A_1{}^k = \frac{1}{1-A_1} = \frac{1}{\Pi_{00}-\Pi_{11}} , \tag{12}$$

because the average number of open descendants of the open vessel in the k position is equal to $A_1{}^k$.

When the vessels do not interact we have

$$\Pi_{11} = p^2, \quad \Pi_{10} = \Pi_{01} = p(1-p), \quad \Pi_{00} = (1-p)^2 \tag{13}$$

where p is the probability for a given vessel to be open. Then

$$A_1 = 2p , \tag{14}$$

$$P = \frac{2p-1}{p^2} . \tag{15}$$

Relations (14) and (15) show that a through flow is possible in the case of non-interacting vessels, if the probability of an open vessel (i.e. a relative density of open vessels) exceeds the threshold value $p_n = 1/2$. This corresponds exactly to the conclusions of the percolation theory /5/. Rewriting (15) in the form $P = 2(p-p_n)/ /p^2$, we obtain for small $p-p_n$:

$$P \cong 8(p-p_n) . \tag{16}$$

Thus, only finite clusters are possible when $\Pi_{00}/\Pi_{11} > 1$ and a combination of finite and infinite clusters is likely to occur when $\Pi_{00}/\Pi_{11} < 1$.

## 3. ANALYSIS OF CIRCULATION REGIMES

The concepts developed above enable us to distinguish two qualitatively different circulation regimes.

The first regime takes place with a through flow in the system when $A_1 > 1$, i.e. when the closing (or the collapse) of the vessels is not too intense. Such a circulation regime is, inevitably, realized in a sound heart muscle, myocardium, because the coronary vascular system feeding it is known to be a dichotomic branching tree /4/. It is characterized by a sharp dependence of the flow probability on the threshold excess of the average number of open descendants of the open vessel. In the case of weakly interacting vessels, this is equivalent to a sharp dependence of the flow probability on the probability of a random vessel to be open, i.e. on the relative density of open vessels. Thus, according to (11) the flow probability $P$ in the absence of coupling varies from 0 to 30% when the density of open vessels increases over a critical value only by 20%, i.e. from $P = 0.5$ to $P = 0.6$. This means that the circulation may be controlled effectively by the variation of the activity of the vessel determining the density of open vessels.

The second regime of circulation is implemented when the bounded spots ($\Pi_{00} > \Pi_{11}$) contact one another, thus making a continuous chain. Each spot in such a chain is born as a result of supplying the tree with the vessels known as vessels-collaterals /11/. Owing to these vessels there is nonzero probability for the descendants of unfilled vessels to be filled. If the size of the spots is limited, the rate of collateral supply to the vessel tree, i.e their density in space, must be rather high for the spots to merge.

Consider now an assembly of merging spots as a new dendritic graph in the original tree. The order of its branching can be evaluated like the number of B vessels within the spot boundaries. This number is equal to

$$B = \sum_{k=0}^{\infty} \Pi_{00} A_1{}^k = \frac{\Pi_{00}}{1-A_1} = \Pi_{00} \cdot N , \tag{17}$$

where N is the average number of vessels in the spot (see Eq. (12)).

Since the critical probability $p_n$ (density) of continuous bonds must be not smaller than $1/n$ (n is the multiplicity of the tree branching) for the flow to be realized /5/, then identifying $p_n$ with the collateral density $C_{crit}^{coll}$ and n with B, we obtain

$$C_{crit}^{coll} \cdot B = \Pi_{00} \cdot NC_{crit}^{coll} = 1 . \tag{18}$$

Since $N = (\Pi_{00} - \Pi_{11})^{-1}$, we finally have

$$C_{crit}^{coll} \cdot \frac{\Pi_{00}}{\Pi_{00} - \Pi_{11}} = 1 . \tag{19}$$

Expression (19) relates the structural characteristics of the circulatory system to to the local characteristics of the interacting vessels, which guarantees the circulation all over the tree. This relationship may be significant for the analysis of the mechanism responsible for the development and germination of the circulatory system in tissues with the adaptation of blood supply in loaded (ischaemia, hypertrophy) and diseased (infractions, tumors) organs.

## 4. DYNAMICS OF PATTERN FORMATION IN ORDERED SYSTEMS

A growing single-root tree must have a connected configuration at each moment of time because the material needed for the growth of new branches is supplied by the available part of the tree. However, the configurations of a growing systems have other, much more rigorous restrictions. Consider, for example, the process of filling a right angle having solid walls with identical heavy squares (Fig.3).

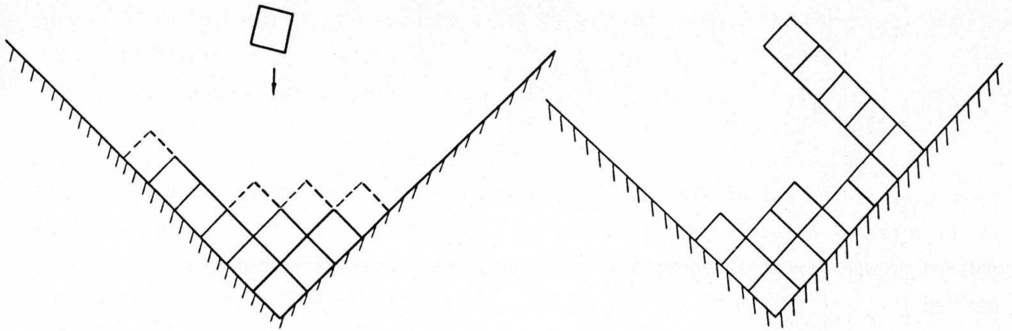

Figure 3. Evolution of configurations in the system of "falling squares".

Figure 4. Simply connected discontinuous configurations.

The dashed curves in this figure denote possible location of a square in accord with the requirements of configuration stability relative to weak impacts. The stable configurations form only a narrow subscale of a set of all connected (even simply connected) configurations, namely, the set of all Young's diagrams.

This process as well as the growth of trees mentioned above are examples of a wide class of growth processes for which the geometric restrictions on the structure and evolution of configurations may be axiomatized on the basis of a partial-order structure.

Assume that the configurations formed in a growing system are parts of some maximal configuration $\Gamma$, for example, of a regular tree or of the filling of part of space by regular polyhedrons. Let there exist some order relation between the elements of the maximal configuration $\Gamma$ (it is known for elements a and b satisfying $a < b$), i.e. a precedes b in the sense that b cannot be part of the configuration in the growth process if there is no element a in it. Apparently, if $a < b$ and $b < c$, then $a < c$, thus $\Gamma$ is a partially ordered set in an ordinary sense. The process will be called an ordered-growth process if all appearing configurations are continuous, i.e. besides its own elements it contains all preceding elements of the maximal configuration $\Gamma$. The continuity requirement is the strongest limitation. For example, the configuration in Fig.4 is connected and simply connected and not continuous.

The authors of /6,12/ and we in the previous section of this paper considered the growth models for dendritic systems the evolution of which is specified by the Langevin equations for random logic variables $x(a,t)$. These variables have the meaning of "truth" if element a enters the configuration of the system at the moment t and "false" in the opposite case. These equations contain the given "external" random fields of the logic variables $\xi(a,t)$, $\eta(a,t)$ which describe the inidividual activity of the elements in the system. In the case of general ordered systems, analogous equations have the form

$$x(a,t+1) = \left[ \bigvee_{b \in C(a)} x(b,t) \vee \overline{\eta}(a,t) \wedge x(a,t) \right] \vee$$

$$\vee \left[ \overline{x}(a,t) \wedge ( \bigwedge_{d \in B(a)} x(d,t)) \wedge \xi(a,t) \right] ,$$

where C(a) is the set of elements following directly after element a and B(a) is the set of elements directly preceding element a. In order to determine the process of ordered growth for these equations, it is necessary and sufficient that truth $\xi(a,t)$ implied truth $\bigwedge_{b \in B(a)} \overline{\eta}(b,t)$ for all a and t. We cannot say that these conditions determine the growth process unambiguously. The results presented in /6,12/ show that these conditions allow for a certain diversity in the behaviour of systems with ordered growth.

## 5. CRITICAL PHENOMENA DUE TO ORDERED GROWTH

The limiting state of the ordered growth of a regular tree under certain restrictions on $\xi$, $\eta$ is a geometrical measure on a set of continuous configurations of the tree /12,13/. Such states can also be determined in the case of arbitrary partially ordered systems /13/. For this purpose let us consider a homogeneous field of random logic variables $\xi(a)$ in $\Gamma$. The probability distribution of this field is the Gibbs distribution for a system of identical noninteracting elements in $\Gamma$. Let us define field $\theta$ by the relation

$$\theta(a) = \bigwedge_{b \leq a} \xi(b) .$$

Field $\theta(a)$ has, apparently, continuous realizations, and therefore its probability distribution is a measure in the set of continuous configurations in $\Gamma$. This measure is obtained as a result of overlapping geometric constraints which are due to the ration of order in $\Gamma$ to the Gibbs state of the system of free particles in $\Gamma$. When $\Gamma$ is a set of natural numbers N, the probability distribution of field $\theta$ is an ordinary geometric probability distribution.

The geometric probability distribution depends on the only parameter $\nu$ which is equal to the probability of true $\xi(a)$. It was found out that the properties of geometric distribution may depend significantly on value $\nu$. Let $m_\Gamma(\nu)$ be the probability of an assembly of all finite continuous configurations in $\Gamma$. It can be shown /13/ that there exists such a number $\nu_\Gamma^*$ that the probability $m_\Gamma(\nu)$ is less than unity when $\nu > \nu_\Gamma^*$, i.e. infinite continuous configurations are possible. The critical value $\nu_\Gamma^*$ is determined by the geometry of the maximal configuration $\Gamma$. In particular, for the trees with the branching multiplicity $N > 1$, we have $\nu_\Gamma^* = 1/N$, while for all regular finite-dimensional lattices we have $\nu_\Gamma^* = 1$. Thus, the abundance of dendritic structures can be explained by purely geometric reasons rather than by the interaction between the elements of the growing system.

Note that parameter $\beta$ of the limit distribution of the ordered tree growth described by (1) is determined by the ratio of the probabilities to be true for the external fields $\xi$ and $\eta$ which can be interpreted as the probability of the birth and death of the elements in the system. Inequality $\nu > \nu_{\Gamma}^{*}$ is fulfilled in this case when relation $\beta$ exceeds value $(N-1)^{N-1}/N^{N}$ (N is the multiplicity of the tree branching), i.e. unlimited growth is possible even under the condition that the probability of birth is much less than the probability of death.

## REFERENCES


1. A.V.Gaponov-Grekhov, M.I.Rabinovich. Nonlinear Physics. Stochasticity and Structures. Preprint No.87, Inst.Appl.Phys., USSR Acad.Sci., Gorky, 1983.
2. L.M.Sander. Scientific American, 1987, 256, 3.
3. S.P.Obukhov. Scaling Models in Problems of Polymer Physics. Preprint, Research Computer Center, Pushchino, USSR, 1985.
4. Architectonics of the Blood Channel. Ed. by V.A.Matyukhin. Nauka, Novosibirsk, 1982.
5. J.W.Essam. Rep.Progr.Phys., 1980, 43, 833-912.
6. V.A.Antonets, M.A.Antonets, I.A.Shereshevsky. In: Collective Dynamics of Excitations and Pattern Formation in Biologic Tissues. Inst.Appl.Phys., USSR Acad.Sci., Gorky, 1988, p.165-177.
7. V.A.Antonets, M.A.Antonets, I.A.Shereshevsky. Stochastic Dynamics of the Blood Flow in a Vascular System - Medical Biomechanics, Riga, USSR, 1986, v.4, 37-47.
8. V.A.Antonets, M.A.Antonets, A.V.Kudryashov. Interacting Markov Processes and Their Application to the Simulations of Biologic Systems, Pushchino, USSR, 1982, 108-118.
9. V.A.Antonets, M.A.Antonets. Flow Structure in the Vascular Network. Preprint No. 177, Inst.Appl.Phys., USSR Acad.Sci., Gorky, USSR, 1987.
10. F.Spitzer. The Annals of Probability, 1975, 3, 3, 387-398.
11. R.Gorlin. Coronary Arteries Diseases. Medicine, Moscow, 1980, p.68.
12. M.A.Antonets, I.A.Shereshevsky. Analysis of a Stochastic Model of Tree Growth. Preprint No.231, Inst.Appl.Phys., USSR Acad.Sci., USSR, 1989.
13. M.A.Antonets, I.A.Shereshevsky. A Critical Phenomenon in a Model of Random Growth. Preprint No.223, Inst.Appl.Phys., USSR Acad.Sci., Gorky, USSR, 1989.

# Chaos and Period Adding Sequence in a Microwave Oscillator with a Ferrite Resonator

*N.F. Rul'kov*

Gorky State University, 23 Gagarin Ave., 603600 Gorky, USSR

A period adding sequence based on a model map is discussed for a microwave oscillator with a ferrite resonator. The regular windows arranged in a parameter space are found to be universal for a period adding sequence.

The appearance of stochastic modes in many dynamical systems is governed by some universal laws with certain scaling properties. These universal laws are typical not only at the onset of chaos, but also within a range of completely developed chaos. An example is a logistic map with the rule of chaotic mode alternation by the regular motions with periodic p-cycles /1,2/. The complex structure of the parameter space splitting in dynamical systems with chaotic behaviour can be explained by studying the sequence rules for regular motion, i.e. the well distinguished windows inside the region of developed chaos and of the motion features inside these windows.

Here the arrangement of r-windows with Pr period (with $P_{r+1}$ = Pr + 1) cycles inside the regions of developed chaos is discussed for a model map. Such a sequence was observed in experimental studies of the Belousov-Zhabotinsky reaction /3/, in a microwave oscillator with a ferrite resonator /4/ and in a periodically driven nonlinear oscillator /5/. Chaos appeared in these systems via the sequence of period doubling bifurcations and further parameter variation was intermitted by the windows of regular motions. Within each window, the appearance of chaos was preceded by Pr-period doubling bifurcations. With the reverse parameter variation chaos appeared via intermittence.

In the experiment with a microwave oscillator with a ferrite resonator /4/, the above scenario of period adding was observed for a relatively low-frequency self-modulation of an output microwave signal ($f_m$ = 10 MHz, $f_h$ = 4 GHz). This means that the bifurcations of periodic motions correspond to the bifurcations of two-dimensional ergodic tori in the phase space of the oscillator. The characteristic property of the r-windows with respect to a parameter variation is their convergence to a critical parameter value when $r \to \infty$. Within a range of supercritical parameter values, the periodic self-modulation of one-revolution cycles was established in the oscillator.

The analysis of a one-dimensional return map for self-modulation obtained in a physical experiment /4/ shows that for the study of bifurcation within the range of parameter values required for the scenario above, one may use a one-hump map governed by the law



Research Reports in Physics Nonlinear Waves 3
Editors: A.V. Gaponov-Grekhov · M.I. Rabinovich · J. Engelbrecht

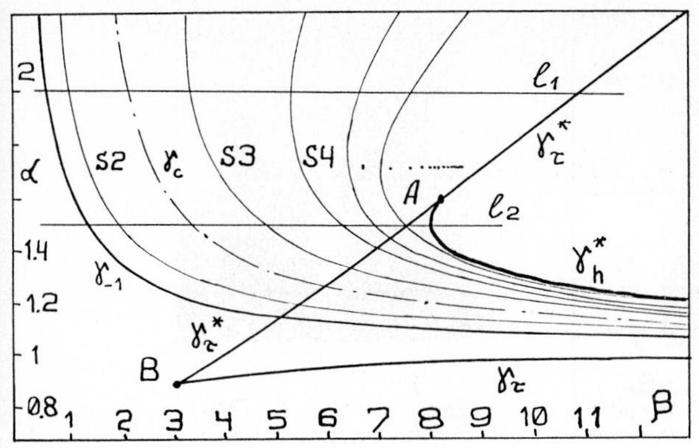


Figure 1. The plane of parameters (β,α).

$$x_{n+1} = F(x_n,\alpha,\beta) \ , \tag{1}$$

where $F(x,\alpha,\beta) = \alpha/\ 1+\beta(x-1)^2$ ,$\alpha > 0$, $\beta > 0$.

The dynamics of system (1) is investigated using the (α,β)-parameter plane (see Fig.1). Within the parameter region bounded by the bifurcation curves $\gamma_\tau$* and $\gamma_\tau$, map (1) has three stationary points $\bar{x}_1$, $\bar{x}_2$, $\bar{x}_3$ ($\bar{x}_1 < \bar{x}_2 < \bar{x}_3$); $\bar{x}_1$ is stable, $\bar{x}_2$ is unstable and $\bar{x}_3$ is stable inside the region below curve $\gamma_{-1}$. At boundary $\gamma_{-1}$ point $\bar{x}_3$ loses the stability as the result of period doubling bifurcation. At $\gamma_\tau$* and $\gamma_\tau$, tangential bifurcations occur resulting in the disappearance of stationary points $\bar{x}_1$, $\bar{x}_2$ and $\bar{x}_2$, $\bar{x}_3$, respectively. All three stationary points merge at point B. Outside the region bounded by curves $\gamma_\tau$* and $\gamma_\tau$ , the map has one stationary point $\bar{x}_3$ which is stable below boundary $\gamma_{-1}$.

The bifurcation curve $\gamma_h$* corresponds to the situation with $F(\alpha,\alpha,\beta) = \bar{x}_2$ and gives the restriction of the parameter value region with a global stability of $\bar{x}_1$. The transient process may be quite complicated and long-lived (metastable chaos). Curve $\gamma_\tau$* is the boundary of the global stability of $\bar{x}_1$ above point A. Hence the regular periodic and chaotic modes in map (1) may be observed only within the region bounded by $\gamma_{-1}$ and $\gamma_h$* curves (below A) and $\gamma_\tau$* (above A).

Let us consider the region of parameter values with possible chaotic modes in more detail. With increasing parameter value of β when $\alpha > 1$, chaos in map (1) arises at boundary $\gamma_c$ as a result of the sequence of period doubling bifurcations. The growth of β is accompanied with the changes of chaotic modes by the windows of periodic motions. The windows, corresponding to successive period increase by 1, are most wide as compared to other kinds of regular motions observed in neighbouring chaotic layers. Inside each r-window the values of the β and α parameters exist corresponding to superstability of the Pr-cycle, i.e., the multiplicator of these cycles turns to zero. In Fig.1 the lines with superstable Pr-cycles are marked by index SPr. With the increase of r, the lines monotonically come closer to the boundary where chaotic modes exist.

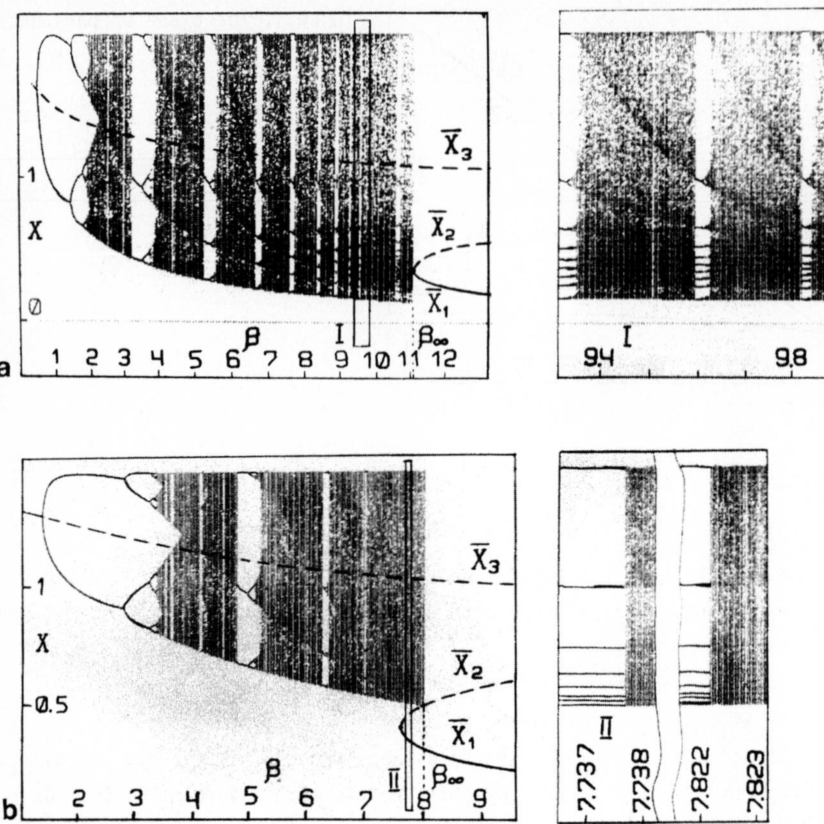


Figure 2. Bifurcation diagrams: a) - along line $\ell_1$; b) - along line $\ell_2$ (see Fig.1). The stationary unstable points $\overline{x}_2$ and $\overline{x}_3$ are marked by dashed lines. Fragments I and II are plotted on the right.

Different scaling properties of SPr, crowding together when $r \to \infty$, are characteristic to mapping (1). So, for boundary $\gamma_\tau{}^*$ above point A the period accumulation is the result of the increasing number of iterations in the vicinity of the point where $\overline{x}_1$ and $\overline{x}_2$ appear (see Fig.2a). Here for $\alpha$ = const , the sequence of $\beta_r$ values corresponding to superstable orbits Pr, tend to $\beta_\infty$ (spaced at the curve $\gamma_\tau{}^*$) under the law /6/:

$$\beta_\infty - \beta_r \sim r^{-2} \, . \tag{2}$$

Figure 3 illustrates the dependence of $-\ell n(\beta_\infty - \beta_r)$ on $\ell n(r)$ for the case $\alpha = 2$. Within the parameter values for the region below point A, a period storage arises as a result of the increasing number of iterations in the vicinity of the stationary unstably point $\overline{x}_2$ (see Fig.2b). Here $\beta_r$ tends to $\beta_\infty$ ( corresponding to $\gamma_h{}^*$) under the law /7/:

$$\beta_\infty - \beta_r \sim z^{-r} \, , \tag{3}$$

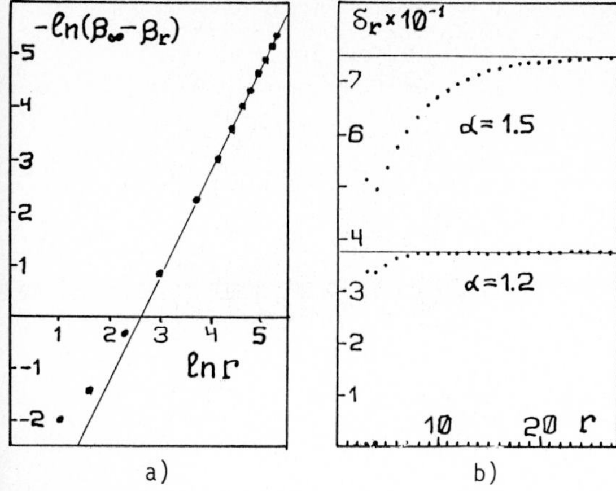


Figure 3. a) - $\beta_\infty$ - $\beta_r$ vs r for the case $\alpha$ = 2; b) - $\delta_r$ vs r for $\alpha$ = 1.5 and $\alpha$ = 1.2.

where z is the multiplicator of the stationary point $\bar{x}_2$, ($z > 1$). Figure 3b displays the laws $\delta_r = (\beta_{r+1}-\beta_r)/(\beta_r-\beta_{r-1})$ vs r.

Cases (2) and (3) discussed above correspond to two different situations in the oscillator phase space. In the first case, the bifurcation of birth of stable and saddle two-dimensional tori takes place in the region of a strange attractor. In the second case, a strange attractor crisis occurs due to the touching of a stable manifold of a saddle two-dimensional torus.


*Acknowledgement*

The author is grateful to I.S.Aranson for useful discussions.


REFERENCES


1. A.N.Sharkovsky, Yu.L.Naistrenko, E.Yu.Romanenko. Difference Equations and Their Applications. Kiev, Naukova Dumka, 1986, 280 (in Russian).
2. R.M.May. Nature, 1976, 261, 559-467.
3. P.Richetti, J.C.Roux, F.Argout, A.Arneodo. J.Chem.Phys., 1987, 86, 3339-3356.
4. I.S.Aranson, N.F.Rul'kov. Sov.Rev.-Tech.Phys., 1988, 58, 1954-1670 (in Russian).
5. R.V.Buskirk, C.Jeffries. Phys.Rev., 1985, A31, 3332-3357.
6. Y.Pomeau, P.Manneuille. Comm.Math.Phys., 1980, 74, 189-197.
7. P.Gaspard , R.Kapral, G.Nicolis. J.Stat.Phys., 1984, 35, 697-727.

# Subject Index

Reference is made to the *first* page of relevant articles

# Index of Contributors

N. G. Chetaev

## *Theoretical Mechanics*

Translated from the Russian by I. Aleksanova

1989. 407 pp. 190 figs. Hardcover DM 68,–
ISBN 3-540-51379-5

This university-level textbook reflects the extensive teaching experience of N. G. Chataev, one of the most influential teachers of theoretical mechanics in the Soviet Union. The mathematically rigorous presentation largely follows the traditional approach, supplemented by material not covered in most other books on the subject. To stimulate active learning numerous carefully selected exercises are provided. Attention is drawn to historical pitfalls and errors that have led to physical misconceptions.
Extensive appendices contain material from additional lectures on optics and mechnics analogies, Poincaré's equation and the special theory of elasticity.

**Distribution rights for the socialist countries, India and Iran: V/O "Mezhdunarodnaya Kniga", Moscow**

**D. Park,** Williams College, Williamstown, MA

## *Classical Dynamics and Its Quantum Analogues*

2nd enl. and updated ed. 1990. IX, 334 pp. 101 figs.
Hardcover DM 78,– ISBN 3-540-51398-1

The primary purpose of this textbook is to introduce students to the principles of classical dynamics of particles, rigid bodies, and continuous systems while showing their relevance to subjects of contemporary interest. Two of these subjects are quantum mechanics and general relativity. The book shows in many examples the relations between quantum and classical mechanics and uses classical methods to derive most of the observational tests of general relativity. A third area of current interest is in nonlinear systems, and there are discussions of instability and of the geometrical methods used to study chaotic behaviour. In the belief that it is most important at this stage of a student's education to develop clear conceptual understanding, the mathematics is for the most part kept rather simple and traditional.
This book devotes some space to important transitions in dynamics: the development of analytical methods in the 18th century and the invention of quantum mechanics.

**A. Hasegawa,** AT & T Bell Laboratories, Murray Hill, NJ

## *Optical Solitons in Fibers*

2nd enl. ed. 1990. XII, 79 pp. 25 figs.
Softcover DM 48,– ISBN 3-540-51747-2

Already after six months high demand made a new edition of this textbook necessary. The most recent developments associated with two topical and very important theoretical and practical subjects are combined: **Solitons** as analytical solutions of nonlinear partial differential equations and as lossless signals in dielectric **fibers.** The practical implications point towards technological advances allowing for an economic and undistorted propagation of signals revolutionizing telecommunications. Starting from an elementary level readily accessible to undergraduates, this pioneer in the field provides a clear and up-to-date exposition of the prominent aspects of the theoretical background and most recent experimental results in this new and rapidly evolving branch of science. This well-written book makes not just easy reading for the researcher but also for the interested physicist, mathematician, and engineer. It is well suited for undergraduate or graduate lecture courses.

**A. G. Sitenko,** Academy of the Ukrainian SSR

## *Scattering Theory*

1990. Approx. 320 pp. 32 figs. (Springer Series in Nuclear and Particle Physics)
Hardcover DM 88,– ISBN 3-540-51953-X

This book is an introduction to nonrelativistic scattering theory. The presentation is mathematically rigorous, but is accessible to upper level undergraduates in physics. The relationship between the scattering matrix and physical observables, i. e. transition probabilities, is discussed in detail. Among the emphasized topics are the stationary formulation of the scattering problem, the inverse scattering problem, dispersion relations, three-particle bound states and their scattering, collisions of particles with spin and polarization phenomena. The analytical properties of the scattering matrix are discussed. Problems round off this volume.

**B. N. Zakhariev,** Moscow; **A. A. Suzko,** Minsk, USSR

## *Direct and Inverse Problems*

### *Potentials in Quantum Scattering*

1990. Approx. 200 pp. 42 figs.
Softcover DM 48,– ISBN 3-540-52484-3

This textbook can almost be viewed as a "how-to" manual for solving quantum inverse problems, that is, for deriving the potential from spectra or scattering data and also, as somewhat of a quantum "picture book" which should enhance the reader's quantum intuition. The formal exposition of inverse methods is paralleled by a discussion of the direct problem. Differential and finite-difference equations are presented side by side. The common features and (dis)advantages of a variety of solution methods are analyzed. To foster a better understanding, the physical meaning of the mathematical quantities are discussed explicitly. Wave confinement in continuum bound states, resonance and collective tunneling, energy shifts and the spectral and phase equivalence of various interactions are some of the physical problems covered.

**P. C. Sabatier,** University of Montpellier (Ed.)

## *Inverse Methods in Action*

Proceedings of the Multicentennials Meeting, Montpellier, November 27th – December 1, 1989

1990. XIV, 636 pp. 125 figs.
Hardcover DM 138,– ISBN 3-540-51994-7

The basic idea of inverse methods is to extract from the evaluation of measured signals the details of the emitting them. The applications range from physics and engineering to geology and medicine (tomography).
Although most contributions are rather theoretical in nature, this volume is of practical value to experimentalists and engineers and as well of interest to mathematicians. The review lectures and contributed papers are grouped into eight chapters dedicated to tomography, distributed parameter inverse problems, spectral and scattering inverse problems (exact theory), wave propagation and scattering (approximations); miscellaneous inverse problems and applications and inverse methods in nonlinear mathematics.